T0212757

An Introduction to Differential Manifolds

Grenoble Sciences

The aim of Grenoble Sciences is twofold:

- to produce works corresponding to a clearly defined project, without the constraints of trends nor curriculum,

- to ensure the utmost scientific and pedagogic quality of the selected works: each project is selected by Grenoble Sciences with the help of anonymous referees. In order to optimize the work, the authors interact for a year (on average) with the members of a reading committee, whose names figure in the front pages of the work, which is then co-published with the most suitable publishing partner.

Contact
Tel.: (33) 4 76 51 46 95
E-mail: grenoble.sciences@ujf-grenoble.fr
Website: *https://grenoble-sciences.ujf-grenoble.fr*

Scientific Director of Grenoble Sciences
Jean Bornarel, Emeritus Professor
at the Joseph Fourier University, Grenoble, France

Grenoble Sciences is a department of the Joseph Fourier University supported by the **ministère de l'Enseignement supérieur et de la Recherche** and the **région Rhône-Alpes**.

An Introduction to Differential Manifolds is a translation of the original book *Introduction aux variétés différentielles* (2nd ed.) by Jacques Lafontaine, EDP Sciences, Grenoble Sciences Series, 2010, ISBN 978 2 7598 0572 3.

The reading committee of the French version included the following members:
- **Pierre Averbuch**, Emeritus CNRS Senior Researcher, Grenoble
- **Pierre Bérard**, Professor, Joseph Fourier University, Grenoble I
- **Gaël Meigniez**, Professor, Bretagne-Sud University
- **Jean-Yves Mérindol**, Professor, President of Sorbonne Paris Cité

Translation from original French version performed by Éric Bahuaud; excerpt from *le Livre de mon ami* by Anatole France translated by Jonathan Upjohn; typesetting: ArchiTeX; figures: Sylvie Bordage; cover illustration: Alice Giraud, after *Klein bottle* by Thomas Banchoff and Jeff Beall (Brown University), *The "figure 8" immersion of the Klein bottle* by Fropuff and Inductiveload (KleinBottle-Figure8-01.png, 2007, *Wikimedia Commons*) and elements provided by the author.

Jacques Lafontaine

An Introduction
to Differential Manifolds

 Springer

Jacques Lafontaine
Département de Mathématiques
Université Montpellier 2
Montpellier, France

Based on a translation from the French language edition:
'Introduction aux variétés différentielles' (2ème édition) by Jacques Lafontaine
Copyright © EDP Sciences, 2010 All Rights Reserved

ISBN 978-3-319-35785-0   ISBN 978-3-319-20735-3   (eBook)
DOI 10.1007/978-3-319-20735-3

Springer Cham Heidelberg New York Dordrecht London
© Springer International Publishing Switzerland 2015
Softcover reprint of the hardcover 1st edition 2015

Printed on acid-free paper

Springer International Publishing AG Switzerland is part of Springer Science+Business Media
(www.springer.com)

Preface

It was at that moment, that Fontanet came up with a third brain-wave. "And what", he exclaimed, "if we did a history of France, in 50 volumes, putting in every single detail?" I found the suggestion fabulous, clapping my hands and shouting for joy in approbation.

We were sent off to bed. But, this sublime idea of a 50-volume history of France, containing every single detail, kindled such excitement that, for a full quarter of an hour, I lay there, unable to sleep.

And so we launched out into this history. To tell the truth, I am no longer quite sure just why we began with King Teutobochus. But, we had to; it was what the project demanded. Our first chapter, then, brought us face to face with King Teutobochus who, as the measurement of his bones (which, incidentally, were discovered by accident) can testify, was 30 feet tall.[1] To be confronted by such a giant, right from the start! Even Fontanet was taken aback.

"We'll have to skip Teutobochus", he said to me. But, I just didn't have the courage.

And so it was that the 50-volume history of France came to an end at Teutobochus.

Anatole France, My Friend's Book

This charming little lesson of methodology applies admirably to the subject of this book. It is for the reader to judge what I have made of it. The first steps in the theory of manifolds can, if one follows Fontanet's footsteps, have dire consequences; there is the danger of demotivation, of being discouraged by the subject before realizing that the real difficulties lie elsewhere.

---

1. Anatole France (1844–1924, Nobel Prize 1921) is a French writer who has, unfortunately, somewhat fallen into oblivion. He was a pacifist and a defender of human rights. In this story he is referring to the period before the development of palaeontology, when the bones of prehistoric animals were taken to be the remains of monsters or giants. The interested reader is referred to the articles "Anatole France" and "Teutobochus" in *Wikipedia*.

Smooth manifolds are the natural generalization of curves and surfaces. The idea of a manifold appeared for the first time (and without discussion!) in 1851, in Riemann's inaugural lecture, and allowed him to construct a satisfactory solution of the problem of analytic continuation of holomorphic functions.

It took some 50 years for a precise definition to emerge. It is a question of conceptualising, not the parts of a space $\mathbf{R}^n$ with large n, defined by a certain number of equations, but, in a more abstract way, objects which, "*a priori*", are not within the "ordinary" space of dimension n, for which the notion of smooth function still makes sense.

There are numerous reasons to be interested in "higher" dimensions. Perhaps one of the more evident comes from classical mechanics. Describing the space of configurations of a mechanical system rapidly depends on more than three parameters: one already needs six for a solid.

The fact that it is not always desirable to consider objects as subsets of $\mathbf{R}^n$ is more subtle. For example, the set of directions in three-dimensional space depends on two real parameters, and naturally forms a manifold of dimension two, called the projective plane. This manifold admits numerous realizations as a subspace of Euclidean space, but these realizations are not immediately obvious and it is not clear how to select a "natural" one amongst them.

These "abstract" manifolds furnish the natural mathematical setting for classical mechanics (both configuration and phase space), but also for general relativity and particle physics.

I wanted to write a text which introduced manifolds in the most direct way possible and principally explores their topological properties, while remaining elementary. In this way a sphere stretched and dented remains a sphere, and in the same setting as curves and surfaces. We will mostly be interested in topological and differential properties over metric properties (length, curvature, etc.).

The reader is expected to have a good knowledge of the basics of differential calculus and a little point-set topology. Certain remarks, always enclosed with ** will require a more elaborate foundation. The first chapter is dedicated to classical differential calculus discussed in a way that will extend easily to the manifold setting.

Our proper study of manifolds starts in the second and third chapters. I tried to give significant examples and results as rapidly as possible.

One class of examples – Lie groups their homogeneous spaces – struck me as deserving its own chapter. Chapters 5, 6 and 7 are devoted to differential forms and their relationship to the topology of manifolds. Each chapter depends on its predecessors with one exception: if Lie groups (Chapter 4) arise in subsequent chapters, it's only through examples and occasionally in exercises. Finally the last chapter treats the Gauss-Bonnet theorem for surfaces. One attractive feature of this result is the variety of techniques it brings into play. Above all it illuminates a phenomena that has never ceased to fascinate me through the years: the appearance of integers (perhaps we could say a quantization?) in geometry.

Each chapter starts with a relatively detailed introduction in which I give motivations and an informal description of the contents appealing to the reader's geometric intuition. A section entitled "Comments" gives possible extensions on the subjects introduced.

As I explained above, I decided mostly to limit myself to discussing differential structures. Except in the last chapter, metric structure is discussed infrequently, and symplectic structure is omitted. I make up a little for this in the "Comments" section and the annotated bibliography.

The numerous exercises (more than 150) are for the most part easy. Those labelled with a star are a little more delicate for beginners. Those labelled with two stars are not necessarily technical but of the "sit and think" style. Many exercises can be thought of as complementary material to the book. For this reason I have included the solutions to many of them.

Throughout the years that I taught the course in differential geometry at Montpellier, I benefited from an agreeable, attentive and demanding audience that would leave no question behind. Their attitude deeply encouraged me as I was preparing the notes which became the first version of this book.

After this book was submitted to Grenoble Sciences, I benefited from numerous remarks and stimulating suggestions from the reading committee. I benefited greatly from the broad scientific perspective and temperaments of these colleagues, and it was they who encouraged me to write the detailed introductions I described above.

I wish also to thank Thomas Banchoff and Jeff Beall for allowing the publisher to reproduce their beautiful realization of the Klein bottle on the front cover.

Last, but not least, I have been profoundly influenced by my mentor Marcel Berger.

The translation into English was performed by Eric Bahuaud. I wish to thank him for an excellent coordination. Moreover, he pointed out and helped me to correct several bugs in the French version. I am of course responsible for the remaining ones!

Our job was supervised by Stéphanie Trine with efficiency and *bonne humeur*.

How to use this book

The first chapter and a good part of Chapters 5 and 6 give a relatively complete discussion of classical differential calculus, from the beginning to Stokes's theorem.

The ambition of Chapter 2 is to explain what smooth manifolds are and how to use them to those that might find this notion too abstract or too technical.

Chapter 3 is more technical, precisely because it explains techniques that too often pass without mention.

The final two chapters can be directly read as soon as one masters a little bit of the notions of manifolds and differential forms.

We also note that a reader who, starting with the word "holomorphic" in the index, completes all of the exercises referred to, will get a sense of the different world of complex manifolds.

The book is self-contained as far as differential calculus goes. However it would be in vain to discuss manifolds without a little topology. We very briefly discuss simple connectedness and covering maps, and I have given precise and usable statements of results, but at the expense of the book being self-contained (as far as topology is concerned).

The French version is provided with electronic complements:

https://grenoble-sciences.ujf-grenoble.fr/pap-ebooks/lafontaine/home.

You will find in particular:

- some perquisites: connexity, proper maps;
- more exercises, more solutions;
- Poincaré-Hopf theorem in any dimension;
- sporadic isomorphisms between small-dimensional Lie groups.

Jacques Lafontaine
October 2014

Contents

List of Figures

Notations

References are made to the subsection (or failing that, the section) where the notation first appears. In this text, we place arrows on only vectors determined by two points: $\overrightarrow{ab}$ denotes the vector from a to b.

$\langle\,,\,\rangle$	Euclidean scalar product	
$\otimes$	tensor product (of vectors or forms)	5.2.1
$\bigotimes$	tensor product (of vector spaces or vector bundles)	5.2.1
$\wedge$	exterior product	5.2.2
$\bigwedge^k$	k-th exterior power	5.2.2
$[\,,\,]$	Lie bracket of two vector fields	3.4
$[x]$	point of $P^n K$ in homogeneous coordinates x	2.5
$\alpha^\sharp$	vector associated to the form α	5.3.1
$v^\flat$	form associated to the vector v	5.3.1
$\coprod$	disjoint union	3.5.1
Ad	adjoint representation	4.3
$B(0,r)$	open ball with center 0 and radius r	
$C^k(M)$	functions of class C^k on M	1.2.3
$C^\infty(M)$	smooth functions on M	1.2.3
$C^\infty(E)$	smooth sections of the vector bundle E	3.5.2
$C^\infty(TM)$	smooth vector fields on M	3.5.2
d	exterior derivative	5.4
$\deg(f)$	degree of the map f	7.4.1
dim	dimension	
dist	distance	
div	divergence	5.5
E^*	vector space dual to the vector space E	5.2.1
$E(C,C')$	linking number of two curves C and C'	7.6

Γ	Gauss map	8.6
δ	derivation	3.3
∂D	boundary of the domain D	6.4.2
∂_k	derivation with respect to the k-th variable	1.2.1
θ_ϕ	isomorphism of $T_x M$ to $\mathbf{R}^n$ defined by ϕ	2.6.1
$\Lambda^k(E^*)$	alternating k-forms on E	5.2.2
ϖ	canonical volume form on the sphere	8.6
φ_t^X	flow of the vector field X	3.6
$\chi(M)$	Euler-Poincaré characteristic of the manifold M	8.2
Ω_g	curvature form of the Riemannian metric g	8.3
$\Omega^p(M)$	differential forms of degree p on M	5.3.2
$\Omega(M)$	differential forms on M	5.3.2
$\Omega_0(M)$	compactly supported differential forms on M	6.3

Chapter 1

Differential Calculus

1.1. Introduction

In this chapter, we review and reinforce the basics of differential calculus in preparation for our subsequent study of manifolds.

The majority of the concepts and results studied are generalization of concepts and results from linear algebra. We have a veritable dictionary:

smooth function	—	linear map
local diffeomorphism	—	invertible linear map
submanifold	—	vector subspace

It's necessary to understand and make this dictionary explicit.

1.1.1. What Is Differential Calculus?

Roughly speaking, a function defined on an open set of Euclidean space is differentiable at a point if we can approximate it in a neighborhood of this point by a linear map, which is called its differential (or total derivative). This differential can be of course expressed by partial derivatives, but it is the differential and not the partial derivatives that plays the central role.

The basic result, aptly called the "chain rule" assures that the differential of a composition of differentiable functions is the composition of differentials. This result gives, amongst other things, a convenient and transparent way to compute partial derivatives of compositions, but for us this will not be essential.

A fundamental notion is that of a *diffeomorphism*. By this we mean a differentiable function that admits a differentiable inverse. By the chain rule, the differential at every point of a diffeomorphism is an invertible linear map.

This property, which is an "evident" remark has an even stronger converse: if the differential of a C^1 function is invertible *at a point*, then it is a diffeomorphism in a neighborhood of this point to its image (this is the inverse function theorem, see 1.13).

This central result, suitably exploited, gives "normal forms" to certain mathematical objects. Suppose for example $f : \mathbf{R}^n \to \mathbf{R}$ is of class C^1, whose differential at a point a is nonzero (this is to say that at least one of the partial derivatives at this point is nonzero). By a local change of variables, this function can be written in a neighborhood of the point in question as a linear map, we can even use the map $(x^1, \ldots, x^n) \mapsto x^1$ (see 1.18 for the precise statement). Put differently, we can find a diffeomorphism ϕ from a neighborhood of 0 to a neighborhood of a such that

$$f\big(\phi(x^1, \ldots, x^n)\big) = x^1 + f(a).$$

This result admits a geometric interpretation: suppose S is the set of points in $\mathbf{R}^n$ that satisfies the equation $f(x) = f(a)$. Then there exists, under the same conditions, a diffeomorphism from a neighborhood U of a, that sends $U \cap S$ to a piece of a hyperplane (see 1.20 and 1.21).

We can also ask what happens when the differential vanishes. We then look at the second order Taylor polynomial, which is a quadratic form. If it is of maximum rank, then after a change of variables, the function can be written in a neighborhood of a as this quadratic form. This is the Morse lemma, proved in Exercise 11. See also Lemma 3.44.

These results have the following points in common:

1. They are consequences (relatively immediate in the first case, slightly disguised in the case of the Morse lemma) of the inverse function theorem.

2. They apply because a certain associated algebraic object is non-degenerate.

3. They are *local* results: the normal form obtained for the mathematical object studied is valid in a neighborhood of a point. Its necessary to keep a simple example in mind: a little piece of the circle is homeomorphic and even diffeomorphic to an interval, but this is not the case for the entire circle.

There are other examples of results of this type in differential calculus, for example the *rank theorem* (see Exercise 10). Looking ahead a little, we mention also that a vector field which is nonvanishing at a point can be written as a constant vector field (see Exercise 16 in Chapter 3), and a symplectic form is locally equivalent to an alternating bilinear form of maximum rank (Darboux's theorem, see Exercise 14 and 17 in Chapter 5).

1.1.2. In This Chapter

Sections 1.2 to 1.5 recall the basics of differential calculus in a way that will extend to situations more general than classical vector calculus. We take up this generalization in the next chapter. Several classical results will not be revisited, for these one can consult [Lang 86] or [Hörmander 90] for example: the Clairaut/Schwarz theorem on symmetry of mixed partial derivatives, Taylor's formula, and sequences and series of differentiable functions. Our goal is to start our study of manifolds rapidly, and these results, while important, enter less into this study.

Section 1.6 is devoted to a result that is not often part of a standard exposition on differential calculus: if h is a continuous function from $\mathbf{R}$ into the group of invertible $n \times n$ matrices such that $h(t + t') = h(t) h(t')$, then $h(t)$ is of the form $\exp tA$. In dimension 1, this is simply a classical characterization of exponential functions. In higher dimensions, one must use the inverse function theorem: it allows us to find a nonzero t_0 such that $h(t_0) = \exp B$ for a suitable matrix B, and we subsequently proceed more or less as in dimension 1.

Critical points are introduced in Section 1.7. The equation which characterizes them is often interesting in itself. We cannot resist the temptation to give the following example. If C is a closed curve in the plane, which we assume to be convex for simplicity, imagine the inscribed polygons for which two consecutive sides satisfy the Descartes/Snell law (polygons formed by light trajectories for physicists, billiard trajectories for mathematicians). If the perimeter function

$$(m_1, \ldots, m_n) \longmapsto \sum_{i=1}^{n} \left\| \overrightarrow{m_i m_{i+1}} \right\| \quad \text{with the convention } m_{n+1} = m_1$$

admits a critical point $(M_1, \ldots, M_n)$, the points M_i are the vertices of a light polygon, by Fermat's principle or Section 1.2.2. The perimeter function, being a function on the Cartesian product C^n, admits a maximum which is realized by compactness. Knowing that the points where a function admits a maximum are critical points, we have in principle a method of showing the existence of these polygons. All of this works very well for triangles (try it!). For $n = 4$ a difficulty occurs: if A and B are two points such that $\mathrm{diam}(C) = \|\overrightarrow{AB}\|$, the quadruplet (A, B, A, B) realizes the maximum perimeter, and it corresponds to a degenerate polygon, with the diameter traversed four times! Moral: more sophisticated methods of finding critical points are needed, for which we refer for example to the excellent [Tabachnikov 95].

A function can admit many critical points. In the extreme case of a constant function, every point of the domain is critical. But there is only one critical value, the constant in question. This extreme case illustrates the fact that critical values are never very numerous. Sard's theorem (Theorem 1.41) confirms this intuition: the set of critical values has measure zero. This is the subject of Section 1.8. This result, whose extension to manifolds is straight-forward, is used twice in this book. The first time is in Chapter 3. After showing that every compact manifold of dimension n is embedded in $\mathbf{R}^N$ for some non-explicit and poorly controlled N, Sard's theorem will allow us to lower the dimension down to $2n+1$. The second time Sard's theorem appears will be in a much more fundamental way, in Chapter 7, to show that there are always regular values (this is to say non-critical values). One does not know how to "explicitly" find a regular value, but we know that nearly every point is a regular value. This is a classical ruse in mathematics.

Finally, Section 1.9 is devoted to differential calculus in infinite dimensions. This subject was very much in fashion in the 1960s. Generalization for its own sake was fashionable and in the spirit of the times. However, it was remarked by mathematicians working on dynamical systems that the inverse function theorem in infinite dimensions gave an efficient proof of a basic result on the existence and uniqueness of systems of differential equations (see the Theorem 1.44). This method gives the smooth dependence of solutions with respect to initial conditions for free, which is not so easy to obtain using classical methods.

1.2. Differentials

1.2.1. Definition and Basic Properties

Definition 1.1. *A function f from an open subset U in $\mathbf{R}^p$ with values in $\mathbf{R}^q$ is differentiable at a point a in U if there exists a linear map L from $\mathbf{R}^p$ to $\mathbf{R}^q$ such that*

$$f(a+h) = f(a) + L \cdot h + o(h).$$

We say L is the *differential* of f at a, or the *total derivative* of f at a.

The notation $L \cdot h$ instead of $L(h)$ is chosen to emphasize the linearity. We designate by $h \mapsto o(h)$ a map from an open set in $\mathbf{R}^p$ with values in $\mathbf{R}^q$ such that for norms $\| \ \|_1$ and $\| \ \|_2$ on the domain and range, we have

$$\lim_{h \to 0} \frac{\|o(h)\|_2}{\|h\|_1} = 0.$$

This property does not depend on the choice of norms used in the formulation above (this will no longer be case when we study differentiability in infinite dimensions, see Section 1.9).

Remark. We can rewrite the definition in the form

$$\overrightarrow{f(a)f(x)} = L \cdot \overrightarrow{ax} + o(\overrightarrow{ax}).$$

In fact, $\mathbf{R}^p$ and $\mathbf{R}^q$ are considered simultaneously as *affine spaces*, where the *points* x and $f(x)$ live, and *vector spaces* where we find the *vectors* $\overrightarrow{ax}$ and $\overrightarrow{f(a)f(x)}$. This drives the reformulation of the definition by allowing us to replace $\mathbf{R}^p$ and $\mathbf{R}^q$ by *affine spaces* E and F of dimensions p and q respectively; in the equation above, L then denotes a linear map between the vector spaces $\overrightarrow{E}$ and $\overrightarrow{F}$ associated to E and F. The affine point of view is explained for example in the first chapter of [Audin 03].

Example: parametric curves

Consider the case of maps from $\mathbf{R}$ to an affine space F, and let $\overrightarrow{F}$ denote the associated vector space. Every linear map from $\mathbf{R}$ to $\overrightarrow{F}$ is of the form $h \mapsto hv$ where v is a vector in $\overrightarrow{F}$. By dividing by the real number h, we see that differentiability at a is equivalent to the existence of a vector $v \in \overrightarrow{F}$ such that

$$\frac{f(a+h) - f(a)}{h} = v + \epsilon(h), \quad \text{where} \quad \lim_{h \to 0} \epsilon(h) = 0.$$

In other words, f is differentiable at a and the vector v is equal to $f'(a)$. It is common to give maps from $\mathbf{R}$ to an affine space the name "parametric curves". The vector $f'(a)$ is then called the *tangent vector* to the curve at $f(a)$.

We will see in a moment that there are important changes when we move from $\mathbf{R}$ to a space with more than one dimension, *i.e.*, from single variable to multiple variable functions.

In any case, we have the following property:

Proposition 1.2. *The map L is unique.*

PROOF. Suppose L' is a second linear map satisfying the same property. Choose $h \in \mathbf{R}^p$, and consider an increment of the form th, where t is a nonzero real number. We have

$$f(a) + L \cdot th + o(th) = f(a) + L' \cdot th + o(th),$$

where

$$L \cdot th - L' \cdot th = t(L \cdot h - L' \cdot h) = o(th).$$

By dividing by t, we see that

$$L \cdot h - L' \cdot h = \frac{o(th)}{t},$$

and taking the limit as $t \to 0$ gives $L = L'$. $\qquad\square$

$$\boxed{\textbf{WE DENOTE THE DIFFERENTIAL OF } f \textbf{ AT } a \textbf{ BY } df_a}$$

Remark. In the previous argument we could restrict to positive values of t. We deduce that a *positive homogeneous function f of degree 1* (this is to say a map from E to F such that $f(tx) = tf(x)$ for all positive real numbers t) that is differentiable at 0 is necessarily *linear*. In particular a norm is *never* differentiable at the origin.

Uniqueness of the differential can also be seen from its explicit expression as a function of partial derivatives of coordinate functions.

Proposition 1.3. *If a map f from an open subset U in $\mathbf{R}^p$ to $\mathbf{R}$ is differentiable at $a \in U$, then the first partial derivatives of f at a are defined and*

$$df_a \cdot h = \sum_{i=1}^{p} \partial_i f(a) h^i \quad (\text{if } h = (h^1, \ldots, h^p)).$$

PROOF. *A priori*, we may write $df_a \cdot h = \sum_{i=1}^{p} u_i h^i$, where the real numbers u_i are to be determined. In writing the property of differentiability for an increment of the form

$$h = (0, \ldots, t, \ldots, 0) \quad (t \text{ in the } i\text{-th place}),$$

we see that the function of one real variable

$$t \longmapsto f(a^1, \ldots, a^i + t, \ldots, a^p)$$

is differentiable, and therefore differentiable at 0, its derivative being u_i. $\square$

This result is easily generalized.

Proposition 1.4. *If a map from an open subset U in $\mathbf{R}^p$ to $\mathbf{R}^q$ is differentiable at $a \in U$, then the partial derivative of each component f^i of f at a exists, and the matrix of differentials with respect to the canonical basis of the domain and range is*

$$\left(\partial_j f^i(a)\right)_{1 \leqslant i \leqslant q, \ 1 \leqslant j \leqslant p}.$$

PROOF. In expressing the property of differentiability component by component, we see that f is differentiable if and only if each component f^i is. It therefore suffices to apply the preceding proposition to f^i. $\square$

Definition 1.5. *The matrix $(\partial_j f^i)_{1 \leqslant i \leqslant q, \ 1 \leqslant j \leqslant p}$ is called the* Jacobian matrix *of f.*

By convention upper indices (superscripts) denote rows, and lower indices (subscripts) denote columns. This is rooted in the *Einstein summation convention*, which will be explained and justified in Section 5.2.

The rank of f at a is by definition the rank of $d_a f$ (or the rank of the Jacobian matrix).

The determinant of the Jacobian matrix (if $p = q!$), is called the Jacobian of f, and will be denoted $J(f)$.

If we replace $\mathbf{R}^p$ and $\mathbf{R}^q$ by vector spaces of dimension p and q represented by the bases $(e_i)_{1 \leqslant j \leqslant p}$ and $(e'_i)_{1 \leqslant i \leqslant q}$, the matrix of the differential represented in these bases can be written in the same way, where we denote the i-th component of f by f^i, and the derivative at $t = 0$ of the function $t \mapsto f^i(a + t e_j)$ by $\partial_j f^i(a)$. If $E = F$, the Jacobian of f is the determinant of the endomorphism df_a.

To not weigh down this exposition, we will most often use the spaces $\mathbf{R}^n$ equipped with their canonical basis. However, there exist many situations (for example when we work with spaces of linear maps) where this is not the natural thing to do.

1.2.2. Three Fundamental Examples

Length

In a Euclidean space, the length function $(x, y) \mapsto \|\overrightarrow{xy}\|$ when (x, y) satisfies $x \neq y$ has differential

$$(u, v) \longmapsto \left\langle \frac{\overrightarrow{xy}}{\|\overrightarrow{xy}\|}, v \right\rangle - \left\langle \frac{\overrightarrow{xy}}{\|\overrightarrow{xy}\|}, u \right\rangle.$$

Cauchy-Riemann equations

A function from $\mathbf{C}$ to $\mathbf{C}$ is *holomorphic* if and only if it is $\mathbf{C}$-differentiable (*i.e.*, we replace $\mathbf{R}$ by $\mathbf{C}$ in Definition 1.1 and we require that the differential be $\mathbf{C}$-linear).

Following the example of functions of a real variable, we have

$$f(z + h) = f(z) + Ah + o(h), \quad \text{with } A = \lim_{h \to 0} \frac{f(z + h) - f(z)}{h},$$

which allows us to write $A = f'(z)$. Regarded as a map from $\mathbf{R}^2$ to $\mathbf{R}^2$, the map f is differentiable, and the Jacobian matrix is of the form

$$\begin{pmatrix} a & -b \\ b & a \end{pmatrix} \quad (\text{if } f'(z) = a + ib).$$

We can obtain the Cauchy-Riemann equations as follows. View f as a differentiable map from $\mathbf{R}^2$ to $\mathbf{R}^2$ with components P and Q. We have

$$\partial_1 P = \partial_2 Q \quad \text{and} \quad \partial_2 P = -\partial_1 Q.$$

In particular, its Jacobian determinant is $|f'(z)|^2$.

There are situations where it is preferable to calculate the differential without appealing to coordinates.

Inverse of a matrix; determinant and trace

We will calculate the differentials

a) *of a map* $\varphi : A \mapsto A^{-1}$ *of* $Gl(\mathbf{R}^n)$ *to itself*. (Note that in the process we verify that $Gl(n, \mathbf{R})$ is open in $\mathrm{End}(\mathbf{R}^n)$.)

Choose a norm on $\mathbf{R}^n$, and equip $\mathrm{End}(\mathbf{R}^n)$ with the associated operator norm:

$$\|A\| = \sup_{\|x\| \leqslant 1} \|Ax\|.$$

If A and B are two endomorphisms, we have $\|AB\| \leqslant \|A\|\,\|B\|$. Then if $\|H\| < 1$, the series

$$\sum_{k=0}^{\infty} (-1)^k H^k$$

is convergent in norm, and therefore convergent. Let S be the sum and note that S satisfies

$$S(I + H) = (I + H)S = I.$$

Thus $I + H$ is invertible, with inverse S and the series expansion of $(I + H)^{-1}$ just obtained gives

$$\|(I + H)^{-1} - I - H\| = \left\| \sum_{k=2}^{\infty} (-1)^k H^k \right\| \leqslant \sum_{k=2}^{\infty} \|H\|^k = \frac{\|H\|^2}{1 - \|H\|}.$$

For $\|H\| < 1/2$, we have

$$(I + H)^{-1} = I - H + r(H), \quad \text{where } \|r(H)\| < 2\|H\|^2,$$

which shows the differential of φ at I is the map $H \mapsto -H$.

To pass to the general case, we write

$$(A + H)^{-1} = \left(A(I + A^{-1}H) \right)^{-1} = (I + A^{-1}H)^{-1} A^{-1}.$$

In using the same series expansion, we see that $Gl(\mathbf{R}^n)$ is *open* in $\text{End}(\mathbf{R}^n)$ (if $\|H\| < 1/\|A^{-1}\|$, then $A + H$ is invertible), and the differential of φ at A is the linear map

$$H \longmapsto -A^{-1}HA^{-1}.$$

Note the analogy with the derivative of the function $1/x$!

b) *of the map $A \mapsto \det A$ from $\text{End}(\mathbf{R}^n)$ to $\mathbf{R}$.*

We leave to the reader to verify, using multilinearity, that if $a_1, \ldots, a_n$ (resp. $h_1, \ldots, h_n$) denotes the column vectors of the matrix A (resp. of the matrix H), we have

$$d\det_{A} \cdot H = \sum_{k=1}^{n} \det(a_1, \ldots, a_{k-1}, h_k, a_{k+1}, \ldots, a_n).$$

However, there is a much more striking intrinsic expression.

We first examine the case where $A = I$. Now

$$\det(I + H) = 1 + \sum_{k=1}^{n} h_k^k + \text{terms of degree} \geqslant 2 \text{ with respect to } h_k^l,$$

which shows that the differential at I is none other than the map

$$H \longmapsto \text{tr}(H).$$

Now if A is invertible, we can write

$$A + H = A(I + A^{-1}H),$$

as in a), and deduce that the differential of det at A is given by

$$H \longmapsto \det(A)\,\text{tr}(A^{-1}H).$$

To pass to the general case, we remark that

$$\det(A)\,\text{tr}(A^{-1}H) = \text{tr}(\widetilde{A}H),$$

where $\widetilde{A}$ is the matrix of cofactors of A. As det is clearly a smooth function, its differential in the general case is therefore

$$H \longmapsto \text{tr}(\widetilde{A}H).$$

This formula is also a direct consequence of our initial calculation.

1.2.3. Functions of Class C^p

We have seen that the components of a differentiable function are differentiable. Conversely, a function defined on an open subset of $\mathbf{R}^p$ that admits partial derivatives, *i.e.*, such that the partial functions

$$t \longmapsto f(a^1, \ldots, a^i + t, \ldots, a^p)$$

are differentiable, need not be differentiable if $p > 1$. A simple counterexample is given by the following function of two variables

$$f(x,y) = \frac{xy}{x^2 + y^2} \quad \text{if } (x,y) \neq (0,0), \text{ and } f(0,0) = 0,$$

which is not continuous at the origin but has partial derivatives at each point. On the other hand, we have the following fundamental result.

Theorem 1.6. *Let f be a map from an open subset U in $\mathbf{R}^p$ to $\mathbf{R}$. If f has partial derivatives on U that are continuous at a, then f is differentiable at a.*

PROOF. Suppose that $p = 2$ to lighten the notation. The general case is treated in the same fashion. Write $a = (b, c)$. We have

$$f(b+h, c+k) - f(b,c) = f(b+h, c+k) - f(b+h, c) + f(b+h, c) - f(b,c).$$

On one hand we have

$$f(b+h, c) - f(b,c) = \partial_1 f(b,c) h + o(h).$$

On the other hand we can apply the mean value theorem to the function $t \mapsto f(b+h, t)$,

$$f(b+h, c+k) - f(b+h, c) = \partial_2 f(b+h, c+\theta k) k \quad (0 < \theta < 1).$$

However, because of the continuity of $\partial_2 f$ at a,

$$\partial_2 f(b+h, c+\theta k) = \partial_2 f(b,c) + o(h,k). \qquad \square$$

This is the most practical and frequently used criteria for differentiability. It gives rise to the following definition:

Definition 1.7. *A map from an open subset U of $\mathbf{R}^p$ to $\mathbf{R}^q$ is of class C^1 (or continuously differentiable) if all of its partial derivatives of order 1 exist and are continuous on all of U, of class C^p (or p-times continuously differentiable) if its partial derivatives are of class C^{p-1}, and finally C^∞ (we also say smooth) if it is of class C^p for all p. Finally note that we often say that a map is a C^p map (respectively C^∞ map) map if it is of class C^p (respectively of class C^∞).*

It is clear that the sum of two functions of class C^p remains of class C^p. The same holds for the product of two real valued functions.

Remark. Differentiability is defined *at a point* (more precisely it depends only on the behavior of a function in an arbitrary neighborhood of the point considered). By contrast, the property of being a C^p map only makes sense on an open subset.

1.3. The Chain Rule

Theorem 1.8. *Suppose f is a map from an open subset U of $\mathbf{R}^m$ to $\mathbf{R}^n$, and g is a map from an open subset V of $\mathbf{R}^n$ to $\mathbf{R}^p$. Suppose that f is differentiable at $a \in U$, with $f(a) \in V$, and further suppose that g is differentiable at $f(a)$. Then $g \circ f$ is differentiable at a, and*

$$d(g \circ f)_a = dg_{f(a)} \circ df_a.$$

In other words, the differential of a composition is the composition of the differentials.

PROOF. We begin by remarking that since f is continuous at a, $f^{-1}(V)$ is a neighborhood of a, and therefore $g \circ f$ is defined on an open subset U' containing a. If $a + h \in U'$, we have

$$f(a + h) = f(a) + L \cdot h + o(h).$$

Write $k = L \cdot h + o(h)$. Then

$$g\big(f(a + h)\big) = g(f(a) + k) = g\big(f(a)\big) + M \cdot k + o(k)$$
$$= g\big(f(a)\big) + M \cdot L \cdot h + o(h). \qquad \square$$

Remarks

a) At the level of Jacobian matrices of g and f, this result yields the formula

$$\partial_j (g \circ f)^i = \sum_{k=1}^{n} \partial_k g^i \big(f(a)\big) \cdot \partial_j f^k(a),$$

which may also be written

$$\big[d(g \circ f)_a\big]^i_j = \sum_{k=1}^{n} \big[dg_{f(a)}\big]^i_k \big[df_a\big]^k_j.$$

b) Implicit usage of the chain rule is very common. For example, suppose $(E, \langle \, , \, \rangle)$ is an inner product space. If u and v are two maps from an open subset U of $\mathbf{R}^n$ to E, the differential of $f = \langle u, v \rangle$ is given by

$$df_a \cdot h = \langle du_a \cdot h, v(a) \rangle + \langle u(a), dv_a \cdot h \rangle.$$

We can either verify this directly, or consider f as a composition of maps $x \mapsto \big(u(x), v(x)\big)$ from U to $E \times E$ and $(y, z) \mapsto \langle y, z \rangle$ from $E \times E$ to $\mathbf{R}$.

Example: inversions

Suppose E is n-dimensional inner product space. *Inversion with center p and power k* is the map $I_{p,k}$ from $E \smallsetminus \{p\}$ to itself defined by

$$\overrightarrow{pI_{p,k}}(x) = k \frac{\overrightarrow{px}}{\|\overrightarrow{px}\|^2}.$$

This map is clearly smooth.

Take $p = 0$, $k = 1$ and set $I_{0,1} = I$. The differential of I at a is therefore

$$dI_a \cdot h = \frac{h}{\|a\|^2} - 2 \frac{\langle a, h \rangle}{\|a\|^4} a = \frac{1}{\|a\|^2} S_a \cdot h,$$

where we write

$$S_a \cdot h = h - 2 \frac{\langle a, h \rangle}{\|a\|^2} a.$$

It is clear that $S_a \cdot a = -a$ and $S_a \cdot h = h$ if h is an element of the hyperplane orthogonal to a. Therefore S_a is the orthogonal reflection with respect to this hyperplane. It is an isometry, and dI_a is a (indirect) similarity.

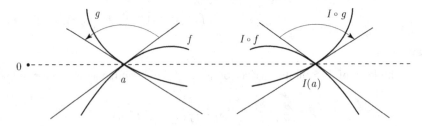

Figure 1.1: Inversion

Now let $t \mapsto f(t)$ and $t \mapsto g(t)$ be two parametric curves such that $f(0) = g(0) = a$. By the chain rule, the tangent vectors at $I(a)$ to the image curves under I are $dI_a \cdot f'(0)$ and $dI_a \cdot g'(0)$. Therefore the preceding discussion shows their angle (in absolute value) is the same as the one between $f'(0)$ and $g'(0)$. In other words, I preserves angles. We say such maps are *conformal*. A theorem of Liouville states that if $n \geqslant 3$, all conformal maps from an open subset of Euclidean space of dimension $n \geqslant 3$ to another are the restriction of a product of inversions. (For a proof, see [Berger 87, Chapter 9].) For $n = 2$ the situation is very different: we can see from the Cauchy-Riemann equations (see Subsection 1.2.2) that f is conformal if and only if it is holomorphic or antiholomorphic, while the products of inversions (called *Möbius transformations*) are fewer in number (they form a finite-dimensional group, see Exercise 16 in Chapter 2).

We return to the general discussion with an immediate consequence of the chain rule.

Corollary 1.9. *Every composition of maps of class C^p ($1 \leqslant p \leqslant \infty$) is itself of class C^p.*

Differential notation. This is justified by the chain rule. Starting from the (obvious) remark that a linear map is differentiable and equal to its differential, we denote (to distinguish the two if we want) dt as the differential of the identity map from $\mathbf{R}$ to $\mathbf{R}$, and dx^i the differential of the i-th coordinate of a vector x in $\mathbf{R}^p$. Let f be a differentiable map from $\mathbf{R}^p$ to $\mathbf{R}$. Denoting h^i the i-th component of the vector h we have

$$df_a \cdot h = \sum_{i=1}^{n} \partial_i f(a) h^i.$$

This gives us the *value* of the linear form df_a for the vector h. As $dx^i(h) = h^i$, we may write

$$df_a = \sum_{i=1}^{n} \partial_i f(a) \, dx^i.$$

In other words, the differential of f is a linear combination of coordinate differentials, with coefficients being the partial derivatives.

Remark. If we simply write df, this can signify:

a) either that we consider the differential of f at a point implied by the context;

b) or we consider the map $x \mapsto df_x$.

Such ambiguity is frequent in differential calculus.

Now if g is a differentiable map from $\mathbf{R}$ to $\mathbf{R}^n$, the chain rule tells us that the differential of $f \circ g$ is obtained by replacing the dx^i by the dg^i, the differentials of the components of g, in the expression for df. We then write

$$df_x = \sum_{i=1}^{n} \partial_i f(x^1, x^2, \dots, x^n)\, dx^i,$$

and

$$d(f \circ g)_t = \sum_{i=1}^{n} \partial_i f(g^1(t), g^2(t), \dots, g^n(t))\, dg^i$$
$$= \left(\sum_{i=1}^{n} \partial_i f(g^1(t), g^2(t), \dots, g^n(t))\, g'^i(t) \right) dt.$$

From this we deduce that the derivative of $f \circ g$ at t is equal to

$$\sum_{i=1}^{n} \partial_i f(g^1(t), g^2(t), \dots, g^n(t))\, g^{i\prime}(t).$$

Remark. We will see two very different generalizations of the differential. Next chapter we will see that the notion of a smooth function has meaning in the more general setting of functions between manifolds (say for the moment between curves and surfaces); we will call this the linear tangent map, and denote it by $T_a f$ (see Section 2.6).

Afterward we will see the differential of functions extends to a linear operator defined on differential forms (see Section 5.4), still denoted by d.

1.4. Local Invertibility

1.4.1. Diffeomorphisms

Definition 1.10. *A map f from an open subset U of $\mathbf{R}^p$ to an open subset V in $\mathbf{R}^q$ is a C^k diffeomorphism if it admits a C^k inverse. We say that U and V are diffeomorphic.*

Denote the inverse map by g. The chain rule applied to $f \circ g$ and $g \circ f$ tells us that if $a \in U$, the linear maps df_a and $dg_{f(a)}$ are mutual inverses. In particular, this forces $p = q$.

Remark. It is also true that an open subset of $\mathbf{R}^p$ cannot be homeomorphic to an open subset of $\mathbf{R}^q$ unless $p = q$. This result, called the *invariance of domain*, is distinctly more difficult to prove, and appeals to algebraic topology (for a proof, see for example [Karoubi-Leruste 87, Chapter V] or [Dugundji 65, Chapter XVII, no. 3]).

Examples: balls and the product of intervals

a) *All open intervals in* **R** *are mutually diffeomorphic, and diffeomorphic to* **R**.

It is clear that all bounded open intervals are mutually diffeomorphic, as are all intervals of the form (a, ∞) or $(-\infty, b)$. On the other hand we have diffeomorphisms $t \mapsto e^t$ from $(0, \infty)$ to **R** and $t \mapsto \frac{t}{1-t^2}$ from $(-1, 1)$ to **R** (for example).

b) *All open balls in* $\mathbf{R}^n$ *(under the Euclidean norm) are diffeomorphic to* $\mathbf{R}^n$.

Using a), we see that

$$x \longmapsto \frac{x}{1 - \|x\|^2}$$

is a diffeomorphism of the open ball $B(0, 1)$ to $\mathbf{R}^n$.

c) *In* $\mathbf{R}^2$, *the interior of a square is diffeomorphic to an open disk.*

It suffices to remark that the map

$$(x, y) \longmapsto \left(\frac{x}{1 - x^2}, \frac{y}{1 - y^2} \right)$$

is a diffeomorphism of the square $(-1, 1) \times (-1, 1)$ to $\mathbf{R}^2$.

Of course there are analogous statements in every dimension. Later we will see that $\mathbf{R}^n$ and $\mathbf{R}^n \smallsetminus \{0\}$ are not diffeomorphic.

Warning. The example $t \mapsto t^3$ from **R** to **R** shows that a smooth homeomorphism may not be a diffeomorphism. In fact its differential at 0 is not invertible, as it vanishes.

Conversely:

Proposition 1.11. *Suppose f is a homeomorphism from an open subset U to an open subset V in $\mathbf{R}^p$. If f is of class C^k, and if df is invertible at every point, then f is a C^k diffeomorphism and*

$$\left(df_{f(x)} \right)^{-1} = (df_x)^{-1}.$$

PROOF. We appeal to an easy but useful lemma whose proof is left as an exercise.

Lemma 1.12. *If A is a bijective linear map between finite-dimensional normed vector spaces, then there exists strictly positive constants m and M such that*

$$\forall x \neq 0, \quad m\|x\| < \|A \cdot x\| < M\|x\|.$$

Let g denote the inverse of f. Suppose $a \in U$ and $b = f(a)$. We first show that g is differentiable at b. Since g is continuous

$$g(b + h) = g(b) + \Delta(h), \quad \text{where } \|\Delta(h)\| = o(1).$$

Composing this equation with f, we obtain

$$b + h = b + df_a \cdot \Delta(h) + o(\Delta(h))$$

and

$$\Delta(h) = (df_a)^{-1} \cdot h + (df_a)^{-1} \cdot o(\Delta(h)).$$

Applying the lemma, $\Delta(h) = O(h)$ therefore the relation above gives

$$\Delta(h) = (df_a)^{-1} \cdot h + o(h).$$

Therefore g is differentiable at b, and $dg_b = \left(df_{g(b)}\right)^{-1}$.

The fact that g is C^k if f is C^k follows from the chain rule. □

A much stronger result is true.

1.4.2. Local Diffeomorphisms

Theorem 1.13 (Inverse function theorem). *Suppose f is a C^k map ($k \geqslant 1$) from an open subset U in $\mathbf{R}^p$ to $\mathbf{R}^p$, and a is a point of U where the differential df_a is invertible. Then there exists an open subset V contained in U and containing a such that $f : V \to f(V)$ is a C^k diffeomorphism.*

In other words, if the differential of f at a is an isomorphism as a linear map, f is itself an isomorphism as a C^k map, provided we remain close to a.

PROOF. The proof rests on a classical result of topology, the fixed point theorem for contraction mappings, and we review the statement now. Note that it's necessary to use a version "with parameters", that is easily obtained in adapting the classical proof.

Theorem 1.14. *Suppose (X, d) is a complete metric space, Y is a topological space, and $F : X \times Y \to X$ is a continuous map. Suppose that F is uniformly contracting, this is to say that there exists a positive real number $k < 1$ such that*

$$d\big(F(x, y), F(x', y)\big) \leqslant kd(x, x')$$

for all x and x' in X and y in Y.

Then, for all $y \in Y$, the equation $F(x, y) = x$ has a unique solution. Let $\varphi(y)$ denote this solution. Then the map $y \mapsto \varphi(y)$ is continuous.

Returning to the proof of the inverse function theorem, by pre and post composing with translations, and precomposing again with df_a^{-1}, we can consider the case where $a = f(a) = 0$ and $df_0 = Id$. By continuity of the map $x \mapsto df_x$, there exists a closed ball $\overline{B}(0, r) \subset U$ on which $\|I - df_x\| \leqslant \frac{1}{2}$. Therefore by the mean value theorem:

- the restriction of f to $\overline{B}(0, r)$ is Lipschitz with constant $\frac{3}{2}$;
- the continuous map $F(x, y) = x - f(x) + y$ sends $\overline{B}(0, r) \times \overline{B}(0, \frac{r}{2})$ to $\overline{B}(0, r)$;
- for all x and x' with norm less than r,

$$\|F(x, y) - F(x', y)\| \leqslant \frac{1}{2}\|x - x'\|.$$

Therefore, by the fixed point theorem, Theorem 1.14, for $y \in \overline{B}(0, \frac{r}{2})$, there exists a unique x in $\overline{B}(0, r)$ such that $F(x, y) = x$, which is to say $f(x) = y$, and the map $g : y \mapsto x$ just defined is continuous. From this we deduce the existence of open subset U' and V' containing 0 such that

$$g \circ f_{|U'} = Id_{|U'} \quad \text{and} \quad f \circ g_{|V'} = Id_{|V'}.$$

As a result f is a homeomorphism from $U' \cap g^{-1}(V')$ to $V' \cap f^{-1}(U')$. Applying Proposition 1.11, we see in fact that f is a diffeomorphism. □

Corollary 1.15. *Suppose f is a C^k map from an open subset $U \subset \mathbf{R}^m$ to $\mathbf{R}^n$. If the differential of f is invertible everywhere, then for all open subsets V of U, $f(V)$ is open in $\mathbf{R}^m$.*

PROOF. The key idea is that every point in V is contained in an open set on which f is a diffeomorphism. □

Example: the "square root" of a endomorphism near the identity

There exists two open subset U and V containing I in $Gl(n, \mathbf{R})$ such that for every matrix $B \in V$ there exists a unique matrix $A \in U$ with square B. To see this, note that

$$(A + H)^2 = A^2 + AH + HA + H^2,$$

and so the map $f : A \mapsto A^2$ is differentiable, with its differential at A given by

$$H \longmapsto AH + HA.$$

However f is C^1, and it suffices to apply the inverse function theorem at $A = I$.

Definition 1.16. *A local diffeomorphism is a C^k map $(k \geqslant 1)$ from an open subset U in $\mathbf{R}^p$ to $\mathbf{R}^p$ whose differential is invertible at every point.*

By the inverse function theorem, it is equivalent to say that every point in U is contained in an further open subset V such that $f_{|V}$ is a diffeomorphism to its image.

We also note that a C^k map ($k \geqslant 1$) whose differential at a point is invertible is a local diffeomorphism from a neighborhood of this point to its image.

There is no reason for a local diffeomorphism to be injective; conversely, by Proposition 1.11 every *bijective* local diffeomorphism is a diffeomorphism. We note finally that a local diffeomorphism is an *open* map (which is to say that the image of every open subset is open) by Corollary 1.15.

Examples

a) The map $(r, \theta) \mapsto (r \cos \theta, r \sin \theta)$ is a local diffeomorphism $(0, \infty) \times \mathbf{R}$ to $\mathbf{R}^2 \smallsetminus \{0\}$.

b) If we identify $\mathbf{C}$ with $\mathbf{R}^2$, the map $z \mapsto z^2$ is a local diffeomorphism of $\mathbf{R}^2 \smallsetminus \{0\}$ to itself, and the map $z \mapsto e^z$ is a local diffeomorphism from $\mathbf{R}^2$ to $\mathbf{R}^2 \smallsetminus \{0\}$.

1.4.3. Immersions, Submersions

It is remarkable that we can still obtain "local" information about f supposing only that its differential at a point a is injective or surjective. To simplify the statements, we suppose that $a = 0$ and $f(a) = 0$, as the general case is deduced without difficulty by performing translations.

Theorem 1.17. *Let f be a C^1 map from an open subset U of $\mathbf{R}^p$ to $\mathbf{R}^q$. Suppose that $0 \in U$ and the differential df_0 is injective. Then there exists an open subset V in $\mathbf{R}^q$ containing 0 and an open subset U' contained in U such that $f(U') \subset V$, and a diffeomorphism φ of V to its image such that*

$$\varphi\big(f(x^1, \ldots, x^p)\big) = (x^1, \ldots, x^p, 0, \ldots, 0).$$

PROOF. Necessarily $p \leqslant q$. Suppose $f^1, \ldots, f^q$ are the components of f. The hypothesis means that the Jacobian matrix of f is of rank p. After permuting the coordinates in the domain space if necessary, we can suppose that the matrix

$$A = \big(\partial_j f^i(0)\big)_{1 \leqslant i \leqslant p, \ 1 \leqslant j \leqslant p}$$

is invertible. We define a map g from $U \times \mathbf{R}^{q-p}$ to $\mathbf{R}^q$ by

$$g\big(x^1, \ldots, x^p, y^1, \ldots, y^{q-p}\big) = \big(f^1(x), \ldots, f^p(x), y^1 + f^{p+1}(x), \ldots, y^{q-p} + f^q(x)\big).$$

The Jacobian matrix of g is of the form

$$\begin{pmatrix} A & 0 \\ & I \end{pmatrix}.$$

This is invertible, therefore there exists an open subset W containing 0 such that $g_{|W}$ is a diffeomorphism to its image. The required diffeomorphism is $\varphi = g^{-1}$.   □

Remark. An immediate consequence of this theorem is the existence of a local left inverse of f, this is to say a map from an open subset of $\mathbf{R}^q$ containing 0 to an open subset of $\mathbf{R}^p$ such that $f_1 \circ f = Id_{\mathbf{R}^p}$: it suffices to take $f_1 = (\varphi^1, \ldots, \varphi^q)$, in other words to keep only the first q components of φ.

Example. If $p = 1$ the image of f is a curve, and this theorem tells us every sufficiently small piece of this curve can be transformed to a segment of a straight line by a diffeomorphism.

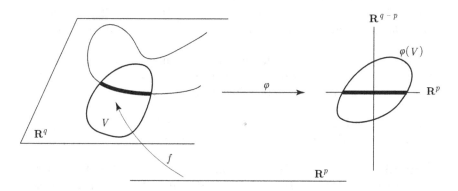

Figure 1.2: Straightening a curve

A dual result concerns maps whose differential is surjective. This time, we compose with a diffeomorphism on the domain side to obtain a linear map.

Theorem 1.18. *Suppose f is a C^1 map from an open subset U of $\mathbf{R}^p$ to $\mathbf{R}^q$. Suppose that $0 \in U$ and that the differential df_0 is surjective. Then there exists an open subset V in $\mathbf{R}^p$ containing 0 and a diffeomorphism ψ of W to its image such that $\psi(W) \subset U$ and*

$$f\big(\psi(x^1, \ldots, x^p)\big) = (x^1, \ldots, x^q).$$

PROOF. Necessarily $p \geqslant q$. This time by permuting the x^i coordinates, we can suppose that the matrix

$$B = \big(\partial_j f^i(0)\big)_{1 \leqslant i \leqslant q,\, 1 \leqslant j \leqslant q}$$

is invertible, and we define a map h from U to $\mathbf{R}^p$ by

$$h(x) = \big(f^1(x), \ldots, f^q(x), x^{q+1}, \ldots, x^p\big).$$

The Jacobian matrix of h at zero is of the form

$$\begin{pmatrix} B & * \\ 0 & I \end{pmatrix}$$

and therefore there exists an open subset W in $\mathbf{R}^p$ containing 0 such that $h_{|W}$ is a diffeomorphism to its image. If ψ is its inverse we find

$$f\big(\psi(x^1,\dots,x^p)\big) = (x^1,\dots,x^q).$$

Indeed, if x is of the form $h(u) = (f(u), u^{q+1},\dots,u^q)$, we have $\psi(x) = u$, and therefore $f\big(\psi(x)\big) = f(u)$. $\square$

Remark. In the same way as before, we deduce a theorem on the existence of *a local right inverse for* f, this is to say a smooth map f_1 from an open subset of $\mathbf{R}^q$ containing 0 to an open subset of $\mathbf{R}^p$ containing 0 such that $f \circ f_1 = Id_{\mathbf{R}^q}$: it suffices to take $f_1(x^1,\dots,x^q) = \psi(x^1,\dots,x^q,0,\dots,0)$.

Example. If $q = 1$, this result implies that, modulo a local diffeomorphism of the domain space, this is to say a "change of variables", a scalar function with a nonzero differential is expressible as a linear form.

Remark. There is a more general statement, that includes both of the two preceding results, the rank theorem, see Exercise 10.

Definitions 1.19. *A C^k immersion from an open subset $U \subset \mathbf{R}^p$ to $\mathbf{R}^q$ is C^k map from U to $\mathbf{R}^q$ with injective differential at each point. A C^k submersion is a C^k map from U in $\mathbf{R}^q$ with surjective differential at each point.*

With this notation, we note that $p \leqslant q$ if f is an immersion, and $p \geqslant q$ if f is a submersion. Of course a map that is both an immersion and submersion is a local diffeomorphism.

Remarks

a) If the differential at point a is injective (resp. surjective) there exists an open subset containing a for which this property subsists. To see this, we can use the preceding theorems by remarking that these properties are equivalent to the nonvanishing of a determinant of order p (resp. of order q) extracted from the Jacobian matrix. This condition is an "open" condition.

b) $\star\star$Theorems 1.17 and 1.18 naturally lead to notions of continuous immersion and submersion: a continuous map f from an open subset U in $\mathbf{R}^p$ to $\mathbf{R}^q$ is a C^0 immersion (resp. a C^0 submersion) if after composition on the range side (resp. domain side) with a suitable homeomorphism, it becomes a injective (resp. surjective) linear map.$\star\star$

> **HEREAFTER, UNLESS OTHERWISE MENTIONED,**
> **WE ASSUME ALL MAPS ARE SMOOTH**

1.5. Submanifolds

1.5.1. Basic Properties

Intuitively, a submanifold of dimension p in $\mathbf{R}^n$ is a union of small pieces each of which can each be straightened in a way to form open subsets of $\mathbf{R}^p$. One can convince oneself for a circle that two pieces are necessary (and sufficient!).

Definition 1.20. *A subset $M \subset \mathbf{R}^n$ is a p-dimensional submanifold of $\mathbf{R}^n$ if for all x in M, there exists open neighborhoods U and V of x and 0 in $\mathbf{R}^n$ respectively, and a diffeomorphism*

$$f : \ U \longrightarrow V \ \ such \ that \ \ f(U \cap M) = V \cap (\mathbf{R}^p \times \{0\}).$$

We then say that M is of codimension $n - p$ *in $\mathbf{R}^n$.*

This definition is better understood with Figure 1.3 kept in mind. We note that p is unique, in other words that M is not a manifold of dimension $p_1 \neq p$. The verification of this is left to the reader, unless they cannot wait until they read the next chapter, where this question will be elucidated in a more general setting.

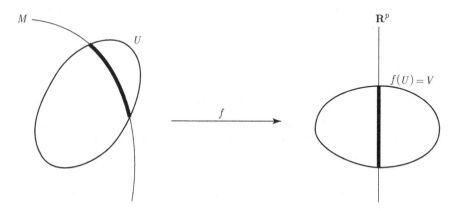

Figure 1.3: Submanifold

Remark. In this definition we can of course replace 0 and $\mathbf{R}^p \times \{0\}$ by any point and any affine subspace of dimension p.

In practice, a submanifold can be locally defined by systems of equations, or by parametric representations. Loosely speaking, if the number of real parameters is equal to the dimension of the ambient space, then a p-dimensional submanifold will be formed by $n - p$ equations. As in linear algebra we call $n - p$ the *codimension*. We now formulate this remark carefully.

Theorem 1.21. *Suppose M is a subset of $\mathbf{R}^n$. The following properties are equivalent:*

i) *M is a submanifold of dimension p of $\mathbf{R}^n$;*

ii) *for all a in M, there exists an open subset U of $\mathbf{R}^n$ containing a and a submersion $g : U \to \mathbf{R}^{n-p}$ such that $U \cap M = g^{-1}(0)$;*

iii) *for all a in M, there exists an open subset U in $\mathbf{R}^n$ containing a, an open subset Ω in $\mathbf{R}^p$ containing 0, and a map $h : \Omega \to \mathbf{R}^n$ which is simultaneously an* immersion *in $\mathbf{R}^n$ and a homeomorphism of Ω on $U \cap M$;*

iv) *for all a in M, there exists an open subset U in $\mathbf{R}^n$ containing a, an open subset V in $\mathbf{R}^p$ containing $(a^1, \ldots, a^p)$ and a smooth map G from V to $\mathbf{R}^{n-p}$ such that, after permuting the coordinates, $U \cap M$ equals the graph of G.*

PROOF. We first show that i) implies ii) and iii). Let f be the diffeomorphism defined on a neighborhood U of $a \in M$, as assured by i). Then f^{-1} is a diffeomorphism of $f(U)$ to U. Its restriction to $\mathbf{R}^p \times \{0\} \cap f(U)$ is an immersion from this open set of $\mathbf{R}^p$ to $\mathbf{R}^n$, and a homeomorphisms on $U \cap M$, giving iii).

To see that i) implies ii), we consider the components $(f^i)_{1 \leqslant i \leqslant n}$ of f. By hypothesis, their differentials are linearly independent at every point of U. Set

$$g = (f^{p+1}, \ldots, f^n).$$

We then have a submersion of U to $\mathbf{R}^{n-p}$ such that $M \cap U = g^{-1}(0)$.

Now suppose that iii) is true. By Theorem 1.17, we may replace Ω by a smaller open subset and find a diffeomorphism φ from an open subset U containing $h(0) = a$ to $\mathbf{R}^n$, such that

$$(\varphi \circ h)(x_1, \ldots, x_p) = (x_1, \ldots, x_p, 0, \ldots, 0).$$

Then

$$\varphi(U \cap M) = \varphi(h(\Omega)) = \varphi(U) \cap (\mathbf{R}^p \times \{0\}).$$

Implication ii) $\Rightarrow$ i) is proved in the same way using Theorem 1.18.

We now show the equivalence of ii) and iv). The fact that iv) implies ii) is elementary: if M is locally the graph of a function $G : V \to \mathbf{R}^{n-p}$ as in the statement, the components $G^1, \ldots, G^{n-p}$, of the map

$$g : \quad x \longmapsto \left(x^{i+p} - G^i(x^1, \ldots, x^p) \right)_{1 \leqslant i \leqslant n-p}$$

is a submersion which satisfies ii) upon restricting its domain of definition. Conversely, given such a submersion, we can suppose after permuting the coordinates as in the proof of Theorem 1.18 that the matrix

$$\left(\partial_{i+p} g^j(a) \right)_{1 \leqslant i, \, j \leqslant n-p}$$

is invertible. We therefore apply the inverse function theorem to the function

$$F : \quad x \longmapsto \left(x^1, \ldots, x^p, g^1(x), \ldots, g^{n-p}(x) \right).$$

This function has a local inverse of the form

$$F^{-1} : \quad x \longmapsto \left(x^1, \ldots, x^p, \gamma^1(x), \ldots, \gamma^{n-p}(x) \right),$$

and thus M is locally the graph of

$$G : \quad (x^1, \ldots, x^p) \longmapsto \left(\gamma^j(x^1, \ldots, x^p, 0 \ldots, 0) \right)_{1 \leqslant j \leqslant n-p}. \qquad \square$$

Remarks

a) The implication ii) $\Rightarrow$ iv) is known as the *implicit function theorem.*

b) Suppose g is a smooth map from an open subset U of $\mathbf{R}^n$ to $\mathbf{R}^p$, and let a in $\mathbf{R}^p$ be such that $g^{-1}(a) \neq \emptyset$. Then for $g^{-1}(a)$ to be a manifold, it suffices to know that the differential of g is surjective at every point of $g^{-1}(a)$. Indeed if this property is true at a point x, it is also true in a neighborhood of x (for example because the surjectivity is equivalent to the nonvanishing of a certain determinant of order p extracted from the Jacobian matrix). This argument is very common.

1.5.2. Examples: Spheres, Tori, and the Orthogonal Group

a) The sphere S^n defined by

$$S^n = \left\{ x = (x_0, \ldots, x_n) \in \mathbf{R}^{n+1} : \ x_0^2 + \cdots + x_n^2 - 1 = 0 \right\}$$

is a submanifold of dimension n (and class C^∞) in $\mathbf{R}^{n+1}$. The map $f : \mathbf{R}^{n+1} \to \mathbf{R}$ defined above is of course a submersion at every point of S^n, and its differential at x is $df_x = (2x_0, \ldots, 2x_n)$.

b) The *torus* T^n of dimension n defined by

$$T^n = \left\{ z = (z_1, \ldots, z_n) \in \mathbf{C}^n \ : \ |z_1|^2 = \cdots = |z_n|^2 = 1 \right\}$$

or

$$\left\{ x = (x_1, \ldots, x_{2n}) \in \mathbf{R}^{2n} \ : \ (x_1^2 + x_2^2 - 1, \ldots, x_{2n-1}^2 + x_{2n}^2 - 1) = 0 \right\}$$

is a submanifold of $\mathbf{R}^{2n} \simeq \mathbf{C}^n$. Here we can also apply the criterion iii) by introducing the map

$$h : (t_1, \ldots, t_n) \longmapsto (e^{it_1}, \ldots, e^{it_n}) \quad \text{from } \mathbf{R}^n \text{ to } \mathbf{C}^n.$$

The torus of revolution (cf. Exercise 13) is easier to see geometrically, but is a little harder to manipulate.

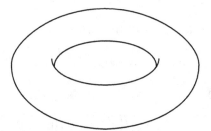

Figure 1.4: Torus of revolution

c) The orthogonal group

$$O(n) = \left\{ A \in M_n(\mathbf{R}) \ : \ {}^t\!A A = Id \right\}$$

is a submanifold of dimension $\frac{n(n-1)}{2}$ on $M_n(\mathbf{R}) \simeq \mathbf{R}^{n^2}$. We define

$$f : M_n(\mathbf{R}) \longrightarrow \text{Sym}(n) \quad \text{by} \quad f(A) = {}^t\!A A - Id,$$

where $\text{Sym}(n)$ is the vector space of $n \times n$ symmetric matrices. Then $O(n) = f^{-1}(0)$ and f is a submersion at every point of $O(n)$. We note then that

$$df_A \cdot H = {}^t\!A H + {}^t\!H A.$$

In particular if S is symmetric and A orthogonal we have

$$df_A \left(\frac{AS}{2} \right) = S.$$

We will see a proof using iii) in Section 1.6.

d) The subset of $\mathbf{R}^3$ defined by the equation

$$x^2 + y^2 - z^2 = 0 \quad \text{(cone of revolution)}$$

is not a submanifold. *A priori*, it seems criteria ii) doesn't work, but this is not the reason! On the contrary, note the straight line with equation $x - y = 0$ is also a solution of $x^3 - y^3 = 0$. The most convenient route is to show that iii) is not true: see the figure below.

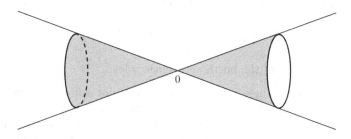

Figure 1.5: A cone with its vertex removed is no longer connected

1.5.3. Parametrizations

Definitions 1.22

a) *A* parametrization *of a p-dimensional submanifold M of $\mathbf{R}^n$ is a map from an open subset Ω in $\mathbf{R}^p$ to $\mathbf{R}^n$ that is simultaneously an immersion in $\mathbf{R}^n$ and a homeomorphism of Ω to an open subset of M.*

b) *A* local parametrization *is a map from Ω to $\mathbf{R}^n$ that induces a parametrization in a neighborhood of every point of Ω.*

By Theorem 1.21 every submanifold can be covered by open subsets that are images of parametrizations.

Examples

a) The map $t \mapsto (\cos t, \sin t)$ from $\mathbf{R}$ to $\mathbf{R}^2$ is a local parametrization of the circle $x^2 + y^2 = 1$. Similarly, the map

$$(u, v) \longmapsto (\cos u, \sin u, \cos v, \sin v)$$

from $\mathbf{R}^2$ to $\mathbf{R}^4$ is a local parametrization of the torus T^2.

b) The image of the map g from $\mathbf{R}^2$ to $\mathbf{R}^3$ defined by

$$g(u, v) = (\sin u \cos v, \sin u \sin v, \cos u)$$

is the sphere S^2. However, g is not a local parametrization unless we remove the lines $u \equiv \frac{\pi}{2} \bmod \pi$. In this case, we obtain a local parametrization of the sphere with its two poles removed (recall that u and v are spherical coordinates – latitude and longitude – on S^2). To show that S^2 is a submanifold using this point of view, one must add other parametrizations to g, for example

$$(x, y) \longmapsto \left(x, y, \pm\sqrt{1 - x^2 - y^2} \right)$$

in a neighborhood of the north and south poles.

It is clear that a single parametrization will not suffice, because S^2 is a compact space, and cannot be homeomorphic to an open subset of $\mathbf{R}^2$. This need to recourse to many parametrizations for such a simple submanifold is one of the reasons justifying the introduction of manifolds.

Note. The lack of symmetry between criteria ii) and iii) in Theorem 1.21 is not accidental. *It is not true that the image of an open subset of $\mathbf{R}^p$ under an immersion is always a submanifold.* One reason is of course because an immersion need not be injective (there can be double points). But even then this is not true for injective immersions.

Counterexample. The map

$$g: \ t \longmapsto \left(\cos t, \sin t, \cos \sqrt{2}\,t, \sin \sqrt{2}\,t \right)$$

is an immersion from $\mathbf{R}$ to $\mathbf{R}^4$, but $g(\mathbf{R})$ is not a submanifold of $\mathbf{R}^4$.

On the one hand, for every irrational number α, the set $\mathbf{Z} + \alpha\mathbf{Z}$ is dense in $\mathbf{R}$ (compare to Theorem 4.40), which implies that $g(\mathbf{R})$ is dense in the torus T^2. On the other hand, it follows from the definition of submanifold that this set is a *locally closed* subset (this is to say open subsets of their closure) of the ambient space.

We will see more of the details on this question in the next chapter.

1.5.4. Tangent Vectors, Tangent Space

Definition 1.23. *Suppose $A \subset \mathbf{R}^p$ and a is an element of A. We say that a vector v is tangent to A at a if there exists a differentiable map $c: (-\epsilon, \epsilon) \to \mathbf{R}^p$ such that $c((-\epsilon, \epsilon)) \subset A$, $c(0) = a$ and $c'(0) = v$.*

Note. This definition, in contrast to one which consists in taking right-sided derivatives of maps defined on $[0, \epsilon)$, is very restrictive, as the following example shows.

Example. The only tangent vector to the origin to the curve C which is the image of $\mathbf{R}$ under the map $t \mapsto (t^2, t^3)$ is the zero vector: if $u \mapsto \big(c_1(u), c_2(u)\big)$ has image on C, then $c_1'(0) = 0$, because $c_1(u)$ is always positive. Since $c_2 = (c_1)^{3/2}$, we also have $c_2'(0) = 0$. In contrast, the map $u \mapsto (u, u^{3/2})$ from $[0, 1)$ to $\mathbf{R}^2$ has image contained in C and a nonzero right derivative at the origin.

In particular, this curve is not a submanifold because of the following property.

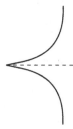

Figure 1.6: A curve with a cusp is not a submanifold

Proposition 1.24. *The tangent vectors at a point of a submanifold of dimension p in $\mathbf{R}^n$ form a vector space of dimension p.*

PROOF. Suppose a is a point of a submanifold M, and f is a diffeomorphism defined on an open subset U containing a and such that $f(U \cap M) = f(U) \cap (\mathbf{R}^p \times \{0\})$. We can suppose that $f(a) = 0$. Now, if v is tangent at a, the chain rule applies to $f \circ c$ and shows that $df(a) \cdot v \in \mathbf{R}^p \times \{0\}$.

Conversely, if $w \in \mathbf{R}^p \times \{0\}$, and choosing ϵ in a way that

$$\forall t, \ |t| < \epsilon, \ tw \in f(U)$$

we see that the curve $t \mapsto f^{-1}(tw)$ defines a tangent vector to M at a, namely $df_0^{-1} \cdot w$. Put differently, the set of tangent vectors is identified with the *image* under the linear map df_0^{-1} of the vector subspace $\mathbf{R}^p \times \{0\}$ of $\mathbf{R}^n$. □

Definition 1.25. The tangent space *of a submanifold M of $\mathbf{R}^n$ at a point a,* denoted $T_a M$, *is the set of points m in $\mathbf{R}^n$ such that the vector $\overrightarrow{am}$ is tangent to M at a.*

By the preceding definition, the tangent space at a is a affine p-dimensional subspace of the ambient space. It becomes a vector space with origin a from which we regard the tails of tangent vectors. An important question, which will be discussed below, is the position of a submanifold with respect to the tangent space at a point. However once we pass to manifolds, the vector point of view will be the pertinent one.

Arguments analogous to those found in the proof of Proposition 1.24 permit us to write the tangent space of a submanifold given by a submersion or parametrization. If g is a submersion defined on an open set U of a and such that $U \cap M = g^{-1}\big(g(a)\big)$, then the tangent space at a is the *kernel* of the linear map dg_a. Consider a curve $t \mapsto c(t)$ defining a tangent vector, and note that we have $g\big(c(t)\big) = g(a)$ and therefore $dg_a \cdot v = 0$. Thus $\operatorname{Ker} dg_a$ is contained in $T_a M$, and these two spaces are equal as they have the same dimension; the $n - p$ components of g give a system of $n - p$ linear equations in n unknowns by differentiation which are of maximum rank, and whose solutions are tangent vectors.

Similarly, if M is defined in a neighborhood of a by a parametrization (such that $g(0) = a$ for example), the tangent space at a is the *image* in $\mathbf{R}^p$ of the linear map dg_0.

Example: surfaces in $\mathbf{R}^3$

We explain all of this for 2-dimensional submanifolds of $\mathbf{R}^3$. If such a submanifold S (as in "surface"!) is given in a neighborhood of a point (a, b, c) by the equation $f = 0$ (where we suppose that f is a submersion), then the equation of the tangent plane at (a, b, c) can be written

$$(x - a)\partial_1 f(a, b, c) + (y - b)\partial_2 f(a, b, c) + (z - c)\partial_3 f(a, b, c) = 0.$$

If S is given in a neighborhood of the same point by a parametrization

$$(u, v) \longmapsto \big(g(u, v), h(u, v), k(u, v)\big),$$

with for example $(a, b, c) = \big(g(0, 0), h(0, 0), k(0, 0)\big)$, this same tangent plane will be given by the parametric representation

$$(u, v) \longmapsto \begin{pmatrix} a + \partial_1 g(0,0)u + \partial_2 g(0,0)v \\ b + \partial_1 h(0,0)u + \partial_2 h(0,0)v \\ c + \partial_1 k(0,0)u + \partial_2 k(0,0)v \end{pmatrix}.$$

To know the position of a surface with respect to a tangent plane, we can always return to the case where S has tangent plane $z = 0$ at 0. The surface is always the graph of a function $(x, y) \mapsto G(x, y)$ in a neighborhood of 0, whose differential at 0 is zero. By composition with a diffeomorphism of the domain space, we can "often" return to the case where

G is a non-degenerate quadratic form (see the Morse lemma, Exercise 11). Then in this neighborhood of 0, either S is on the same side of its tangent plane at 0 as this quadratic form – or, what amounts to the same thing, the quadratic form defined by the second derivatives of G at 0 – is of type $++$ or $--$, or S *crosses its tangent plane* if this quadratic form is of type $\pm, \mp$ (see Figure 1.7 and Exercise 22).

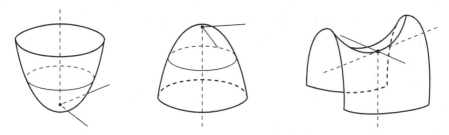

Figure 1.7: Minimum, maximum, saddle point

1.6. One-Parameter Subgroups of the Linear Group

Recall that for a field K, we denote the multiplicative group of invertible matrices from K^n to itself by $Gl(n, K)$. The cases that interest us here are $K = \mathbf{R}$ and $K = \mathbf{C}$. If K^n is equipped with a norm, recall that there is an associated norm on $\operatorname{End}(K^n)$ defined by

$$\forall A \in \operatorname{End}(K^n), \quad \|A\| = \sup_{\|x\| \leqslant 1} \|Ax\|.$$

Definition 1.26. *The exponential of an endomorphism $A \in \operatorname{End}(K^n)$ ($K = \mathbf{R}$ or $\mathbf{C}$) is the endomorphism defined by*

$$\exp A = \sum_{k=0}^{\infty} \frac{A^k}{k!}.$$

It is clear that this series converges: appealing to the properties of a norm of a linear map,

$$\left\| \frac{A^k}{k!} \right\| \leqslant \frac{\|A\|^k}{k!},$$

and we have a *norm convergent* series in a finite-dimensional normed vector space. We also see that

$$\| \exp A \| \leqslant e^{\|A\|}.$$

Moreover, we have the following properties:

Lemma 1.27

 i) exp *is continuous.*

 ii) *If A and B commute,*

$$\exp(A + B) = (\exp A)(\exp B).$$

   In particular, $\exp A$ *is invertible with inverse* $\exp(-A)$.

 iii) *If P is invertible,* $\exp(P^{-1}AP) = P^{-1}(\exp A)P.$

 iv) $\det(\exp A) = e^{\mathrm{tr}(A)}.$

 v) $\exp {}^{t}A = {}^{t}\exp A,\ \exp \overline{A} = \overline{\exp A}.$

PROOF

 i) is immediate: by the inequality above, we have a uniformly convergent series on every compact subset.

 ii) Since both sides are norm convergent series, the proof is the same as the classical proof of the identity $e^{z+z'} = e^{z}e^{z'}$ for z and z' in $\mathbf{C}$. We use the identity

$$(A + B)^{k} = \sum_{i=0}^{k} \binom{k}{i} A^{i}B^{k-i}$$

which is true if A and B commute.

 iii) is immediate from i) after passing to the limit, as for every integer k, $P^{-1}A^{k}P = (P^{-1}AP)^{k}.$

 iv) The property is evident for diagonal matrices, and by iii) for diagonalizable matrices (and even for real matrices that are $\mathbf{C}$-diagonalizable), which form a dense subset of $\mathrm{End}(K^{n})$. Therefore the property holds for all matrices by continuity of exp and det (see Exercise 24 for another proof).

 v) is clear.   □

It is clear from the definition that the exponential is differentiable at 0, and the differential is the identity map. To be able to apply the inverse function theorem, one must verify that exp is C^{1}. The classical criterion of continuity and differentiability applies easily, but we can say much more by less pedestrian means, so long as we are willing to do a little complex analysis.

Theorem 1.28. *The exponential is smooth.*

PROOF. For every matrix whose spectrum is contained in the open disk $D(r) = \{z : |z| < r\}$ of the complex plane, we have

$$\exp A = \frac{1}{2i\pi} \int_{C(0,r)} (zI - A)^{-1}e^{z}\, dz.$$

Indeed, we provisionally call the right hand side $f(A)$. As $f(P^{-1}AP) = P^{-1}f(A)P$, it suffices to use the same density argument from the previous lemma, and to show that $f(D) = \exp D$ for diagonal matrices D, in which case we return to Cauchy's integral formula.

The integrand of the right hand side is smooth (in fact real analytic) on the (open!) set of matrices whose spectrum is contained in $D(r)$, and we conclude by applying the theorem on the differentiation of integrals depending on parameters.[1]   □

Remark. We can give an elementary but pedestrian proof of this result, by using the usual theorem on the term-by-term differentiation of series of functions.

Corollary 1.29. *There exists an open subset U of* $\mathrm{End}(\mathbf{R}^n)$ *containing 0 such that the exponential is a diffeomorphism from U to its image.*

We now deduce the following fundamental result.

Theorem 1.30. *For every* continuous *group homomorphism f of the additive group $(\mathbf{R}, +)$ to $Gl(n, \mathbf{R})$, there exists a unique endomorphism A such that*
$$\forall t \in \mathbf{R}, \quad f(t) = \exp tA.$$

Remark. Conversely, by Lemma 1.27, $t \mapsto \exp tA$ is a continuous morphism of $(\mathbf{R}, +)$ to $Gl(n, \mathbf{R})$.

PROOF. We can suppose f is not constant (if it is take $A = 0$!). Let U be an open ball with center 0 in $\mathrm{End}(\mathbf{R}^n)$ such that $\exp$ restricts to the ball of double the radius (which we denote $2U$) as a diffeomorphism. By the continuity of f, there exists an interval I containing 0 such that $f(I) \subset \exp U$. Since f is nonconstant there exists $c \in I$ such that $f(c) = A \in \exp U$, with $A \neq I$. By the choice of U, there exists nonzero $B \in U$, such that $\exp B = A$. Now it follows $\exp B/2 = f(c/2)$.

Furthermore, given the choice of U, we have
$$f\left(\frac{c}{2}\right) = \exp B', \quad \text{with } B' \in U.$$
Then
$$\exp 2B' = f(c) = \exp B,$$
while the endomorphisms $2B'$ and B are both in the open subset $2U$, on which $\exp$ is a diffeomorphism, and in particular an injection. Therefore $B' = B/2$.

---

1. This elegant argument was communicated to me by Max Karoubi on the terrace of a Parisian café.

The same reasoning proves that for every integer p,

$$\exp \frac{B}{2^p} = f\left(\frac{c}{2^p}\right).$$

Therefore by the algebraic properties of f, we have

$$\exp \frac{kB}{2^p} = f\left(\frac{kc}{2^p}\right)$$

for all integers k and p. But the real numbers of the form $\frac{k}{2^p}$ are dense in $\mathbf{R}$, and using continuity once more we see that

$$\forall t \in \mathbf{R}, \quad \exp tB = f(tc). \qquad \square$$

An analogous argument allows us to show that the exponential furnishes a parametrization of certain subgroups of $Gl(n, K)$.

Definitions 1.31

a) *The* special linear *group, denoted $Sl(n, K)$ is the subgroup of $Gl(n, K)$ of endomorphisms of determinant 1.*

b) *The* orthogonal *group, denoted $O(n)$, is the subgroup of $Gl(n, \mathbf{R})$ of endomorphisms such that ${}^t A A = I$.*

c) *The* special orthogonal *group, denoted $SO(n)$, is the subgroup $Sl(n, \mathbf{R}) \cap O(n)$.*

We have seen that $O(n)$ is a submanifold of $\mathrm{End}(\mathbf{R}^n)$. In the same way, it is an immediate result from the end of Section 1.2.2 that $Sl(n, \mathbf{R})$ is also a submanifold. On the other hand the formulas of Lemma 1.27 show that the exponential of a *trace-free* endomorphism is in $Sl(n, \mathbf{R})$, and that the exponential of an antisymmetric endomorphism is in $O(n)$.

Proposition 1.32. *There exists an open subset V containing 0 in the vector space of trace-free endomorphisms (resp. in the vector space of antisymmetric endomorphisms) such that $\exp|_U$ is a parametrization of $Sl(n, \mathbf{R})$ (resp. of $O(n)$).*

PROOF. Let us look at the second case (the first one is straightforward). To start we choose an open subset U containing 0 in $\mathrm{End}(\mathbf{R}^n)$ such that $\exp : U \mapsto \exp(U)$ is a diffeomorphism, and is stable under the maps $A \mapsto -A$ and $A \mapsto {}^t A$.

Let $B \in \exp(U) \cap O(n)$. There exists a unique $A \in U$ such that $B = \exp A$. Therefore we have

$$B^{-1} = \exp(-A) \quad \text{and} \quad {}^t B = \exp {}^t A,$$

therefore $-A = {}^t A$ by the choice of U. It then suffices to take V as the intersection of U with the space of antisymmetric endomorphisms. $\qquad \square$

1.7. Critical Points

Definitions 1.33

a) *If f is a smooth map from an open subset $U \subset \mathbf{R}^m$ to $\mathbf{R}^n$, a point $x \in U$ is called* critical *if*

$$\operatorname{rank}(d_x f) < n.$$

b) *A point is called* regular *if it is not critical.*

If $m = n$, the regular points are those where the differential is invertible, *i.e.*, those where the inverse function theorem applies.

If the range is one dimensional, a point x is critical if and only if $d_x f = 0$. Here is an important example.

Proposition 1.34. *Let f be a C^1 map from an open subset U in $\mathbf{R}^m$ to $\mathbf{R}$. If f admits a local maximum (resp. minimum) at a, i.e., if there exists a further open subset V containing a such that*

$$\forall x \in V, \quad f(x) \leqslant f(a) \quad (\text{resp. } f(x) \geqslant f(a)),$$

then a is a critical point.

PROOF. Suppose by way of contradiction that v is a vector such that $d_a f \cdot v \neq 0$. If t is a sufficiently small real number, $f(a + tv) - f(a) = d_a f \cdot tv + o(tv)$ is nonzero and of the same sign as $d_a f \cdot tv$. By choosing two such t of opposite sign results in a contradiction.   □

To take full advantage of this result, it is best to be assured of the existence of extrema, for example by using compactness.

This is why this property takes its interest when applied not only to open subsets of $\mathbf{R}^m$ but to (possibly compact) submanifolds (and also to manifolds, as we will see later).

Proposition 1.35. *Let S be a submanifold of $\mathbf{R}^m$, and f a real-valued function defined on an open subset containing S. If the restriction of f to S admits a local maximum or minimum at a point a of S, then $d_a f$ annihilates the tangent space at a.*

PROOF. Let U be an open subset containing a such that there exists a diffeomorphism g on an open subset $g(U)$ containing 0 such that $g(U \cap S) = g(U) \cap (\mathbf{R}^p \times \{0\})$ (see Definition 1.20; note here p is the dimension of S). By Proposition 1.34, the differential of $f \circ g^{-1}$ at 0 is annihilated on $\mathbf{R}^p$. The statement now follows by the definition of the tangent space.   □

Remark. In practice, we principally use this result when S is given in a neighborhood of a by a submersion. If $h_1, \ldots, h_{n-p}$ are the coordinates of such a submersion, the proposition above says precisely that df_a is a linear combination of the $(dh_i)_a$. The coefficients of this linear combination are called the *Lagrange multipliers*. Here is a sample application.

Theorem 1.36. *Every self-adjoint endomorphism of an inner product space is diagonalizable and admits an orthogonal basis of eigenvectors.*

PROOF. We proceed by induction on the dimension of E. The property is clear in dimension 1. Let u be such an endomorphism. By definition, for every x and y in E, $\langle u(x), y \rangle = \langle x, u(y) \rangle$. The unit sphere S is compact, so the function $f(x) = \langle u(x), x \rangle$ realizes its maximum at a point a in S. Moreover, the function f is smooth, and its differential is

$$d_a f \cdot h = \langle u(h), a \rangle + \langle u(a), h \rangle = 2 \langle u(a), h \rangle,$$

so the tangent space of S at a is orthogonal to $u(a)$. By elementary linear algebra, $u(a)$ is co-linear to a, which is to say a is an eigenvector. As u is self-adjoint, it fixes the subspace $\mathbf{R}a^{\perp}$, which gives the inductive step. □

Remarks

a) We have the same result in the hermitian case.

b) This type of argument is important in functional analysis. See for example [Brezis 83, VI.4]. A typical example is the spectral decomposition for the Laplacian. If D is a compact domain in $\mathbf{R}^n$ with smooth boundary, the spectrum of the Dirichlet problem is the set of λ for which there exists a function $f \in C^2(D)$ not identically zero such that

$$\Delta f = -\lambda f \quad \text{with} \quad f_{|\partial D} = 0.$$

One shows that there is a sequence $\lambda_1 \leqslant \lambda_2 \leqslant \cdots \leqslant \lambda_p \leqslant \cdots$, with $\lim_{p \to +\infty} \lambda_p = +\infty$, and that $L^2(D)$ admits an orthonormal basis of eigenfunctions. We note in passing that Δ is "formally[2] self-adjoint" for the inner product, which is to say that if f and g are C^2 and vanish on the boundary,

$$\int_D f \Delta g \, dx^1 \ldots dx^n = \int_D g \Delta f \, dx^1 \ldots dx^n.$$

See Exercise 14 in Chapter 6, where one also shows, with the same hypothesis on f that

$$\int_D f \Delta f \, dx^1 \ldots dx^n = \int_D |df_x|^2 \, dx^1 \ldots dx^n.$$

---

2. "Formally" because the operator is not defined on the whole space.

This spectral decomposition start index check theorem is also a perfect analogue of Theorem 1.36, and its proof rests, modulo adequate functional analysis, on a study of the extrema of the functional $\int_D |df_x|^2 \, dx^1 \ldots dx^n$ on the unit sphere of $L^2(D)$. See for example [Attouch-Buttazzo-Michaille 06, Chapter 7].

This variational method also gives a way to control the volume of parallelepipeds in Euclidean space. Recall that the Lebesgue measure is normalized by giving the value 1 to a parallelepiped (and therefore to all parallelepipeds) defined by an orthonormal basis. If

$$P = \left\{ \sum_{i=1}^{n} t_i a_i, \ 0 \leqslant t_i \leqslant 1 \right\}$$

is the parallelepiped defined by n vectors $a_1, \ldots, a_n$,

$$\mathrm{vol}(P) = |\det(a_1, \ldots, a_n)|,$$

with the determinant being taken with respect to an orthonormal basis. In two dimensions, we know that $\mathrm{vol}(P) = \|a_1\| \|a_2\| |\sin \alpha|$, where α denotes the angle between vectors a_1 and a_2, and it is clear that $\mathrm{vol}(P) \leqslant \|a_1\| \|a_2\|$, and that equality is realized if and only if the vectors are orthogonal. An analogous result is true in all dimensions.

Theorem 1.37 (Hadamard inequality). *Suppose E is a Euclidean vector space of dimension n, and P is the parallelepiped defined by n vectors $a_1, \ldots, a_n$. Then*

$$\mathrm{vol}(P) \leqslant \prod_{i=1}^{n} \|a_i\|,$$

and equality is realized if and only if the a_i are mutually orthogonal.

PROOF. It suffices to prove the result when the a_i are of norm 1. Suppose $(v_1, \ldots, v_n)$ is an n-tuple of vectors for which the continuous map

$$(a_1, \ldots, a_n) \longmapsto |\det(a_1, \ldots, a_n)|$$

attains it maximum. In a neighborhood of this n-tuple, this map is equal to the determinant (or its negative).

By Proposition 1.35 applied to the partial functions

$$a_i \longmapsto \det(a_1, \ldots, a_n),$$

for all i, the determinant $\det(v_1, \ldots, v_{i-1}, h, v_{i+1}, \ldots, v_n)$ is zero once $\langle v_i, h \rangle = 0$, i.e.,

$$(v_i)^{\perp} \subset \mathrm{span}(v_1, \ldots, v_{i-1}, v_{i+1}, \ldots, v_n).$$

Since these two subspaces are of codimension 1, they are therefore equal, and v_i is orthogonal to each v_k for $k \neq i$. The system $(v_1, \ldots, v_n)$ is orthonormal, and the maximum is equal to 1. Conversely, if equality is realized, the volume is 1 which is maximal and the system is orthonormal. □

1.8. Critical Values

Phenomena like "Peano's curve" (a continuous surjection of $[0,1]$ onto $[0,1] \times [0,1]$) are not produced with differentiable functions. Here we make this precise. We denote Lebesgue measure by μ.

Proposition 1.38. *Suppose $M \subset \mathbf{R}^n$ is a submanifold of dimension $p < n$. Then M is of measure zero in $\mathbf{R}^n$.*

It suffices to show that every $x \in M$ is contained in an open subset U such that $U \cap M$ is of measure zero. We directly apply the definition of submanifolds, and we consider an open subset U containing x and a diffeomorphism $f : U \mapsto V$ such that

$$f(U \cap M) = V \cap (\mathbf{R}^p \times \{0\}).$$

The right hand side is clearly of measure zero, and the argument is reduced to the following property:

Lemma 1.39. *Suppose $U \subset \mathbf{R}^n$ is an open subset and $f : U \mapsto \mathbf{R}^n$ is a C^1 map. The image under f of a set of measure zero is of measure zero.*

PROOF. Suppose E is such a set. It suffices to show that for every closed ball $B \subset U$, $f(E \cap B)$ is of measure zero. If

$$K = \sup_{x \in B} \|f'(x)\|,$$

the mean value theorem shows that f is K-Lipschitz on B, therefore f transforms every cube of measure δ to a set of smaller measure $K^n \delta$. Also, if $C \supset E \cap B$ is a union of cubes such that $\mu(C) < \epsilon$, we have

$$\mu\big(f(E \cap B)\big) \leqslant \mu\big(f(C)\big) < K^n \epsilon. \qquad \square$$

In fact we can say much more. We begin with another definition.

Definition 1.40. *If f is a smooth map of an open subset $U \subset \mathbf{R}^m$ to $\mathbf{R}^n$, a point $y \in \mathbf{R}^n$ is a* critical value *if there exists a critical point x such that $y = f(x)$. A point that is not a critical value is called a* regular value.

Remark. By the definition, a point that is not in the image of f is a regular value. This convention is perfectly coherent with all of the results concerning regular values.

Theorem 1.41 (Sard's theorem). *The set of critical values of a smooth map from an open subset $U \subset \mathbf{R}^m$ to $\mathbf{R}^n$ is of measure zero.*

We will only give the proof in the easiest case, when $m \leqslant n$. It suffices to suppose that f is of class C^1. We first note that the result is clear if $m < n$. Every point is critical, and applying Lemma 1.39 to the map $f_1 : U \times \mathbf{R}^{n-m} \to \mathbf{R}^n$ defined by $f_1(x,y) = f(x)$ we see that $f(U)$ is of measure zero.

In the case $m = n$, the proof rests on the following lemma, which states that a function of class C^1 is "uniformly differentiable" on every compact subset. To lighten the notation, we denote the differential of f at x by $f'(x)$ (instead of df_x).

Lemma 1.42. *If f is of class C^1 on U, then for every connected compact $K \subset U$, there exists a real number $\alpha > 0$ and a function $\lambda : [0, \alpha] \mapsto \mathbf{R}^+$ such that*

$$\|f(y) - f(x) - f'(x) \cdot (x - y)\| < \lambda(\|x - y\|)\|x - y\|, \quad \text{with} \quad \lim_{t \to 0} \lambda(t) = 0,$$

for all points $x, y \in K$ such that $\|x - y\| < \alpha$.

PROOF. We have

$$f(y) - f(x) = \int_0^1 f'\big(x + t(y - x)\big) \cdot (y - x)\, dt,$$

where

$$\left\| \int_0^1 f'\big(x + t(y - x)\big) \cdot (y - x) - f'(x) \cdot (y - x)\, dt \right\|$$

$$\leqslant \int_0^1 \left\| f'\big(x + t(y - x)\big) \cdot (y - x) - f'(x) \cdot (y - x) \right\| dt.$$

The result is therefore a consequence of the uniform continuity of f' on K. □

PROOF OF THEOREM 1.41. Let C be the set of critical points. It suffices to show that $f(C \cap A)$ is of measure zero for every cube A. We first note that if $x \in C$, the vector space $\operatorname{Im} f'(x)$ is contained in a hyperplane $H \subset \mathbf{R}^n$. Let $r > 0$, and y be such that $\|y - x\| < r$.

By the lemma, the distance from $f(y)$ to the affine hyperplane H' parallel to H and containing $f(x)$ is less than $\lambda(r)$. On the other hand, if $K = \sup_{x \in B} \|f'(x)\|$, we have $\|f(y) - f(x)\| < Kr$. Also, $f(B(x,r))$ is contained in a cylinder of base $H' \cap B(f(x), Kr)$ and height $2r\lambda(r)$. Moreover,

$$\mu\big(f\big(B(x,r)\big)\big) \leqslant 2^n K^{n-1} r^n \lambda(r).$$

Now, the cube A is included in at most $(ak)^n$ cubes with side $\frac{1}{k}$, where we have denoted the side length of A by a.

Each cube that meets C can be enclosed in a ball $B\big(x, 2\frac{\sqrt{n}}{k}\big)$, where $x \in C$. Finally if $\omega_n r^n$ denotes the volume of a ball of radius r, we find that

$$\mu\big(f(A \cap C)\big) \leqslant (ak)^n 2^n K^{n-1} \omega_n \left(2\frac{\sqrt{n}}{k}\right)^n \lambda\left(2\frac{\sqrt{n}}{k}\right) \leqslant C(n,a,K)\lambda\left(2\frac{\sqrt{n}}{k}\right),$$

from which the conclusion follows upon allowing k to tend to infinity. $\square$

For the (more difficult) proof in the case where $m > n$, see for example [Hirsch 76] or [Golubitsky-Guillemin 73].

1.9. Differential Calculus in Infinite Dimensions

The notion of differential and the chain rule extends word for word to the case of normed vector spaces on the condition that we require the differential L to be a *continuous* linear map. The inverse function theorem extends as well to Banach spaces (we must of course suppose that the differential has a continuous inverse[3]). Indeed the proof rests essentially on the fixed point theorem for contraction mappings, valid for all *complete* metric spaces.

We will see this generalization is not straightforward and will shed light on the situation in finite dimensions. We start with the following property, which is the most elementary variant of what analysts call the "ω lemma".

Proposition 1.43. *Let I be a compact interval, U an open subset of $\mathbf{R}^n$, and $f : U \to \mathbf{R}^n$ a C^1 function. Then the set of continuous functions $g : I \to U$ is an open subset of $C^0(I, \mathbf{R}^n)$ (equipped with the uniform norm) on which the map $g \mapsto f \circ g$ is continuously differentiable.*

---

3.   We can invoke Banach's theorem that ensures that the inverse of a bijective continuous linear map from one Banach space to another is continuous. In fact, in all of the examples we are confronted with this continuity is very easily verified.

PROOF. The first part is classical: suppose g_0 is a function such that $g_0(I) \subset U$. As $g_0(I)$ is compact, its distance to the complement of U is a strictly positive number α, and if $\|g - g_0\| < \alpha$, we have $g(I) \subset U$.

For the second part, we apply Lemma 1.42 to a suitable compact neighborhood of $g_0(I)$. We have, in the notation of this lemma

$$\left\| f\big(g(t)\big) - f\big(g_0(t)\big) - df_{g_0(t)} \cdot \big(g(t) - g_0(t)\big) \right\| \leqslant \lambda(\|g(t) - g_0(t)\|)\, \|g(t) - g_0(t)\|,$$

which gives precisely the differentiability of $g \mapsto f \circ g$, with the domain $C^0(I, \mathbf{R}^n)$ and range $C^0(I, \mathbf{R})$ equipped with the uniform norm. The differential of g_0 is clearly the continuous linear map which assigns to $h \in C^0(I, \mathbf{R}^n)$ the real-valued function

$$t \longmapsto f'\big(g_0(t)\big) \cdot h(t).$$

This map depends continuously on g_0 by the uniform continuity of $x \mapsto f'(x)$ on compact subsets. $\qquad\qquad\qquad\qquad\qquad\qquad\qquad\qquad\qquad\qquad\qquad\square$

It was remarked by Joel Robbin during the 1970s (see [Robbin 68]) that one can show existence results for differential equations by using the Banach inverse function theorem.

We consider the equation

$$x' = f(t, x),$$

where f is (say) a function of class C^1 on $I \times U$, where I is an interval and U is an open subset of $\mathbf{R}^n$.

If we impose the initial condition $x(t_0) = x_0$ on the unknown function $x : J \to U$, (where J is a subinterval of I) this equation is equivalent to an integral equation

$$x(t) = x_0 + \int_{t_0}^{t} f\big(u, x(u)\big)\, du.$$

The traditional proof of the existence and uniqueness theorem consists of showing the map defined on the left hand side is a contraction on suitable function space and admits a fixed point (for this approach, see for example [Lang 86, Chapter 18], or [Hirsch-Smale-Devaney 03, Chapter 17]).

It is possible to use the inverse function theorem by proceeding as follows. If $[t_0 - \epsilon, t_0 + \epsilon]$ is an (unknown) interval of existence, we pose

$$t = t_0 + \epsilon u \quad \text{and} \quad y(u) = x(t_0 + \epsilon u) - x_0.$$

The differential equation is therefore equivalent to

$$y'(u) = \epsilon f\big(t_0 + \epsilon u, x_0 + y(u)\big).$$

Let E be the Banach space of C^1 functions on $[-1, 1]$, with values in $\mathbf{R}^n$ and zero at 0, equipped with the supremum norm of the derivative and $F = C^0([-1, 1], \mathbf{R}^n)$ equipped with the uniform norm. We define a map φ from $E \times \mathbf{R} \times \mathbf{R}$ to $F \times \mathbf{R} \times \mathbf{R}$ by setting

$$\varphi(y, x_0, \epsilon) = \left(y'(u) - \epsilon f\left(t_0 + \epsilon u, x_0 + y(u)\right), x_0, \epsilon\right).$$

We see that φ is of class C^1. It is now a consequence of Proposition 1.43 and an immediate generalization of Theorem 1.6 that assures that function defined on a product space is C^1 if each factor is separately C^1.

The differential of φ at $(0, x_0, 0)$ is the linear map

$$(Y, X, \mathcal{E}) \longmapsto (Y' - \mathcal{E}f(t_0, x_0), X, \mathcal{E}) \quad \text{from } E \times \mathbf{R} \times \mathbf{R} \text{ to } F \times \mathbf{R} \times \mathbf{R}$$

whose inverse is evidently

$$(Z, X, \mathcal{E}) \longmapsto \left(\int_0^u (\mathcal{E}f(t_0, x_0) + Z(v))\, dv, X, \mathcal{E}\right).$$

The inverse function theorem directly gives the local existence and uniqueness of the solution, which is equal to $\varphi^{-1}(0, x_0, \epsilon)$ for $\epsilon > 0$ sufficiently small. We obtain as a bonus the differentiable dependence with respect to the initial condition x_0.

We have therefore proved the following result:

Theorem 1.44. *There exists an open subset $\Omega \subset I \times U$ and a function $\phi : \Omega \to U$ of class C^1 such that*

i) *for all $x_0 \in U$, $I \times \{x_0\} \cap \Omega$ is an open interval containing t_0;*

ii) *$f(t_0, x_0) = x_0$;*

iii) *$\partial_t \phi(t, x_0) = f\left(t, \phi(t, x_0)\right)$;*

iv) *if $u : J \to U$ is a C^1 function defined on an open interval $J \subset I$ containing t_0, such that $u(t_0) = x_0$ and $u'(t) = f\left(t, u(t)\right)$ for $t \in J$, then $u(t) = \phi(t, x_0)$ for all $t \in J \cap I_{x_0}$.*

1.10. Comments

Local models of maps

We have seen that every scalar function with nonvanishing differential at a point a can be transformed by a local change of variables of the domain to a linear function. Based on the strength of this success, we can ask ourselves the more general question: if f is a smooth map on an open subset containing $a \in \mathbf{R}^n$, can we use a local diffeomorphism at a to bring us to the most simple model possible? This is exactly what was done in the case of maps whose differential was injective, surjective, or more generally of constant rank in a neighborhood of a (see Exercise 10).

For example, if the differential of f is zero, we ask if we can transform f to a quadratic form. The answer is yes if the second order Taylor polynomial of f is a *non-degenerate* quadratic form (see the Morse lemma, cf. Exercise 11). Beyond that, things truly get complicated. This is the theory of singularities, founded by Hassler Whitney and developed by René Thom, which is still lively today, for which one may for example consult [Demazure 00] (lively, and accessible), [Golubitsky-Guillemin 73] and [Arnold 78] (70 rich pages).

Of course this theory studies the local models of maps from $\mathbf{R}^p$ to $\mathbf{R}^q$ beyond the simple case where the differential is of constant rank in a neighborhood of a.

Transversality

Two submanifolds X and Y in $\mathbf{R}^n$ are said to be *transverse* if $X \cap Y = \emptyset$ or if for all $p \in X \cap Y$, $T_pX + T_pY = \mathbf{R}^n$ (we do not assume that this is a direct sum). It is then easy to show that $X \cap Y$ is also a submanifold. We really want to say that if we move two transverse submanifolds a little, they remain transverse.

To see this, (and to give a precise meaning to this assertion) it is preferable to generalize the definition in the following way: we consider a smooth map f from an open subset of $\mathbf{R}^m$ to $\mathbf{R}^n$ and a submanifold X in $\mathbf{R}^n$. Then we say f is *transverse* to X at $p \in \mathbf{R}^m$ if $T_{f(p)}X + \mathrm{Im}(T_pf) = \mathbf{R}^n$ or if $f(p) \notin X$, and that f is transverse to X if this is true for all p. For example, a submersion is transverse at every point of the range space, and Theorem 1.18 admits a natural and easy generalization that we will see next chapter (Theorem 2.30) in the more natural case of manifolds:

If f is transverse to X, then $f^{-1}(X)$ is a submanifold of $\mathbf{R}^m$ of the same codimension as X.

Thom's theorem on transversality ensures notably that the set of transverse maps to a closed submanifold is an open dense subset of the space of smooth maps from $\mathbf{R}^m$ to $\mathbf{R}^n$.

For a more precise and general statement (it is necessary to define an adequate topology on $C^\infty(\mathbf{R}^m, \mathbf{R}^n)$, and it proves to be useful to replace $\mathbf{R}^m$ and $\mathbf{R}^n$ by manifolds) see [Golubitsky-Guillemin 73, Chapter II]. See also Exercise 29 of Chapter 2 for a weaker form of this statement.

Weakening of the regularity hypotheses

In this book we essentially work with smooth functions. It is not an easy task to relieve ourselves of this hypothesis.

It turns out that the space of smooth functions cannot be equipped with a norm (which takes into account the convergence of all derivatives). Thus they do not easily lend themselves to functional analysis.

In the majority of problems of analysis on domains of $\mathbf{R}^n$ or manifolds, we frequently work with L^2 spaces and Sobolev spaces (for example the space of f such that f and $\|df\|$ are L^2), and then prove regularity theorems. For example, this is the case in the Dirichlet problem mentioned after Theorem 1.36, where after showing the existence of eigenfunctions in an suitable Sobolev space, one then proves that they are smooth.

We also mention Section 8.5.3; there we have piecewise C^1 vector field, and we need to use a theorem which requires that the field should be C^2.

The analytic case

In place of working with smooth functions we can also work with analytic functions. The inverse function theorem and the existence theorem for differential equations are both valid in this setting (see [Chaperon 08]). We will not tackle these questions for reasons of internal coherence, given for example our intensive use of "bump functions" (see Section 3.2)

1.11. Exercises

1. Let E and F be two vector spaces. Write the differential of a bilinear map ϕ from $E \times E$ to F.

2. *Laplacian and isometries*

Suppose f is a C^2 function from an open subset of $\mathbf{R}^n$ to $\mathbf{R}$. The *Laplacian* of f, denoted Δf, is defined by

$$\Delta f = \sum_{i=1}^{n} \partial_i^2 f.$$

a) Suppose that f is defined on all of $\mathbf{R}^n$ (or on $\mathbf{R}^n \setminus \{0\}$), and depends only on the distance to the origin, in other words there exists a function ϕ defined on $\mathbf{R}^+$ (or $\mathbf{R}^+ - 0$) such that

$$f(x) = \phi(\|x\|).$$

Show that ϕ is of class C^2 and calculate the derivatives in terms of ϕ.

b*) Characterize the linear maps A from $\mathbf{R}^n$ to $\mathbf{R}^n$ such that for all C^∞ functions f, we have

$$\Delta f \circ A = \Delta(f \circ A).$$

c*) Suppose T is a C^2 map from $\mathbf{R}^n$ to $\mathbf{R}^n$ satisfying the property above. Show that the differential of T is an orthogonal linear map. What can we deduce about T?

3. *Differentiation and integration*

a) Suppose a, b, f are functions defined on $\mathbf{R}$. Suppose that a and b are differentiable, and f is continuous. Show that the function

$$F(x) = \int_{a(x)}^{b(x)} f(t)\, dt$$

is differentiable and calculate its derivative.

b) Repeat the question above replacing $f(t)$ by $h(t, x)$, where h is continuous on $\mathbf{R}^2$ and continuously differentiable with respect to the second variable.

4. Suppose f is a differentiable map from $\mathbf{R}^n \setminus \{0\}$ to $\mathbf{R}$ such that for all $t > 0$ we have

$$f(tx) = t^\alpha f(x) \quad \text{(where } \alpha \text{ is real)}.$$

We then say that f is homogeneous of degree α. Show that

$$d_x f \cdot x = \alpha f(x) \quad \text{(Euler's identity)}.$$

5. Show that $\mathbf{R}^2 \setminus \{0\}$ is diffeomorphic to the complement of a closed ball in $\mathbf{R}^2$.

6. *Cusps of the "second kind" and diffeomorphisms*

Show that the map $f : (x, y) \mapsto (x, y - x^2)$ is a local diffeomorphism in a neighborhood of 0. Sketch the curve $t \mapsto (t^2, t^4 + t^5)$ and its transformation under f. What can you say?

7. Sketch the image of each of the following types of simple closed curve (*i.e.*, a curve without double points) under the map $z \mapsto z^2$ of $\mathbf{C}$ to $\mathbf{C}$,

a) not surrounding the origin;

b) surrounding the origin;

c) passing through the origin.

8*. *Cartan's decomposition of the linear group*

Consider $\mathbf{R}^n$ equipped with an inner product. A symmetric endomorphism S is said to be *positive* if $\langle Sx, x \rangle \geqslant 0$ for all x, and *strictly positive* if $\langle Sx, x \rangle > 0$ for all $x \neq 0$.

a) Show that if S is strictly positive, then S is invertible; show that a symmetric endomorphism is strictly positive if and only if there exists a real number $k > 0$ such that

$$\forall x \in \mathbf{R}^n, \quad \langle Sx, x \rangle \geqslant k\|x\|^2$$

(use diagonalization for symmetric matrices). Deduce that the set of strictly positive endomorphisms is an open subset of $\mathbf{R}^{\frac{n(n+1)}{2}}$.

b) Show that every strictly positive endomorphism S has a unique strictly positive square root, and the map

$$S \longmapsto T$$

is a diffeomorphism.

c) Suppose M is an invertible endomorphism on $\mathbf{R}^n$. Show that there exists an orthogonal endomorphism A and a strictly positive endomorphism S such that $M = AS$. Show that A and S are unique and that the map

$$M \longmapsto (A, S)$$

is differentiable. Show that we have analogous results for a decomposition of the form $M = S'A'$.

Note. This decomposition is also called the *polar decomposition*.

9*. For invertible $S \in \mathrm{Sym}(n)$, define a map f_S from $M_n(\mathbf{R})$ to $\mathrm{Sym}(n)$ by $f_S(A) = {}^tASA$. Show there exists an open subset U of $\mathrm{Sym}(n)$ containing S, an open subset V of $M_n(\mathbf{R})$ containing I, and a smooth map g from U to V such that $T = {}^tg(T)Sg(T)$. What property can you deduce for the quadratic forms $q_S(x) = {}^txSx$ and $q_T(x) = {}^txTx$ defined by the matrices S and T?

10*. *Rank theorem*

This result encompasses Theorems 1.17 and 1.18 as is stated as follows:

Let Ω be an open subset of $\mathbf{R}^n$ and $f : \Omega \to \mathbf{R}^m$ by a differentiable map of constant rank r (which is to say that the rank of the differential is constant). Then for all $x_0 \in \Omega$, there exists on the one hand a diffeomorphism φ of an open subset U containing 0 in $\mathbf{R}^n$ to an open subset U' containing x_0 in Ω, with $\varphi(0) = x_0$, and on the other hand, a diffeomorphism ψ from an open subset V' containing $y_0 = f(x_0)$ on an open subset $V \ni 0$ in $\mathbf{R}^m$, with $\psi(y_0) = 0$, such that the map $\psi \circ f \circ \varphi : U \to V$ coincides with the restriction to U of a linear map of $\mathbf{R}^n$ to $\mathbf{R}^m$.

a) Show that it suffices to examine the case where $x_0 = 0$, $f(x_0) = 0$, and where $f'(0)$ is of the form

$$(x^1, \ldots, x^n) \longmapsto (x^1, \ldots, x^r, 0, \ldots, 0).$$

b) These conditions being satisfied, let g be the map from Ω to $\mathbf{R}^m$ defined by

$$g(x^1, \ldots, x^n) = (f^1(x), \ldots, f^r(x), x^{r+1}, \ldots, x^n)$$

where f^j denotes the j-th component of f. Show that g defines a diffeomorphism between two open neighborhoods of 0 in $\mathbf{R}^n$, and that the map f_1 defined on an appropriate open neighborhood of 0 in $\mathbf{R}^n$ by $f_1(g(x)) = f(x)$ is of the form

$$f_1(x^1, \ldots, x^n) = \left(x^1, \ldots, x^r, k^1(x^1, \ldots, x^r), \ldots, k^{m-r}(x^1, \ldots, x^r)\right)$$

where k is a differentiable map from an open neighborhood of 0 in $\mathbf{R}^r$ to $\mathbf{R}^{m-r}$, satisfying $k(0) = 0$ and $k'(0) = 0$.

c) On an appropriate open neighborhood of 0 in $\mathbf{R}^m$, we define a map h with values in $\mathbf{R}^m$ by

$$h(y^1, \ldots, y^m) = \left(y^1, \ldots, y^r, y^{r+1} + k^1(y^1, \ldots, y^r), y^m + k^{m-r}(y^1, \ldots, y^r)\right).$$

Show that h defines a diffeomorphism between two open neighborhoods of 0 in $\mathbf{R}^m$, and that the map f_2 defined in a neighborhood of 0 of $\mathbf{R}^m$ by $h(f_2(x)) = f_1(x)$ is of the form

$$f_2(x^1, \ldots, x^n) = (x^1, \ldots, x^r, 0, \ldots, 0).$$

Deduce for all $y_0 \in f(\mathbf{R}^n)$, the preimage $f^{-1}(y_0)$ is a submanifold of dimension $n - r$ in $\mathbf{R}^n$.

Application. By considering the map f defined by $f(M) = {}^t M M$, show that

$$O(n) = \left\{M \in GL(n, \mathbf{R}) : \; {}^t M = M^{-1}\right\}$$

is a submanifold of dimension $\frac{n(n-1)}{2}$ in $GL(n, \mathbf{R})$.

What difference is there with the proof given in 1.5.2?

11**. *Morse lemma*

Let $f : U \mapsto \mathbf{R}$ be a smooth function on an open subset U of $\mathbf{R}^n$. Suppose that $0 \in U$ is a *non-degenerate critical point*. This means that $df_0 = 0$ and that the quadratic form defined by the matrix

$$S = \left(\partial_{ij}^2 f(0)\right)_{1 \leqslant i, \, j \leqslant n}$$

is non-degenerate. Show that there exists a diffeomorphism ϕ from an open subset containing 0 to another such that

$$f(\phi^{-1}(x)) = f(0) + \sum_{i=1}^{p} x_i^2 - \sum_{i=p+1}^{n} x_i^2,$$

where $(p, n - p)$ is the signature of the quadratic form to associated to S.

Hint. First show by applying Lemma 3.12 twice that there exists a smooth map h defined on an open subset of $\mathbf{R}^n$ containing 0 with values in $\mathrm{Sym}(n)$, such that $f(x) = f(0) + {}^t x h(x) x$ and $h(0) = S$. Then use Exercise 9.

12. Show that the graph of a smooth map from $\mathbf{R}^p$ to $\mathbf{R}^q$ is a p-dimensional submanifold of $\mathbf{R}^{p+q}$.

13. Write the equation of the surface in $\mathbf{R}^3$ given by rotating the circle with center $(a, 0, 0)$ and radius r in the plane xOy about the Oy axis. Show that this is a submanifold if and only if $a > r$ (a torus of revolution). Show, by using a well chosen parametrization that this submanifold is homeomorphic (and even diffeomorphic by using notions of the next chapter) to $S^1 \times S^1$.

14. Show that the graph of the function $x \mapsto |x|$ is not a submanifold of $\mathbf{R}^2$.

15. Show that if X is a submanifold of $\mathbf{R}^n$, the set of ordered pairs $(x, v) \in \mathbf{R}^n \times \mathbf{R}^n$ such that $x \in X$ and v is tangent to X at x is a submanifold of $\mathbf{R}^n \times \mathbf{R}^n$.

16. Consider the circular helix with parametric equation

$$t \longmapsto (a \cos t, a \sin t, bt)$$

in an orthonormal frame. Show that the surface given by the set of straight lines which meet both the helix and the axis Oz orthogonally (*the right helicoid*) is a submanifold of $\mathbf{R}^3$.

17. Show that

$$f : t \longmapsto \left(\frac{\sin 2t}{1 + \cos^2 t}, \frac{2 \sin t}{1 + \cos^2 t} \right)$$

is a periodic immersion of $\mathbf{R}$ to $\mathbf{R}^2$ with period 2π. The image of each interval of length less than π is a submanifold. On the other hand, $f(\mathbf{R})$ is not a submanifold (it is a closed "figure eight" curve, called *Bernoulli's lemniscate*).

18*. Is the intersection of the unit sphere $x^2 + y^2 + z^2 = 1$ and the cylinder with equation $x^2 + y^2 - x = 0$ a submanifold?

19. *Pseudo-orthogonal group*

Let Q be the quadratic form in $\mathbf{R}^n$ defined by

$$Q(x) = \sum_{i=1}^{p} x_i^2 - \sum_{i=p+1}^{n} x_i^2.$$

Show that the set of matrices $A \in M_n \mathbf{R}$ such that $Q(Ax) = Q(x)$ for all x is a submanifold of dimension $n(n-1)/2$.

20. Let P be a homogeneous polynomial in $n + 1$ variables on $\mathbf{R}$ such that the $\partial_i P$ have no zero in common other than 0 (for example $P(x) = \sum_{i=0}^{n} x_i^k$). Show that the intersection of the unit sphere and of $P^{-1}(0)$ is a submanifold.

21*. *Sphere with two holes*

Show that the subset of $\mathbf{R}^3$ defined by the equation

$$\left(4x^2(1-x^2) - y^2\right)^2 + z^2 - \frac{1}{4} = 0$$

is a submanifold of dimension 2. Consider the intersections by the planes $z = $ constant, and deduce that this submanifold is a sphere with two holes, in order words a connected sum of two tori.

22*. *Position of a hypersurface with respect to a tangent plane*

Denote the coordinates of $\mathbf{R}^{n+1}$ by $(x^0, \ldots, x^n)$. Let S be a submanifold of codimension 1 that contains 0 and has the hyperplane $x^0 = 0$ as the tangent plane at the origin.

a) Show that S can be defined in a neighborhood of 0 by the graph of a function $(x^1, \ldots, x^n) \mapsto f(x^1, \ldots, x^n)$ where 0 is a critical point.

b) Suppose that the quadratic form Q defined by

$$(x^1, \ldots, x^n) \longmapsto \sum_{i,j=1}^{n} (\partial_{ij}^2 f(0)) x^i x^j$$

is non-degenerate. Show that if Q is positive definite (resp. negative definite), then f admits a local minimum (resp. maximum) at 0.

c) Now suppose that Q has signature $(p, n - p)$, with $0 < p < n$. Show that every neighborhood of 0 simultaneously contains points of S situated both above and below $T_0 S$. Show that there exists an open subset U containing 0 such that $(U \cap S \cap T_0 S) \smallsetminus \{0\}$ is a submanifold of dimension $n - 1$ of $T_0 S$ that we specified. What happens if we add 0? (Start with the case $n = 2$.)

23. *The exponential is not a group morphism*

a) Compute $\exp(A + B)$ and $(\exp A)(\exp B)$ for the matrices

$$A = \begin{pmatrix} 0 & 1 \\ 0 & 0 \end{pmatrix} \quad \text{and} \quad B = \begin{pmatrix} 0 & 0 \\ 1 & 0 \end{pmatrix}.$$

b) Show that if $\exp t(A + B) = (\exp tA)(\exp tB)$ for every real t, then A and B commute.

24. Justify the details of the argument in Lemma 1.27 iv); give another proof by calculating the derivative of the map $t \mapsto \det(\exp tA)$ at 0.

25. Let $f : D(0,r) \to \mathbf{C}$ be a nonconstant $\mathbf{C}$-differentiable function such that $f(0) = 0$. Show that there exists an integer $k > 0$, a disk $D(0,r')$ and a diffeomorphism $\phi : D(0,r') \to \phi\big(D(0,r')\big) \subset D(0,r)$ such that $f\big(\phi(u)\big) = u^k$.

Chapter 2

Manifolds: The Basics

2.1. Introduction

"The notion of a manifold is hard to define precisely." This is the famous opening of Chapter III of *Leçons sur la Géométrie des espaces de Riemann* by Elie Cartan. It is followed by a stimulating heuristic discussion on the notion of manifold which can still be read with pleasure. For additional historic perspective we also mention Riemann's inaugural lecture, translated with annotations for the modern reader in [Spivak 79].

2.1.1. A Typical Example: The Set of Lines in the Plane

To see how things go, let us start with a simple – yet not too simple – example, the set of straight lines in the plane. A straight line depends on 2 real parameters. We can formalize this somewhat vague idea by choosing a system of Cartesian coordinates and representing each line by its equation relative to this frame. All non-vertical lines are represented by a unique equation of the form $y = ax + b$, and so we may encode this by the ordered pair (a, b) of real numbers. In the same way, all non-horizontal lines are represented by a unique equation of the form $x = cy + d$, which we may again encode by the corresponding pair (c, d). Lines that are neither vertical nor horizontal have two possible encodings. We can pass easily from one code to the other by the formulas

$$c = \frac{1}{a} \qquad d = -\frac{b}{a} \tag{2.1}$$

which clearly show that passing from one code to the other is a diffeomorphism.

In fact, it is not too difficult to see that the set of straight lines can be viewed as a topological space in which the non-vertical and non-horizontal lines are two open subsets homeomorphic to $\mathbf{R}^2$, with the transformations above being homeomorphisms.

We have just seen that the straight lines of the plane form a smooth manifold of dimension 2. The preceding open subsets, with the homeomorphisms of $\mathbf{R}^2$ we have written, are called *charts*, and the formula (2.1) is called a *transition map*.

Amongst other things, this point of view lets us define the notion of a smooth function on the set of straight lines: it will be a function such that the restriction to non-vertical lines is smooth in the variables (a, b), and the restriction to horizontal lines is smooth in (c, d). The formulas in 2.1 show that this point of view is consistent and well-defined: these two properties are equivalent for lines that are neither vertical nor horizontal.

There are other ways of representing lines in the plane. To avoid distinguishing between the "non-horizontal" and "non-vertical" cases, we can write the equation of a line in the form $ux + vy + w = 0$ (with $(u, v) \neq (0, 0)$). But another difficultly appears, because now a line is not specified by a unique triple as any nonzero multiple of (u, v, w) yields the same line. In fact, as will be explained in Section 2.5, by identifying proportional triples (u, v, w) (here $(u, v, w) \neq (0, 0, 0)$), we form a 2-dimensional manifold, called the projective plane. The set of straight lines then appears as the projective plane with a point removed, this point corresponding to the forbidden triples $(0, 0, w)$.

Using the structure of the Euclidean plane, we can also represent straight lines by their direction and their distance from the origin: a line with direction vector $(\cos\theta, \sin\theta)$ has an equation of the form

$$(-\sin\theta)x + (\cos\theta)y - p = 0$$

where p is the (signed) distance from the line to the origin, measured in the direction orthogonal to the line. Now be careful: we have just written a bijection between the set of *oriented* lines and the cylinder $S^1 \times \mathbf{R}$. The latter object is a manifold: we can realize it as a submanifold of $\mathbf{R}^3$. We can also say that it is a product of two manifolds of dimension 1.

The equations of oriented lines

$$(-\sin\theta)x + (\cos\theta)y - p = 0 \quad \text{and} \quad (-\sin\theta')x + (\cos\theta')y - p' = 0$$

represent the same line if and only if

$$p' = -p \quad \text{and} \quad \theta' = \theta + \pi \quad (\text{modulo } 2\pi).$$

This relation defines an action of $\mathbf{Z}/2\mathbf{Z}$ on the cylinder $S^1 \times \mathbf{R}$: the set of lines appears also as the quotient of the cylinder by this action (the reader may recognize a ghost of the Möbius strip). This example shows that it is important to define a notion of quotient for manifolds. The details are given in Section 2.7. The simplest case is of a quotient by a finite group action without fixed points.

2.1.2. In This Chapter

The notions of manifolds and smooth maps between manifolds are discussed in a general setting.

This setting allows us to give a proof of the fundamental theorem of algebra due to J. Milnor, which is more or less as follows.

A polynomial P with complex coefficients is a smooth map from the complex plane to itself. Its extension to the Riemann sphere by continuity is again smooth (with the identification of S^2 with $\mathbf{C} \cup \{\infty\}$ – to be made precise below). If f is this extension, it can have only a finite number of singular points (the zeros of P' and the point at infinity) and therefore singular values. An argument mixing topology and differential calculus shows that the cardinality of $f^{-1}(y)$ is finite and locally constant when y varies over the regular values. But this set is the complement of a finite subset of the sphere S^2 and is therefore connected, which shows that the cardinality of $f^{-1}(y)$ is in fact constant. Finally, this constant is nonzero since P is not constant.

We then derive two particularly important examples of manifolds, the real and complex projective spaces.

Following this, we extend the notions from the previous chapter: immersions, submersions, tangent space, and submanifolds from the vector calculus setting to the more general setting of manifolds. The examples of manifolds that we give along the way will also motivate two fundamental notions which are difficult to define if you restrict yourself to submanifolds of $\mathbf{R}^n$, that of a *fibration*, and that of a *covering space*.

Some definitions, that of a smooth map and most of all that of a tangent vector, may seem painful upon introduction. Happily we have a toolbox (composition, restriction to the domain or the range) which we can often use directly (after all, the rigorous notion of a real number is not easy, but the everyday work of the analyst rarely needs this!).

2.2. Charts, Atlases

2.2.1. From Topological to Smooth Manifolds

Definition 2.1. *An* n-*dimensional* topological manifold *is a Hausdorff*[1] *topological space where every point is contained in an open subset homeomorphic to an open subset of* $\mathbf{R}^n$.

---

1. We can always suppose that the topology is defined by a metric; the price we pay is a little more complication when we define certain constructions (tangent bundle, quotients).

We obtain the same definition by taking smaller open subsets in supposing that M admits a refinement of the covering by open subsets homeomorphic to open balls in $\mathbf{R}^n$. The *invariance of domain* (see for example [Karoubi-Leruste 87, 5.3] or [Dugundji 65, XVII.3]) assures that if two open subsets of $\mathbf{R}^n$ and $\mathbf{R}^m$ are homeomorphic, then $m = n$. Therefore two homeomorphic manifolds have the same dimension.

Examples. From their definition, the n-dimensional submanifolds of a vector space are n-dimensional topological manifolds. The graph of the function $x \mapsto |x|$ (or of any continuous function from $\mathbf{R}$ to $\mathbf{R}$) is a topological manifold of dimension 1, because the projection onto the first factor is a homeomorphism to $\mathbf{R}$.

On the other hand, the union X of solutions to the straight-line equations $y = x$ and $y = -x$ in $\mathbf{R}^2$ is not a topological manifold. Indeed, the complement of $(0,0)$ in any of its neighborhoods has at least four connected components, which precludes the existence of an open subset of X containing $(0,0)$ and homeomorphic to an interval.

Definitions 2.2

a) *A chart of a topological manifold X is an ordered pair (U, φ) consisting of an open subset U of X (the* domain *of the chart) and a homeomorphism φ from U to an open subset of $\mathbf{R}^n$.*

b) *An* atlas *of X is a family $(U_i, \varphi_i)_{i \in I}$ (not necessarily finite) of charts, such that the domains U_i cover X.*

Sometimes we will not mention the domain of the chart. The expression *local coordinate system* is a common synonym for a chart.

This terminology speaks for itself: the surface of the Earth is a sphere S^2 that we can consider as a manifold of dimension 2. The charts are flat representations, necessarily partial (a compact space cannot be homeomorphic to an open subset of $\mathbf{R}^n$), and an atlas is necessary if we want to represent the entire Earth.

A point can clearly belong to the domain of many charts. Then the following property is clear:

If two charts (U, φ) and (V, ψ) are such that $U \cap V \neq \emptyset$, then the map

$$\psi \circ \varphi^{-1} : \ \varphi(U \cap V) \longrightarrow \psi(U \cap V)$$

is a homeomorphism.

In the case of a submanifold of Euclidean space, we can say much more: the map above is a diffeomorphism. This is an immediate consequence of the following proposition.

Proposition 2.3. *Let $M \subset \mathbf{R}^n$ be a submanifold of dimension p, and let (Ω_1, g_1) and (Ω_2, g_2) be two parametrizations.*

Then

$$g_2^{-1} \circ g_1 : \; \Omega_1 \cap g_1^{-1}\big(g_2(\Omega_2)\big) \longrightarrow \Omega_2 \cap g_2^{-1}\big(g_1(\Omega_1)\big)$$

is a diffeomorphism.

PROOF. Let $m \in g_1(\Omega_1) \cap g_2(\Omega_2)$ (there is nothing to show if this intersection is empty). By Definition 1.20 there exists an open subset U containing m and a diffeomorphism f from U to $\mathbf{R}^n$ such that $f(U \cap M) = f(U) \cap (\{0\} \times \mathbf{R}^p)$. Then $f \circ g_1$ and $f \circ g_2$ are immersions from Ω_1 and Ω_2 to $\mathbf{R}^n$. Now if we consider these maps as maps with values in $\mathbf{R}^p$, we obtain smooth homeomorphisms with invertible differentials, and therefore these maps are diffeomorphisms. The same argument applies to

$$(f \circ g_2)^{-1} \circ (f \circ g_1) = g_2^{-1} \circ g_1. \qquad \qquad \square$$

As is often the case in mathematics, we take a property verified in its natural setting and elevate it to an axiom.

Definitions 2.4

a) *Two charts (U_1, φ_1) and (U_2, φ_2) of a topological manifold M are compatible to order k $(1 \leqslant k \leqslant \infty)$ if $U_1 \cap U_2 = \emptyset$ or if the map*

$$\varphi_2 \circ \varphi_1^{-1} : \; \varphi_1(U_1 \cap U_2) \longrightarrow \varphi_2(U_1 \cap U_2)$$

(called a transition function*) is a C^k diffeomorphism.*

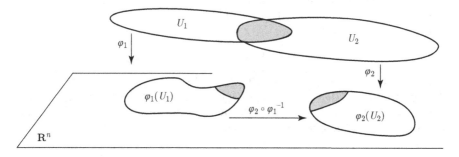

Figure 2.1: Transition function

b) *A C^k atlas of a topological manifold M is an atlas $(U_i, \varphi_i)_{i \in I}$ of M such that any two charts are compatible to order k.*

Take for example a smooth submanifold of codimension 1 in $\mathbf{R}^n$, defined by a submersion $f : \mathbf{R}^n \to \mathbf{R}$. This submanifold admits a smooth atlas with cardinality at most n (where the domains are open subsets U_i of M where the i-th partial derivative of f is nonvanishing). But every smooth parametrization has an inverse compatible with this atlas. This drives the following definitions.

Definitions 2.5

a) *A C^k atlas of a topological manifold M is said to be* maximal *if it contains every chart compatible with the charts in the atlas (one also finds the words "complete" and "saturated" in the literature). Such an atlas also called a C^k* differentiable structure.

b) *A* differentiable manifold *of class C^k is a topological manifold equipped with a differentiable structure of class C^k.*

Every atlas is clearly contained in a unique maximal atlas, obtained by adding all possible compatible charts. For example, by Proposition 2.3 a smooth submanifold of $\mathbf{R}^n$ has a natural smooth structure. This structure is obtained by taking the atlas formed by the inverses of *every* (!) parametrization.

In practice, we define a differentiable structure by taking an atlas that is "not too large": the differentiable structure is given by the corresponding maximal atlas. We already proceeded in this way for submanifolds of $\mathbf{R}^n$.

2.2.2. First Examples

The sphere

The differentiable structure on the sphere S^n defined by the equation $\sum_{i=0}^{n} x_i^2 = 1$ can be defined by an atlas consisting of two charts. Let N and S denote the North and South poles of S^n (this is to say $N = (1, 0, \ldots, 0)$ and $S = (-1, 0, \ldots, 0)$) and set

$$U_1 = S^n \smallsetminus \{N\} \quad \text{and} \quad U_2 = S^n \smallsetminus \{S\}.$$

We obtain homeomorphisms, denoted i_N and i_S, of U_1 and U_2 to $\mathbf{R}^n$, called *stereographic projection* from the North and South pole, by assigning to each $x \in U_i$ the intersection of the straight line passing through x and N ($i = 1$) or S ($i = 2$) with the hyperplane $x_0 = 0$. Explicitly,

$$i_N(x) = \frac{(x_1, \ldots, x_n)}{1 - x_0} \quad \text{and} \quad i_S(x) = \frac{(x_1, \ldots, x_n)}{1 + x_0},$$

Though we will not use this fact again, one verifies easily that i_N (resp. i_S) is the restriction to S^n of the inversion with center N (resp. S) of power 2.

One can also verify that

$$i_N^{-1}(y) = \frac{(|y|^2 - 1, 2y_1, \ldots, 2y_n)}{(|y|^2 + 1)} \quad \text{and} \quad i_S^{-1}(y) = \frac{(-|y|^2 + 1, 2y_1, \ldots, 2y_n)}{(|y|^2 + 1)}.$$

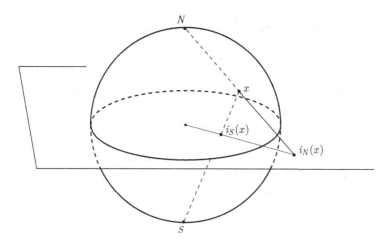

Figure 2.2: The sphere seen as a manifold

Now $i_S \circ i_N^{-1}$ is the diffeomorphism from $\mathbf{R}^n \setminus \{0\}$ to itself given by

$$y \longmapsto \frac{y}{\| y \|^2} \qquad \text{(inversion with center 0 and power 1).}$$

For $n = 1$, there is an even simpler atlas.

The circle S^1 can be equipped with an atlas whose transition functions are translations. Starting with the local parametrization $h : t \mapsto (\cos t, \sin t)$ we remark that $h_{|(0,2\pi)}$ and $h_{|(-\pi,\pi)}$ are homeomorphisms to $U_1 = S^1 \setminus \{(1,0)\}$ and $U_2 = S^1 \setminus \{(-1,0)\}$ respectively. We call ϕ_1 and ϕ_2 the inverse homeomorphisms. We note (this is what makes this example a little puzzling despite its simplicity) that $U_1 \cap U_2$ has two connected components, which we denote U^+ and U^- for the obvious reasons. Now

$$\phi_1(U_1 \cap U_2) = \phi_1(U^+) \cup \phi_1(U^-) = (0, \pi) \cup (\pi, 2\pi)$$

and

$$\varphi_2\big(\varphi_1^{-1}(t)\big) = \begin{cases} t & \text{if } t \in (0, \pi) \\ t - 2\pi & \text{if } t \in (\pi, 2\pi). \end{cases}$$

We now present another example of a manifold which is of interest because it cannot be realized in an obvious way as a submanifold of Euclidean space.

The manifold of affine straight lines of the plane

We saw this example in the introduction! See Section 2.1.1.

2.3. Differentiable Functions; Diffeomorphisms

If we search for a definition of a C^k map between C^k manifolds, it is natural to require that the charts and inverse be C^k, and that a composition of C^k maps be C^k. These requirements impose the following definition.

Definition 2.6. *Let M and N be two C^k manifolds. A continuous map f from M to N is said to be a C^k map if for every $a \in M$, there exists a chart (U, φ) of M, with $a \in U$, and a chart (V, ψ) of N, with $f(a) \in N$, such that the map*

$$\psi \circ f \circ \varphi^{-1} : \ \varphi\left(f^{-1}(V) \cap U\right) \longrightarrow \psi(V)$$

is of class C^k.

Remarks

a) This definition leads to the following commutative diagram

$$
\begin{array}{ccc}
U & \xrightarrow{\ \ f\ \ } & V \\
\varphi \downarrow & & \downarrow \psi \\
\varphi(U) & \xrightarrow{\psi \circ f \circ \varphi^{-1}} & \psi(V)
\end{array}
$$

with a small precaution: one must modify the open subset of the domain (for example consider not U but $U \cap f^{-1}(V)$) so that the compositions are well-defined.

b) It is important to assume that f is continuous to be sure that $\psi \circ f \circ \varphi^{-1}$ is defined on an *open subset* of $\mathbf{R}^m$ (if dim $M = m$).

One of the charts is not necessary if the domain or range manifold is an open subset of Euclidean space. A very important case is that of $N = \mathbf{R}$. A continuous map from M to $\mathbf{R}$ is C^k if $f \circ \varphi^{-1}$ is C^k for every chart (U, φ). It follows that *the sum and the product of two real-valued C^k functions on a manifold is C^k.*

c) It suffices to verify this property for a C^k atlas that defines the differentiable structures on M and N.

Example. Suppose that M is a p-dimensional submanifold of $\mathbf{R}^n$, and $f : \mathbf{R}^n \to N$ is a C^k map. Then the restriction of f to M is of class C^k: by taking Definition 1.20, we return to the clear case where $M = U \cap \mathbf{R}^p \times \{0\}$, for an open subset $U \subset \mathbf{R}^n$.

Example: extension "to infinity" for polynomials

For $z \in \mathbf{C}$, let
$$P(z) = a_0 z^n + a_1 z^{n-1} + \cdots + a_n,$$

where the a_k are complex constants, with $a_0 \neq 0$, and $n > 0$. We consider the sphere S^2 as a subset of $\mathbf{C} \times \mathbf{R}$. Denote the stereographic projection from the North and South poles by i_N and i_S respectively (see Figure 2.2). We define $f : S^2 \to S^2$ by

$$f(x) = i_N^{-1}\big(P(i_N(x))\big) \quad \text{if } x \neq N, \quad \text{and} \quad f(N) = N.$$

(Heuristically, as $\mathbf{C}$ is diffeomorphic to $S^2 \smallsetminus \{N\}$, we may consider S^2 as $\mathbf{C}$ where we add a point at infinity, and then f is the continuous extension of P.)

It is easy to see that f is continuous, and we will see that f is smooth. On $S^2 \smallsetminus \{N\}$ this is true because $i_N \circ f \circ i_N^{-1} = P$ is smooth. It remains therefore to study the situation in the neighborhood of N, which can be done with a chart $(S^2 \smallsetminus \{S\}, i_S)$. In a neighborhood of 0 in $\mathbf{C}$ we have

$$\big(i_S \circ f \circ i_S^{-1}\big)(z) = \begin{cases} \big(i_S \circ i_N^{-1} \circ P \circ i_N \circ i_S^{-1}\big)(z) & \text{if } z \neq 0 \\ 0 & \text{if } z = 0. \end{cases}$$

Knowing that $(i_S \circ i_N^{-1})(z) = 1/\bar{z}$, we see

$$\big(i_S \circ i_N^{-1} \circ P \circ i_N \circ i_S^{-1}\big)(z) = \frac{1}{\overline{P\left(\frac{1}{\bar{z}}\right)}}$$

$$= \frac{z^n}{\overline{a_0} + \overline{a_1} z + \cdots + \overline{a_n} z^n}.$$

Because $n > 0$ and $a_0 \neq 0$, the expression obtained is smooth in a neighborhood of 0, and gives 0 for $z = 0$, where it was not defined *a priori*.

We return to the general theory. We must verify the following property.

Proposition 2.7. *Every composition of C^k maps is C^k.*

PROOF. Consider three manifolds M, N, and P, maps $f \in C^k(M, N)$ and $g \in C^k(N, P)$. For $a \in M$, we take charts (U, φ), (V, ψ), (W, χ) of M, N, P, with open subsets U, V, W containing a, $f(a)$, $g(f(a))$ respectively. The proof can then be read from the commutative diagram

In the second line, we have C^k maps defined on open subsets of vector spaces, where composition is $\chi \circ g \circ f \circ \varphi^{-1}$. Be careful: this map is defined only on $\varphi(U \cap f^{-1}(g^{-1}(W) \cap V))$ (ouch!). This is not the important point, the key point is having an open subset containing a. □

In particular, if f is a C^k map from $\mathbf{R}^m$ to a manifold M, and if N is a submanifold of $\mathbf{R}^m$, the *restriction* of f to N is C^k.

A little less immediate, but also important is the following property.

Proposition 2.8. *If M is a submanifold of $\mathbf{R}^n$ and if f is a C^k map from an open subset $U \subset \mathbf{R}^m$ to $\mathbf{R}^n$ such that $f(U) \subset M$, then f is C^k as a map from U to M.*

PROOF. The property is clear when M is a *vector subspace* of $\mathbf{R}^n$. The definition of submanifold allows us to reduce to this case. Let $a \in U$, $f(a) \in M$, V an open subset $\mathbf{R}^n$ containing $f(a)$ for which there exists a diffeomorphism g of V to its image such that

$$g(V \cap M) = g(V) \cap (\mathbf{R}^p \times \{0\}) \quad (p = \dim M).$$

By the preceding remark, $f \circ g$ is then a C^k map from U to $\mathbf{R}^p$, and the restriction of g to $V \cap M$ is a chart of M. □

Example: the orthogonal group

We saw in Section 1.5 that $O(n)$ is a submanifold of $\text{End}(\mathbf{R}^n)$ (in fact of the open subset $Gl(n, \mathbf{R})$ in $\text{End}(\mathbf{R}^n)$). The maps $A \mapsto A^{-1}$ and $(A, B) \mapsto AB$ are clearly C^k maps from $Gl(n, \mathbf{R})$ to itself and from $Gl(n, \mathbf{R}) \times Gl(n, \mathbf{R})$ to $Gl(n, \mathbf{R})$ respectively. We have just seen that these two maps remain C^k as maps of $O(n)$ to itself and of $O(n) \times O(n)$ to $O(n)$ respectively. The group $O(n)$ appears also as a manifold for which the multiplication and inverse maps are smooth. Such a group is called a *Lie group*. This situation will be studied more systematically in Chapter 4.

The previous remarks motivate the definition of a *product* of two manifolds.

Definition 2.9. *If M and N are two C^k manifolds equipped with the atlas $(U_i, \varphi_i)_{i \in I}$ and $(V_j, \psi_j)_{j \in J}$, the manifold structure on the product $M \times N$ is given by the atlas*

$$(U_i \times V_j, \varphi_i \times \psi_j)_{(i,j) \in I \times J}.$$

It is very easy to verify that the projections $\text{pr}_1 : M \times N \to M$ and $\text{pr}_2 : M \times N \to N$ are of class C^k, and that for fixed $n \in N$ (for example), the restriction of pr_1 to $M \times \{n\}$ is a diffeomorphism onto M.

Definition 2.10. *A C^k diffeomorphism between two manifolds M and N is a C^k bijection with C^k inverse.*

Example: quadrics

In the product $E \times F$ of two inner product spaces of dimensions p and q, the *quadric* Q is the solution of the equation $\|x\|^2 - \|y\|^2 = 1$, and is a submanifold of dimension $p + q - 1$. As a manifold, Q is diffeomorphic to $S^{p-1} \times \mathbf{R}^q$.

The differential of $f(x,y) = \|x\|^2 - \|y\|^2$ is given by

$$df_{(x,y)} \cdot (h,k) = 2\langle x, h \rangle - 2\langle y, k \rangle.$$

This differential is surjective at every point of Q, which is therefore a submanifold of codimension 1 in $\mathbf{R}^{p+q}$. The map

$$(x,y) \longmapsto \left(\frac{x}{\sqrt{1 + \|y\|^2}}, y \right) \quad \text{of} \quad Q \quad \text{to} \quad S^{p-1} \times \mathbf{R}^q$$

is smooth. It admits an inverse map, given by

$$(u,y) \longmapsto \left(\sqrt{1 + \|y\|^2}\, u, y \right) \quad \text{where} \quad u \in S^{p-1},$$

which is smooth for the same reasons. In particular, Q is connected if $p > 1$, and has two connected components if $p = 1$. All this applies as well to

$$S^0 = \left\{ x \in \mathbf{R} : \ x^2 = 1 \right\} = \{-1, 1\}!$$

Warning. Two differentiable structures on the same set can be diffeomorphic but distinct, in other words defined by different maximal atlases. See Exercise 27.

> **FROM THIS POINT ON, UNLESS EXPLICITLY MENTIONED,**
> **ALL MANIFOLDS AND ALL MAPS ARE ASSUMED TO BE SMOOTH**

In using charts in the previous definitions, we see a guiding principle: every definition or property of open subsets of $\mathbf{R}^n$ which is invariant under diffeomorphisms extends to manifolds.

Definition 2.11. *If M and N are two manifolds, a map $f : M \to N$ is a* local diffeomorphism *if every point of M is contained in an open subset U such that $f_{|U}$ is a diffeomorphism to its image.*

At this stage, we can ask if there exists a version of the inverse function theorem adapted to manifolds. The answer is of course yes, on the condition that we define the differential in this setting. We will see this in Section 2.6. Beforehand, we give an example that shows we already have the means to prove significant theorems.

2.4. Fundamental Theorem of Algebra

Theorem 2.12. *Every nonconstant polynomial with complex coefficients has at least one zero in* **C.**

In the proof we will use the following notion as seen in Section 1.40 that generalizes to manifolds.

Definition 2.13. *Let X and Y be two manifolds of the same dimension, and $f : X \to Y$ a smooth map. A point $a \in X$ is said to be a* regular point *of f if f restricts to a local diffeomorphism on a neighborhood of a. A point $b \in Y$ is called a* regular value *of f if its inverse image $f^{-1}(b)$ consists only of regular points.*

In particular, every point $b \notin f(X)$ is a "regular value".

Theorem 2.14. *Let X and Y be manifolds of the same dimension, with X compact. Let $f : X \to Y$ be a smooth map, with y a regular value of f. Then*
i) *$f^{-1}(y)$ is finite;*
ii) *there exists an open set V containing y such that:*

$$\forall z \in V, \quad \mathrm{card}\left\{f^{-1}(z)\right\} = \mathrm{card}\left\{f^{-1}(y)\right\}.$$

PROOF. If $f^{-1}(y) = \emptyset$ it suffices to prove ii). But $f(X)$ is the image of a compact subset X under a continuous map, so it is a compact subset and is therefore closed in Y. It suffices to take $V = Y \smallsetminus f(X)$.

If $f^{-1}(y) \neq \emptyset$ we first note that it is a *compact* subset of X, as it is a closed subset of a compact set. Let $x \in f^{-1}(y)$. By hypothesis, x is a regular point of f: there exists an open set U_x containing x such that $f_{|U_x}$ is a diffeomorphism between U_x and $f(U_x)$. In particular, x is the only point of U_x where $f(x) = y$. The family $(U_x)_{x \in X}$ is an open cover of $f^{-1}(y)$. We can extract a finite subcover $(U_{x_i})_{1 \leqslant i \leqslant p}$, which shows that $f^{-1}(y)$ is a finite subset with p elements (more concisely we can say the topological space $f^{-1}(y)$ is both compact and discrete, and therefore finite).

By replacing each open set U_{x_i} by a possibly smaller open set, we can suppose that the U_{x_i} are mutually disjoint. We set

$$V = f\left(U_{x_1} \cap \cdots \cap f(U_{x_p})\right) \smallsetminus f\left(X - U_{x_1} - \cdots - U_{x_p}\right).$$

This is an intersection of finitely many open sets containing y.

If $z \in V$, then by construction z has a preimage under f in each U_{x_i}, and it has no other preimages since V is disjoint from $f\left(X \smallsetminus \bigcup_{1 \leqslant i \leqslant p} U_{x_i}\right)$. In other words, card $\left(f^{-1}(z)\right) = p$. ☐

PROOF OF THE FUNDAMENTAL THEOREM OF ALGEBRA. Let $P(z) = a_0 z^n + a_1 z^{n-1} + \cdots + a_n$ be a nonconstant polynomial with complex coefficients. We can associate an "extension to infinity" to P, denoted f, defined and studied in Section 2.3. We will see that f has a finite number of non-regular values.

On $S^2 \smallsetminus \{N\}$, f has the same number of non-regular values as P (we composed on the right and left by diffeomorphisms). But the differential of P at a point z is precisely *multiplication by the complex number $P'(z)$, seen as a linear map of $\mathbf{R}^2$ in $\mathbf{R}^2$* (see Section 1.2.2).

The non-regular points of P are those such that P' is zero. There are only finitely many of these, and so P has only a finite number of non-regular values. The same holds for f seen as a map of $S^2 \smallsetminus \{N\}$ to itself, and therefore for f as a map of S^2 in S^2 (if necessary, we adjoin the North pole).

Let F denote the set of these values. Since F is finite, $S^2 \smallsetminus F$ is *connected*. By Theorem 2.14, the function

$$x \longmapsto \mathrm{card}\left\{f^{-1}(x)\right\}$$

is locally constant on $S^2 \smallsetminus F$, therefore constant by connectedness. This constant is nonzero. If not, f, and moreover P have only singular values, which says that P' is identically zero and therefore P is constant, contrary to the hypothesis.

All points in $S^2 \smallsetminus F$, and therefore all the points in S^2 are values taken by f. Returning to the definition of f, we deduce that all of the points of $\mathbf{C}$ are values taken by P. ☐

2.5. Projective Spaces

When we draw in perspective or study geometric optics, we must take "infinity" into account. To do this, we adjoin extra points to the usual plane or space, namely one point for each plane or spatial direction. The model situation for perspective proceeds as follows: we represent $\mathbf{R}^n$ as the (affine!)

subspace $\{1\} \times \mathbf{R}^n$ of $\mathbf{R}^{n+1}$. Perspective with respect to the origin is a bijection between $\{1\} \times \mathbf{R}^n$ and the set of (vectorial) straight lines of $\mathbf{R}^{n+1}$ that are not contained in $\{0\} \times \mathbf{R}^n$ (see Figure 2.3). These are the straight lines of $\{0\} \times \mathbf{R}^n$ which model the points at infinity. Projective space formalizes this situation.

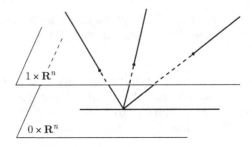

Figure 2.3: A chart of projective space

Definition 2.15. *The real projective space of dimension n, denoted $P^n\mathbf{R}$ is the quotient space of $\mathbf{R}^{n+1} \smallsetminus \{0\}$ by the equivalence relation*

$$x \sim y \quad \text{if and only if } x \text{ and } y \text{ are collinear,}$$

equipped with the quotient topology (for this notion, see the beginning of Section 2.7).

Let $p : \mathbf{R}^{n+1} \smallsetminus \{0\} \to P^n\mathbf{R}$ be the quotient map. Recall that a subset U of $P^n\mathbf{R}$ is open if and only if $p^{-1}(U)$ is open in $\mathbf{R}^{n+1} \smallsetminus \{0\}$ (this is practically the definition of the quotient topology).

We can then consider $P^n\mathbf{R}$ as the set of straight lines in $\mathbf{R}^{n+1}$. Another interpretation is possible: the restriction of $\sim$ to S^n identifies the points x and $-x$, and the real projective space is homeomorphic to the quotient of S^n under this identification. One verifies that $P^n\mathbf{R}$ is *Hausdorff* (see also 2.7.1). From this we deduce that $P^n\mathbf{R}$ is *compact*, being the image of S^n under the continuous map p.

Definition 2.16. *The $(n+1)$-tuple $x = (x_0, \dots, x_n)$ is a homogeneous coordinate system of $p(x)$.*

It is convenient to denote the homogeneous coordinates of x by $[x] = [(x_0, \dots, x_n)]$.

We will equip $P^n\mathbf{R}$ with the atlas $(U_i, \phi_i)_{0 \leqslant i \leqslant n}$ which makes it into a manifold. Set

$$V_i = \big\{ x = (x_0, \dots, x_n) \in \mathbf{R}^{n+1} : x_i \neq 0 \big\} \quad (0 \leqslant i \leqslant n)$$

and define the maps $\Phi_i : V_i \to \mathbf{R}^n$ by

$$\Phi_i(x) = \left(\frac{x_0}{x_i}, \ldots, \frac{\widehat{x_i}}{x_i}, \ldots, \frac{x_n}{x_i} \right)$$

where the symbol $\widehat{}$ signifies that the corresponding term has been omitted. These maps are continuous and

$$\Phi_i(x) = \Phi_i(y) \text{ if and only if } p(x) = p(y).$$

From the properties of quotient topology, $U_i = \Phi_i(V_i)$ is an open subset of $P^n\mathbf{R}$, and Φ_i passes to the quotient and gives a continuous and bijective map ϕ_i from U_i to $\mathbf{R}^n$. Explicitly,

$$\phi_i([x]) = \left(\frac{x_0}{x_i}, \ldots, \frac{\widehat{x_i}}{x_i}, \ldots, \frac{x_n}{x_i} \right).$$

The inverse map is given by

$$\phi_i^{-1}(y_0, \ldots, y_{n-1}) = p(y_0, \ldots, y_{i-1}, 1, y_i, \ldots, y_{n-1}),$$

which shows that ϕ_i is a homeomorphism of U_i to $\mathbf{R}^n$.

The transition functions $\phi_j \circ \phi_i^{-1}$ are diffeomorphisms from $\phi_i(U_i \cap U_j)$ to $\phi_j(U_i \cap U_j)$, because for $y_j \neq 0$ we have

$$(\phi_j \circ \phi_i^{-1})(y_0, \ldots, y_{n-1}) = \left(\frac{y_0}{y_j}, \ldots, \frac{y_{i-1}}{y_j}, \frac{1}{y_j}, \ldots, \frac{\widehat{y_j}}{y_j}, \ldots, \frac{y_{n-1}}{y_j} \right).$$

Thus we have a smooth manifold structure on $P^n\mathbf{R}$.

In an analogous way we can define *complex projective space* $P^n\mathbf{C}$. In everything said above, it is sufficient to replace $\mathbf{R}$ by $\mathbf{C}$, and to remark that a map of the form

$$(z, t) \longmapsto \frac{z}{t} \text{ from } \mathbf{C} \times \mathbf{C} \smallsetminus \{0\} \text{ to } \mathbf{C}$$

is smooth, provided we consider it as a map from $\mathbf{R}^2 \times \mathbf{R}^2 \smallsetminus \{0\}$ to $\mathbf{R}^2$.

Remark. **We can also define *complex (analytic) manifolds*. A complex structure is defined by an atlas with values in $\mathbf{C}^n$ such that the transition functions are complex analytic. A simple examination of the formulas above show this is the case for $P^n\mathbf{C}$. In view of the very particular properties of complex analytic functions (maximum principle, etc.), the world of complex manifolds is very different and we will only mention it occasionally.**

It is instructive to understand the classical notions of complex analysis in terms of manifolds. For example, reviewing the discussion at the beginning of Section 2.3 on the extension of polynomials to infinity, we see that meromorphic functions are holomorphic functions with values in $P^1\mathbf{C}$.

In a manner analogous to the real case, we can consider the restriction of the equivalence relation to the set of vectors in $\mathbf{C}^{n+1}$ of norm 1, which is to say the sphere

$$S^{2n+1} = \left\{ (z_0, z_1, \ldots, z_n) \in \mathbf{C}^{n+1} : \sum_{i=0}^{n+1} |z_i|^2 = 1 \right\}.$$

Denoting the quotient map as always by p, it is clear that $p(S^{2n+1}) = P^n\mathbf{C}$ (which shows in passing the compactness of complex projective space), and that, for $z, z' \in S^{2n+1}$, we have $p(z) = p(z')$ if and only if $z = uz'$, where u is a complex number of unit modulus. In other words, if we consider

$$p: \quad S^{2n+1} \longrightarrow P^n\mathbf{C}$$

the inverse map of p of a point in $P^n\mathbf{C}$ is a great circle of S^{2n+1}. Moreover, we have:

Lemma 2.17. *The open subset $p^{-1}(U_i)$ of S^{2n+1} is diffeomorphic to $S^1 \times U_i$.*

PROOF. We take the case $i = 0$ to lighten the notation. A point $x \in U_0$ has a unique homogeneous coordinate system of the form $(1, \zeta_1, \ldots, \zeta_n)$ (the point $(1, \phi_0^{-1}(x))$ of $\mathbf{C}^{n+1}$). Now a point $(z_0, \ldots, z_n)$ of S^{2n+1} satisfies $p(z) = x$ if and only if

$$(z_0, z_1, \ldots, z_n) = \frac{u}{\sqrt{1 + \sum_{i=1}^{n} |\zeta_i|^2}} (1, \zeta_1, \ldots, \zeta_n).$$

The map $(u, \zeta) \mapsto z$ is a diffeomorphism from $S^1 \times \mathbf{C}^n$ to $p^{-1}(U_0)$, and the required diffeomorphism is obtained by composing with $Id \times \phi_0$ □

We have just seen a particular case of the following situation.

Definition 2.18. *Let E and B be two C^k manifolds $(0 \leqslant k \leqslant +\infty)$. A C^k map p from E to B is a fibration (with base B and total space E), if for every $x \in B$ there exists an open subset U containing x, a C^k manifold F and a diffeomorphism $\phi : U \times F \to p^{-1}(U)$ such that $p(\phi(y, z)) = y$ for all y in U and z in F.*

We also say that E is a *fibered space*. We note that the restriction of p to $p^{-1}(U)$ is equal to $\mathrm{pr}_1 \circ \phi^{-1}$, which proves (looking ahead a little, see 2.6.2) that p is a submersion. For b in B, $p^{-1}(b) = E_b$ is a closed submanifold of E, of dimension $\dim E - \dim B$, that we call the *fiber* of (or over) b. For $b \in U$, E_b is diffeomorphic to F. In fact, if the base B is connected, the manifold F is "always the same".

Lemma 2.19. *If $p : \quad E \to B$ is a fibration with connected base, then the fibers E_b are diffeomorphic.*

PROOF. We choose a point b_0 in B. For all $b \in B$, there exists a continuous path $\gamma : [0,1] \to B$ joining b_0 and b. Every $x \in \gamma([0,1])$ is contained in an open subset U_x satisfying the property of Definition 2.18. By compactness, we can find a finite number of subsets $U_1, \ldots, U_m$ covering $\gamma([0,1])$. We may suppose that $b_0 \in U_1$, $b \in U_m$, and that $U_i \cap U_{i+1} \neq \emptyset$ for $1 \leqslant i \leqslant m-1$. Then the fibers E_x for $x \in U_i$ are mutually diffeomorphic, and diffeomorphic to the fibers E_x for $x \in U_{i+1}$. We conclude that E_{b_0} is diffeomorphic to E_b.   □

This property justifies the name *model fiber* for F. This allows us to see that a submersion is not necessarily a fibration (Theorem 1.18 seems to say the contrary, but one must shrink the open subset of the domain space, which need not be of the form $p^{-1}(U)$ for an open subset U of the range). For an explicit counterexample, see Section 2.9.

Lemma 2.17 says exactly that

$$p : \ S^{2n+1} \longrightarrow P^n \mathbf{C}$$

is a fibration with model fiber S^1. An analogous argument to the one in the lemma shows that

$$p : \ \mathbf{R}^{n+1} \smallsetminus \{0\} \longrightarrow P^n \mathbf{R}$$

is a fibration with fiber $\mathbf{R}^*$.

Definitions 2.20

a) *The* trivial *fibration is the one for which $E = B \times F$ and $p = \mathrm{pr}_1$; we say that a fibration is* trivializable *if there exists a diffeomorphism ϕ from E to $B \times F$ such that $p = \mathrm{pr}_1 \circ \phi$.*

   We say that ϕ is a trivialization.

b) *An* isomorphism *between two fibrations E_1 and E_2 having the same base B is a diffeomorphism $f : E_1 \to E_2$ such that $p_2 \circ f = p_1$.*

For example, a trivializable fibration is isomorphic to the trivial fibration. Definition 2.18 says exactly that every fibration becomes trivializable above sufficiently small open subsets of the base. In other words, every fibration is locally trivial.

Example: the Hopf fibration

For

$$(u,v) \in S^3 = \big\{ (u,v) \in \mathbf{C}^2 : \ |u|^2 + |v|^2 = 1 \big\},$$

we set

$$h(u,v) = \big(2u\bar{v}, |u|^2 - |v|^2 \big).$$

We will see that h is a fibration of S^3 over S^2 with fiber S^1.

It is immediate that h sends S^3 to the sphere S^2 seen as

$$\{(z,t) \in \mathbf{C} \times \mathbf{R} : |z|^2 + t^2 = 1\}.$$

If we set $N = (0,1)$ and $S = (0,-1)$ the calculations of 2.2 show that

$$i_N\big(h(u,v)\big) = \frac{u}{v} \quad \text{and} \quad i_S\big(h(u,v)\big) = \frac{\bar{v}}{\bar{u}}$$

As a result, $S^2 \setminus \{N\}$ are $S^2 \setminus \{S\}$ local trivializations for a fibration whose model fiber is the set of complex numbers of unit modulus (see the proof of Lemma 2.17).

From this we deduce an explicit diffeomorphism between $P^1\mathbf{C}$ and S^2, given by

$$[(u,v)] \longmapsto i_N^{-1}\left(\frac{u}{v}\right) \quad \text{if } v \neq 0 \quad \text{and} \quad [(u,v)] \longmapsto i_S^{-1}\left(\frac{\bar{v}}{\bar{u}}\right) \quad \text{if } u \neq 0.$$

The inverse diffeomorphism may be written

$$(z,t) \longmapsto [(z, 1-t)] \quad \text{if } t \neq 1 \quad \text{and} \quad (z,t) \longmapsto [(1+t, \bar{z})] \quad \text{if } t \neq -1.$$

In the same way, we can see that $P^1\mathbf{R}$ is diffeomorphic to S^1: if we restrict to $\mathbf{R}^2$, we obtain the map

$$h : \ (u,v) \longmapsto (u^2 - v^2, 2uv) \quad \text{from } S^1 \text{ to } S^1;$$

see Figure 2.4.

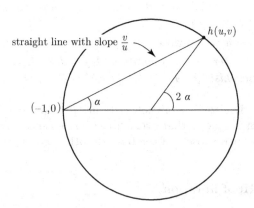

Figure 2.4: From the projective line to the circle

We can also write $z = u + iv$, and note that upon passing to the quotient, the map $z \mapsto z^2$ from S^1 to S^1 gives a diffeomorphism between $S^1/\pm I$ and S^1 (see Section 2.7).

2.6. The Tangent Space; Maps

2.6.1. Tangent Space, Linear Tangent Map

Up to now, we have spoken of smooth maps between manifolds but not their differentials! To define these, we must first define the tangent space to a point m in a manifold M. We take inspiration from what we have done for submanifolds of $\mathbf{R}^n$, by using curves passing through m. We denote $\mathcal{C}_m^M$ (or simply $\mathcal{C}_m$ if there is no ambiguity) the set of smooth curves $c : I \mapsto M$ defined on an open interval I containing 0 and such that $c(0) = m$.

Definition 2.21. *Two curves $c_1 : I_1 \to M$ and $c_2 : I_2 \to M$ in $\mathcal{C}_m$ are tangent at m if $c_1(0) = c_2(0) = m$ and if there exists a chart (U, ϕ), such that $m \in U$ and*
$$(\phi \circ c_1)'(0) = (\phi \circ c_2)'(0).$$
We also say these curves have the same velocity at m.

This condition is independent of the choice of chart: if (V, ψ) is a second chart defined in an open subset containing m, the chain rule gives
$$(\psi \circ c_i)'(0) = d(\psi \circ \phi^{-1})_{\phi(m)} \cdot (\phi \circ c)'(0).$$
We have therefore defined an *equivalence relation* on $\mathcal{C}_m$.

Taking inspiration from the case of submanifolds of $\mathbf{R}^n$, we are driven to make the following definition.

Definition 2.22. *Let M be a smooth manifold and let $m \in M$. A tangent vector to M at m is an equivalence class of the equivalence relation above. The set of tangent vectors at m is denoted $T_m M$.*

Given a chart (U, ϕ) in a neighborhood of m, we define a map θ_ϕ from $T_m M$ to $\mathbf{R}^n$ (if $n = \dim M$), by
$$\theta_\phi(\xi) = (\phi \circ c)'(0)$$
(the right hand side is well defined as it depends only on the equivalence class of c). By the same definition, θ_ϕ is *injective*. Also, θ_ϕ is surjective: a vector $v \in \mathbf{R}^n$ is the image under θ_ϕ of the equivalence class of the curve $t \mapsto \phi^{-1}(tv)$. Therefore the map $\theta_\phi : T_m M \mapsto \mathbf{R}^n$ is a *bijection*.

Now let (V, ψ) be another chart such that $m \in V$, and $v \in \mathbf{R}^n$. Then
$$(\theta_\phi \circ \theta_\psi^{-1})(v) = d(\phi \circ \psi^{-1})_{\psi(m)} \cdot v.$$

We obtain a linear map. This allows us to give $T_m M$ the structure of a vector space over $\mathbf{R}$: if $\xi, \eta \in T_m M$, $\lambda \in \mathbf{R}$, we set
$$\xi + \eta = \theta_\phi^{-1}\big(\theta_\phi(\xi) + \theta_\phi(\eta)\big) \quad \text{and} \quad \lambda\xi = \theta_\phi^{-1}\big(\lambda\theta_\phi(\xi)\big),$$
and the result does not depend on ϕ.

Definition 2.23. *The tangent space to M at m, denoted $T_m M$ is the set of tangent vectors at m equipped with the vector space structure defined above.*

Examples

a) If U if an open subset of M, the preceding construction shows that $T_m U$ is canonically isomorphic to $T_m M$.

b) The vector space tangent to an affine space E at a is identified with the vectorized E_a of E at a: at $v \in E_a$ we associate the equivalence class of the curve $t \mapsto a + tv$.

c) If M and M' are two manifolds, the tangent space at (m, m') to $M \times M'$ is the direct sum $T_m M \oplus T_{m'} M'$.

We are now ready to define the differential of a smooth map between two manifolds.

Definition 2.24. *If X and Y are two manifolds, and $f : X \to Y$ is a smooth map, the linear tangent map at $x \in X$, denoted $T_x f$ is the map obtained by passing to quotient in the map $c \mapsto f \circ c$ from $\mathcal{C}_x^X$ to $\mathcal{C}_{f(x)}^Y$.*

Let (U, ϕ) and (V, ψ) be charts of X and Y, where the domains U and V contain x and $f(x)$ respectively. From the differentiability of $\psi \circ f \circ \phi^{-1}$ (see Definition 2.6) and the fact that $(\psi \circ f \circ \phi^{-1}) \circ (\phi \circ c) = \psi \circ f \circ c$, we see immediately that the images of the two tangent curves are tangent. Moreover we have the following commutative diagram

$$
\begin{array}{ccc}
T_x X & \xrightarrow{\ \ T_x f\ \ } & T_{f(x)} Y \\
\theta_\phi \downarrow & & \downarrow \theta_\psi \\
\mathbf{R}^p & \xrightarrow{\ d_{\phi(x)}(\psi \circ f \circ \phi^{-1})\ } & \mathbf{R}^q
\end{array}
$$

where $p = \dim X$ and $q = \dim Y$.

Examples

a) If M if a submanifold of $\mathbf{R}^n$, and $i : M \to \mathbf{R}^n$ is the natural inclusion, $T_m i$ is an isomorphism between $T_m M$ and the tangent space of Definition 1.25.

b) If ϕ is a chart, $T_x \phi$ is the isomorphism θ_ϕ seen above.

c) If f is a smooth map from M to $\mathbf{R}$, $T_x f$ is a linear map from $T_x M$ to $T_{f(x)} \mathbf{R} \simeq \mathbf{R}$; if (U, ϕ) is a chart whose domain contains m, then $T_x f \circ \theta_\phi^{-1}$ is the differential of $f \circ \phi^{-1} : \phi(U) \to \mathbf{R}$. We continue to denote the differential of a map with values in $\mathbf{R}$ by df_x (we note that certain authors, Bourbaki and Berger-Gostiaux for example, make a distinction between $T_x f$, the linear map from $T_x M$ to $T_x \mathbf{R}$, and df_x, the linear map from $T_x M$ to $\mathbf{R}$).

Proposition 2.25. *If $f : X \to Y$ and $g : Y \to Z$ are two smooth maps between manifolds, then*

$$\forall x \in X, \quad T_x(g \circ f) = T_{f(x)}g \circ T_x f.$$

PROOF. We take charts (U, ϕ), (V, ψ) and (W, χ) in a neighborhood of x, $f(x)$, $g(f(x))$ respectively. The given property is a consequence of the definitions and the ordinary chain rule applied to the functions

$$\phi \circ f \circ \psi^{-1} \quad \text{and} \quad \psi \circ g \circ \chi^{-1}. \qquad \square$$

2.6.2. Local Diffeomorphisms, Immersions, Submersions, Submanifolds

After we establish the results of this section, the results of Section 1.4 concerning the inverse function theorem and its consequences extend word for word to manifolds.

Theorem 2.26. *Let X and Y be two manifolds of dimensions m and n respectively, and let $f : X \to Y$ be a smooth map, and $x \in X$.*

i) *If $T_x f$ is bijective, there exists an open subset U containing x such that $f_{|U}$ is a diffeomorphism to $f(U)$.*

ii) *If $T_x f$ is injective or surjective, there exists open subsets U containing x and V containing $f(x)$, and charts (U, ϕ) and (V, ψ) such that*

$$(\psi \circ f \circ \phi^{-1})(x_1, \ldots, x_n) = \begin{cases} (x_1, \ldots, x_n, 0, \ldots, 0) & \text{if } T_x f \text{ is injective} \\ (x_1, \ldots, x_m) & \text{if } T_x f \text{ is surjective.} \end{cases}$$

PROOF. Write f with the help of charts and apply the results of Section 1.4. $\square$

It is therefore natural to extend the definitions seen in Section 1.4 to manifolds.

Definitions 2.27

a) *A map f from a manifold X to a manifold Y is a immersion (resp. a submersion) if for every $x \in X$ the linear tangent map is injective (resp. surjective).*

b) *A subset M of a manifold X of dimension n is a p-dimensional submanifold of X if for every x in M, there exists open neighborhoods U and V of x in X and 0 in $\mathbf{R}^n$ respectively, and a diffeomorphism*

$$f : U \longrightarrow V \quad \text{such that} \quad f(U \cap M) = V \cap (\mathbf{R}^p \times \{0\}).$$

It is of course the same to say that for every $x \in M$ there exists a chart (U, ϕ), where $x \in U$, such that $\phi(U \cap M)$ is a submanifold of $\mathbf{R}^n$.

c) *A map f from X to Y is an* embedding *if $f(X)$ is a submanifold of Y and if f is a diffeomorphism of X to $f(X)$.*

d) *The* rank *of f at x is the rank of the linear map $T_x f$.*

e) *If $f : X \to Y$ is smooth, a point $x \in X$ is said to be a* critical point *if $\mathrm{rg}(T_x f) < \dim Y$; the image of a critical point is a* critical value.

Example: local extrema

For a map $f : X \to \mathbf{R}$, a point $a \in X$ is critical if and only if the differential df vanishes at a. Proposition 1.34 extends immediately: if $a \in X$ and f admits a local extrema at a then, a is a critical point. This reduces to the case of $f \circ \phi^{-1}$, where ϕ is a chart whose domain contains a.

We note also that in the case where X is a submanifold of $\mathbf{R}^n$, the notion of critical point is equivalent to that of constrained extremum. Indeed suppose X is of codimension p and defined in a neighborhood of a point a by a submersion $g = (g^1, \ldots, g^p)$ in $\mathbf{R}^p$. Let f be a smooth function on $\mathbf{R}^n$. The point a will be critical for $f_{|X}$ if and only if df_a vanishes on $T_a X$, or

$$\bigcap_{i=1}^{p} \operatorname{Ker} dg_a^i \subset \operatorname{Ker} df_a.$$

Since the linear forms dg_a^i are independent by hypothesis, we find this reduces to saying that there exists real numbers $(\lambda_i)_{1 \leqslant i \leqslant p}$ (called *Lagrange multipliers*) such that

$$df_a = \sum_{i=1}^{p} \lambda_i dg_a^i.$$

Example: projective hypersurfaces

Let P be a homogeneous polynomial in $n+1$ variables, whose partial derivatives $\partial_i P$ are never simultaneously vanishing if $x \neq 0$. The subset of $P^n \mathbf{R}$ consisting of points whose homogeneous coordinates satisfy $P(x) = 0$ is a submanifold of $P^n \mathbf{R}$.

Suppose for example that the first homogeneous coordinate x_0 is nonzero, and use the charts seen in Section 2.5. Then

$$\phi_0(U_0 \cap M) = \{ u \in \mathbf{R}^n : P(1, u_1, \ldots, u_n) = 0 \}.$$

By Euler's identity (see Chapter 1, Exercise 4),

$$(\deg P)P(1, u_1, \ldots, u_n) = \partial_0 P(1, u_1, \ldots, u_n) + \sum_{i=1}^{n} u_i \partial_i P(1, u_1, \ldots, u_n).$$

Now, at a point of $\phi_0(U_0 \cap M)$, the derivatives $\partial_i P(1, u_1, \ldots, u_n)$ cannot be simultaneously zero without $\partial_0 P(1, u_1, \ldots, u_n)$ also vanishing.

These considerations also apply to complex projective space.

Proposition 2.28

i) *If $f : X \to Y$ is a submersion, then for all $y \in Y$, the inverse image $f^{-1}(y)$ is a (possibly empty) submanifold of X.*

ii) *If X is compact and if $f : X \to Y$ is an injective immersion, then f is an embedding of X in Y.*

PROOF. Let (U, ϕ) be a chart of X whose domain contains x such that $f(x) = y$, and let (V, ψ) be a chart Y whose domain contains y. Then $\psi \circ f \circ \phi^{-1}$ is again a submersion, and we reduce to the case of open subsets of $\mathbf{R}^n$. (See Theorem 1.21.) For ii), we first note that because we assume X to be compact, f is a homeomorphism to its image. Then if (U, ϕ) is a chart in a neighborhood of $x \in X$, $f \circ \phi^{-1}$ is an immersion and a local homeomorphism, and the arguments of 1.21 apply once more.   □

Warning. This result is false when X is not compact. We have already seen an example of this situation in Section 1.5.3. It is now easier to see this example as the immersion of $\mathbf{R}$ in T^2 given by

$$t \longmapsto \left(\cos t, \sin t, \cos \sqrt{2}\,t, \sin \sqrt{2}\,t\right)$$

(see Figure 2.5). The image of this map is often called the *Kronecker line*. The allusion to the great number theorist Leopold Kronecker comes from the arithmetic character of this example.

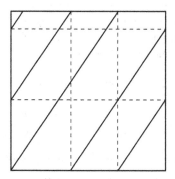

Figure 2.5: Kronecker line

It is not even sufficient to have an injective immersion whose image is closed. To see this, we can take the restriction to $(-\infty, 1)$ of the immersion

$$t \longmapsto \left(\frac{t^2 - 1}{t^2 + 1}, \frac{t(t^2 - 1)}{t^2 + 1} \right) \quad \text{from} \quad (-\infty, 1) \quad \text{to} \quad \mathbf{R}^2,$$

often called "the snake that bites its belly".

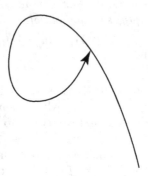

Figure 2.6: The snake that bites its belly

In both of these cases, the induced topology on the image is not locally Euclidean. Otherwise, these two examples do not have much in common, and the second is fairly artificial. Some colleagues in Montpellier suggested the following definition.

Definition 2.29. *An immersion i of a manifold X into a manifold Y is strict if for every manifold Z and every smooth map $f : Z \to Y$ such that $f(Z) \subset i(X)$, the map $i^{-1} \circ f$ is smooth.*

The Kronecker line is given by a strict immersion, but the snake is not.

Remark. When we have an immersion (resp. a strict immersion) from X to Y, we also say that X is a *immersed submanifold* (resp. *strictly immersed*) in Y. This notion is important in the theory of foliations, but we will not have occasion to use it much in the sequel.

We finish this section with an important generalization of Proposition 2.28.

Theorem 2.30. *Let $f : X \to Y$ be a smooth map, and let Z be a smooth submanifold of Y. Suppose that for every x in X such that $f(x) \in Z$ we have*

$$\operatorname{Im}(T_x f) + T_f(x)Z = T_{f(x)}Y \quad \text{(transversality hypothesis)}.$$

Then $f^{-1}(Z)$ is a submanifold of X, with codimension equal to $\operatorname{codim}(Z)$ if it is nonempty.

Remarks

a) This statement still makes sense if $f^{-1}(Z) = \emptyset$.

b) When Z is a point, we recover the first part of Proposition 2.28.

PROOF. Being a local question, by choice of an appropriate chart we can reduce to the case where Y is an open subset of $\mathbf{R}^n$ and $Z = (\mathbf{R}^p \times \{0\}) \cap Y$. By hypothesis, there exists a vector subspace Z' of $\mathrm{Im}(T_x f)$ such that $\mathbf{R}^n = Z' \bigoplus \mathbf{R}^p \times \{0\}$. Let Φ be the projection on Z' parallel to $\mathbf{R}^p \times \{0\}$. Then $f^{-1}(Z) = (\Phi \circ f)^{-1}(0)$. It then suffices to verify that the differential of $\Phi \circ f$ *seen as a function with values in* Z' is surjective. But

$$\mathrm{Im}(\Phi \circ T_x f) \supset \Phi(Z') = Z'.$$

Finally,

$$\mathrm{codim}\left((\Phi \circ f)^{-1}(0)\right) = \dim Z' = n - p = \mathrm{codim}\, Z. \qquad \square$$

2.7. Covering Spaces

We begin by reviewing a few notions related to the quotient topology (for details, see [Dugundji 65, XVI.4], [Munkres 00, 2.22] or [Massey 77, Appendix A]).

If X is a topological space and if $\mathcal{R}$ is an equivalence relation on X, there exists a natural topological structure on $X/\mathcal{R}$. By definition, a subset U of $X/\mathcal{R}$ is open if and only if its inverse image under the quotient map $p : X \mapsto X/\mathcal{R}$ is open in X. This definition is made to ensure the following property: a map f from $X/\mathcal{R}$ to a topological space Z is continuous if and only if $f \circ p : X \to Z$ is continuous. In other words, the continuous maps on $X/\mathcal{R}$ are continuous maps on X that are constant on equivalence classes.

Now if X is a manifold, there is no reason that $X/\mathcal{R}$, equipped with the quotient topology is still a manifold: the topological constraints are too strong. Further, if we want smooth maps on X that pass to the quotient to still be smooth, it is natural to require that if $g = f \circ p$ is such a map, the differential of g determines that of f. As $T_x g = T_{p(x)} f \circ T_x p$, this will be the case if p is a submersion (in which case the equivalence classes will become closed submanifolds).

We will not go much further, and we note simply that the idea of a quotient manifold is not immediate. For more details, see [Godement 05, Chapter 4, § 9]. We will only give this partial consideration, starting with the important case of a quotient of a manifold by a discrete group. Even in this case, some topological prerequisites are indispensable.

2.7.1. Quotient of a Manifold by a Group

Definitions 2.31

a) *A left action by a group Γ on a set X is a map*

$$(\gamma, x) \longmapsto \gamma \cdot x \quad \text{from} \quad \Gamma \times X \quad \text{to} \quad X$$

such that

$$e \cdot x = x \quad \text{and} \quad \gamma_1 \cdot (\gamma_2 \cdot x) = (\gamma_1 \gamma_2) \cdot x$$

for any $x \in X$, $\gamma_1, \gamma_2 \in \Gamma$.

b) *A continuous action of a group Γ on a topological space X is an action such that for all γ the map $x \mapsto \gamma \cdot x$ is continuous.*

c) *A smooth action of a group Γ of a manifold X is an action such that for all γ the map $x \mapsto \gamma \cdot x$ is smooth.*

Remarks

a) In the definition, a) reduces to giving a morphism from Γ to the group of bijections of X.

b) It follows from the definitions that for fixed γ, the map $x \mapsto \gamma \cdot x$ is a homeomorphism in case b), and a diffeomorphism in case c).

c) We can reformulate b) (resp. c)) by saying that map $(\gamma, x) \mapsto \gamma \cdot x$ is a *continuous* map, from $\Gamma \times X$ to X, Γ being equipped with the discrete topology. This allows us to link topological group actions with Lie group actions that we will see in Chapter 4.

To avoid trivialities, we will need the following definition.

Definition 2.32. *A group action is* effective *if for $\gamma \neq e$ the map $x \mapsto \gamma \cdot x$ is distinct from the identity, or in other words if the homeomorphism $\gamma \mapsto (x \mapsto \gamma \cdot x)$ from Γ to the group of bijections of X is injective.*

In this case, we again denote the map $x \mapsto \gamma \cdot x$ by γ.

Example. The action of $\mathbf{Z}$ on $\mathbf{R}$ defined by $(n, x) \mapsto (-1)^n x$ is not effective, but the analogous action of the group $\{\pm 1\}$ is. This is an example of a general situation described in Exercise 18, that justifies why we usually consider effective actions.

Definition 2.33. *If E is a topological space on which a group Γ acts, the* quotient of E by Γ, *denoted E/Γ, is the space of orbits of Γ, or put differently the quotient of E by the equivalence relation*

$$x \simeq y \iff \exists \gamma \in \Gamma, \ y = \gamma \cdot x,$$

equipped with the quotient topology.

We let p denote the quotient map. Recall that the *orbit* of a point x in E is its equivalence class, this is to say the subset $\{\gamma \cdot x\}, \gamma \in \Gamma$ of X. We denote this by $\Gamma \cdot x$.

We have in mind the case where E is a manifold, and we look for sufficient conditions on the action of Γ so that E/Γ will be a manifold. First, we check that E/Γ is locally compact. The essential point is the Hausdorff property.

Definition 2.34. *A discrete group Γ acts* properly *on a locally compact space X if for every pair of compact subsets K and L of X, the set*

$$\{\gamma \in \Gamma : \ \gamma(K) \cap L \neq \emptyset\}$$

is finite.

Note that some references call this a "discontinuous action", which is both questionable and widespread.

Examples and counterexamples

a) Every finite group acts properly.

b) The action of $\mathbf{Z}$ on $\mathbf{R}$ given by

$$(n, x) \longmapsto x + n$$

is proper, but the one given by

$$(n, x) \longmapsto 2^n x$$

is not.

c) The group $\mathbf{Z}^n$ acts properly on $\mathbf{R}^n$ by translations: it suffices to check that every compact subset of $\mathbf{R}^n$ is contained in a sufficiently large cube $\{(x_1, \ldots, x_n) : |x_i| \leqslant A\}$. Conversely, if α is an irrational number, the action of $\mathbf{Z}^2$ on $\mathbf{R}$ given by

$$(m, n, x) \longmapsto x + \alpha m + n$$

is not proper. (We can verify for example that the quotient topology of the space of orbits is the trivial topology.)

For more examples and counterexamples, see Exercise 19.

Theorem 2.35

i) *For a map f from E/Γ to a topological space X to be continuous, it is necessary and sufficient that $f \circ p$ is continuous.*

ii) *The image of an open subset of E under p is an open subset of E/Γ.*

iii) *If E is locally compact and if Γ is a discrete group acting properly, then E/Γ is locally compact.*

PROOF. Recall that by definition of the quotient topology, a subset V of E/Γ is open if and only if $p^{-1}(U)$ is open; i) is then immediate.

For ii), it suffices to remark that

$$p^{-1}(p(U)) = \bigcup_{\gamma \in \Gamma} \gamma(U)$$

is open once U is open, as the γ give homeomorphisms of E.

We move on to iii). To show that E/Γ is Hausdorff, the essential point is that it suffices to see if $y \notin \Gamma \cdot x$, that there exists neighborhoods V and W of x and y respectively such that

$$\gamma(V) \cap \gamma'(W) = \emptyset$$

for all $\gamma, \gamma' \in \Gamma$, or again

$$V \cap \gamma(W) = \emptyset$$

for all $\gamma \in \Gamma$. First, as E is locally compact, x and y have compact disjoint neighborhoods K and L. The action of Γ is proper, and so $\gamma(L)$ only meets K for a finite number of γ, say

$$\gamma_1, \ldots, \gamma_n.$$

However $x \neq \gamma_i(y)$ for any i, therefore by continuity x and y are contained in compact neighborhoods A_i and B_i such that

$$A_i \cap \gamma_i(B_i) = \emptyset.$$

It now suffices to take

$$V = K \cap \left(\bigcap_{i=1}^{n} A_i \right), W = L \cap \left(\bigcap_{i=1}^{n} B_i \right).$$

Local compactness now follows immediately from ii). □

We now examine the case where the quotient space is a manifold. The action $x \mapsto \pm x$ of $\mathbf{Z}/2\mathbf{Z}$ on $\mathbf{R}$ gives a simple example of a proper action whose quotient is not a manifold. We can also remark that if an element γ different from the identity has a fixed point, p is very likely not a submersion: if $\gamma(a) = a$ and if for example $T_a\gamma - I_a$ is invertible, then from the equality $T_a p \circ T_a \gamma = T_a p$, we may deduce by differentiating $p \circ \gamma = p$ at a that $T_a p = 0$! The following definition is made to avoid this situation.

Definition 2.36. *A group Γ acting on a set E is said to be* free *if*

$$\forall \gamma \neq e, \forall x \in E, \quad \gamma \cdot x \neq x.$$

The quotient map of a free and proper discrete group action on a locally compact space has remarkable properties meriting study themselves.

Definition 2.37. *A continuous map $p : X \to B$ is a* covering *map with base B and total space X if every $b \in B$ is contained in an open subset U such that $p^{-1}(U)$ is a union (possibly infinite) of mutually disjoint open subsets $(V_\alpha)_{\alpha \in A}$ of X, such that the restriction of p to each V_α is a homeomorphism on U.*

If X and B are manifolds and p is a smooth map, by replacing homeomorphims in the previous definition with diffeomorphisms, we obtain the notion of smooth covering.

If B is connected, the definition implies that the cardinality of $p^{-1}(x)$ is constant. We call this cardinality the *degree* of the covering.

Remarks

a) A covering is in fact a fibration where the fibers are endowed with the discrete topology (compare to Theorem 2.14).

b) If X is a manifold, we may instead suppose every $b \in B$ is contained in a connected open subset U such that the restriction of p to each connected component of $p^{-1}(U)$ is a diffeomorphism (or a homeomorphism).

Examples

a) If I is any subset equipped with the discrete topology, the projection of $I \times B$ to B is a covering: for each $i \in I$, $\{i\} \times B$ is an open and closed subset of the total space that is homeomorphic to the base. Such a covering is

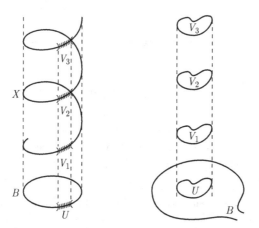

Figure 2.7: Local trivialization

called *trivial*. More generally, a covering $p : X \to B$ is called *trivializable* if there exists a discrete space I and a homeomorphism $f : B \times I \to X$ such that $p \circ f$ is the projection of $B \times I$ to B. We note that the definition of coverings can be reformulated by saying that every point in the base is contained in an open subset U such that $p : p^{-1}(U) \to U$ is trivializable.

We note that a covering whose total space X is connected is not trivializable unless p is a homeomorphism.

b) Take $X = \mathbf{R}$, $B = S^1$, and

$$p(t) = (\cos t, \sin t).$$

Let $\tau \in (0, \pi)$. For each t_0, we consider the open subset

$$U = \{(\cos(t_0 + t), \sin(t_0 + t)),\ t \in (-\tau, \tau)\}.$$

Then

$$p^{-1}(U) = \bigcup_{n \in \mathbf{Z}} (t_0 - \tau + 2n\pi, t_0 + \tau + 2n\pi).$$

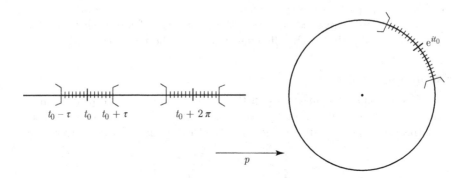

Figure 2.8: Covering of the circle by the line

c) Take $X = B = \mathbf{C}^*$, or $X = B = S^1$, and in both cases

$$p(z) = z^k.$$

Here, the same argument as in b) shows that each $b \in B$ is contained in an open subset U such that $p^{-1}(U)$ is composed of k open subsets homeomorphic to U.

d) If $f : X \to Y$ is a smooth map between two manifolds of the same dimension, and if X is compact, f is a covering of the set of regular points to the set of regular values. The is the content of the statement of Theorem 2.14.

In case b), p passes to the quotient as a homeomorphism (and even a diffeomorphism, as we will soon see) of $\mathbf{R}^2/2\pi\mathbf{Z}$ to S^1. In case c), two points have the same image under p if and only if they are transformed to each other by a rotation about the center 0 and angle $2r\pi/k$. These rotations define a group action $\mathbf{Z}_k = \mathbf{Z}/k\mathbf{Z}$, and from this we deduce that $S^1/\mathbf{Z}_k$ is homeomorphic to S^1.

This is an example of the situation described by the following theorem.

Theorem 2.38

i) *If a discrete group Γ acts freely and properly on a locally compact space X, the map*

$$p: \ X \longrightarrow X/\Gamma$$

is a covering map.

ii) *If in addition X is a manifold on which Γ acts smoothly, there exists a unique smooth structure on X/Γ on which p is a smooth covering map.*

We rely mainly on the following lemma.

Lemma 2.39. *Under the same hypotheses, every $x \in X$ is contained in an open subset V such that the images $\gamma(V)$ are mutually disjoint.*

PROOF. Suppose W is a compact neighborhood of x. By the properness hypothesis, W can only meet a finite number of its images under elements of Γ, say $\gamma_1(W), \gamma_2(W), \ldots, \gamma_p(W)$. For each i between 1 and p, we can find disjoint open subsets W_i' containing x and W_i'' containing $\gamma_i(x)$. It then suffices to take

$$V = W \cap \left(\bigcap_{1 \leqslant i \leqslant p} W_i' \cap \gamma_i^{-1}(W_i'') \right).$$

Then by construction V will not meet $\gamma_i(V)$, and since $V \subset W$, V will not meet the other $\gamma(V)$. $\qquad\square$

PROOF OF THEOREM 2.38. If we write $U = p(V)$, we see that

$$p^{-1}(U) = \bigcup_{\gamma \in \Gamma} \gamma(V).$$

The restriction of p to each $\gamma(V)$ is by construction a bijection on U, which is continuous and open, and therefore a homeomorphism, which shows the first part.

If, in the case where X is a smooth manifold, we want p to be a smooth covering, the manifold structure on the quotient space is imposed. Indeed, let $y \in X/\Gamma$, and let U contain y such that $p^{-1}(U)$ is a union of disjoint open subsets diffeomorphic to U. By i), if V is any one of them, the others are of the form $\gamma(V)$, where γ runs over Γ. By replacing V by a smaller open subset containing x in V with image y, we can always suppose that V (and therefore $\gamma(V)$ for each γ) is diffeomorphic to an open subset of $\mathbf{R}^n$. Then if

$$\phi_{\gamma,V} : \ \gamma(V) \longrightarrow \mathbf{R}^n$$

is a chart, the same must be true of

$$\phi_{\gamma,V} \circ p^{-1}_{|\gamma(V)} : \ U \longrightarrow \mathbf{R}^n.$$

The compatibility condition is satisfied. For the same U, the transition function is

$$\left(\phi_{\gamma,V} \circ p^{-1}_{|\gamma(V)}\right) \circ \left(\phi_{\gamma',V} \circ p^{-1}_{|\gamma'(V)}\right)^{-1} = \phi_{\gamma,V} \circ \left(p^{-1}_{|\gamma(V)} \circ p_{|\gamma'(V)}\right) \circ \phi^{-1}_{\gamma',V}$$

$$= \phi_{\gamma,V} \circ \gamma^{-1} \circ \gamma' \circ \phi^{-1}_{\gamma',V},$$

which is the local expression of a diffeomorphism $\gamma^{-1} \circ \gamma'$. For two distinct open subsets U and U' with nonempty intersection, we proceed in the same way, after we remark that there exists a γ_0 such that

$$p^{-1}(U \cap U') = \bigcup_{\gamma \in \Gamma} \gamma\big(V \cap \gamma_0(V')\big). \qquad \square$$

Corollary 2.40. *Under hypothesis ii) above, for a map f from X/Γ to a manifold Y to be smooth, it is necessary and sufficient that $f \circ p$ be smooth.*

PROOF. It suffices to use the fact that p is a local diffeomorphism. $\qquad \square$

Examples: real projective space and tori

a) The group $\mathbf{Z}/2\mathbf{Z}$ with two elements acts properly and freely on S^n by $x \mapsto \pm x$, and the quotient manifold is diffeomorphic to $P^n\mathbf{R}$. This permits a more geometric view of projective geometry: the projective plane with a disk removed is homeomorphic to the quotient of the sphere with two diametrically opposite disks removed. The quotient can also be obtained by starting with a rectangle and identifying opposite sides after inverting the direction. The space obtained is called the *Möbius strip*.

This is an example of a manifold with boundary (this notion will be introduced in Chapter 6) whose boundary has a single connected component. Projective space itself can be seen as a Möbius strip with a disk adjoined to the boundary. One might suspect that such a realization is impossible in the 3-dimensional space. In fact, there does not exist an embedding of $P^2\mathbf{R}$ in $\mathbf{R}^3$, because $P^2\mathbf{R}$ is not orientable (this assertion will be explained and proved in Section 6.2), while every compact submanifold with codimension 1 in $\mathbf{R}^n$ is orientable.

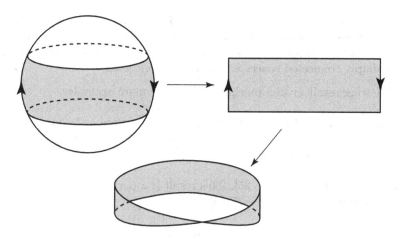

Figure 2.9: From the projective plane to the Möbius strip

b) The action $(n, x) \mapsto x + n$ of $\mathbf{Z}$ on $\mathbf{R}$ is proper and smooth, and the manifold $\mathbf{R}/\mathbf{Z}$ is diffeomorphic to S^1. We obtain such a diffeomorphism by the map $x \mapsto \exp 2i\pi x$ upon passing to the quotient. Similarly, $\mathbf{Z}^n$ acts freely on $\mathbf{R}^n$, and the quotient $\mathbf{R}^n/\mathbf{Z}^n$ is diffeomorphic to the torus T^n.

2.7.2. Simply Connected Spaces

Instead of forming the quotient of a manifold by a group, we can reverse the process and look for coverings of a given manifold. This is the object of elementary homotopy theory (cf. [Dugundji 65], [Stillwell 08] for the essentials, [Fulton 95] and [Massey 77] for further details) and we will state some of the basic results without proof but with sufficient examples to give an idea of what is going on.

The fundamental property is that of a simply connected space. This is, more or less, path connectedness in the space of closed curves.

Definitions 2.41

a) *A* loop *in a topological space X is a continuous map from S^1 to X.*

b) *Two loops in X given by maps f and g are* homotopic *if there exists a continuous map $H : [0,1] \times X \to X$ such that, for every x in X, $H(0, x) = f(x)$ and $H(1, x) = g(x)$*

c) *A path connected space is said to be* simply connected *if every loop is homotopic to a constant map.*

A clear but important example is that of $\mathbf{R}^n$: two loops in $\mathbf{R}^n$ (or more generally a convex subset of $\mathbf{R}^n$) are homotopic. In particular, the convex subsets of $\mathbf{R}^n$ are simply connected. It is also immediate that the product of two simply connected spaces is also simply connected.

The following result gives a machine to produce more examples.

Proposition 2.42.[2] *Let X be a topological space written as $U \cup V$, where U and V simply connected open subsets whose intersection is path connected. Then X is simply connected.*

For a proof, see [Apéry 87, p. 25]. This result is a particular case of the Van Kampen theorem (see [Massey 77, IV.2]).

Corollary 2.43. *The sphere S^n (for $n \geqslant 2$) and complex projective space $P^n\mathbf{C}$ are simply connected.*

PROOF. We view S^n as

$$\left\{ x \in \mathbf{R}^{n+1} : \sum_{i=0}^{n} x_i^2 = 1 \right\}$$

and we apply the proposition to $U = \{x \in S^n : x_0 < \frac{1}{2}\}$ and $V = \{x \in S^n : x_0 > -\frac{1}{2}\}$

For $P^n\mathbf{C}$, we use homogeneous coordinates. By Section 2.5, the open subsets U and V defined by $z_0 \neq 0$ and $z_1 \neq 0$ respectively are diffeomorphic to $\mathbf{C}^n$, and their intersection is diffeomorphic to the complement of a *complex* hyperplane of $\mathbf{C}^n$. □

This argument apples neither to S^1 nor to $P^n\mathbf{R}$: in this case, $U \cap V$ is not connected. In fact, in general it is often easy to prove a space is simply connected, while proving a space is not simply connected is much harder. It is true that S^1 and $P^n\mathbf{R}$ are not simply connected, and this is a consequence of homotopy theory, and more precisely the following fundamental result.

Theorem 2.44.[3] *Every covering of a simply connected manifold is trivializable. In particular, every connected covering of a simply connected manifold is a diffeomorphism.*

So in particular $P^n\mathbf{R}$, which admits a covering of degree 2 by S^n, is not simply connected.

---

2. Stated without proof.
3. Stated without proof.

Corollary 2.45 (monodromy theorem). *Let* $p : Y \to X$ *be a covering of a smooth manifold* X, *and let* X' *be a smooth simply connected manifold, and let* $f : X' \to X$ *be a smooth map. Let* a' *be a point of* X'. *Then for all points* $b \in Y$ *such that* $p(b) = f(a')$, *there exists a unique smooth map* $g : X' \to Y$ *such that* $p \circ g = f$ *and* $g(a') = b$.

PROOF. Introduce

$$Y' = \{(x', y) \in X' \times Y : \ f(x') = p(y)\}.$$

We show this is a submanifold of $X' \times Y$. Indeed, it is the inverse image of the diagonal of $X \times X$ under $(f, p) : X' \times Y \to X \times X$; the image of $T_{(x', y)}(f, p)$ contains $\{0\} \times T_x X$ (here we denote the common value of $f(x')$ and $p(y)$ by x), which is transverse in $T_{(x, x)} X \times X$ to the tangent space to the diagonal. We can therefore apply Theorem 2.30. Let p' and f' be restrictions to Y' of the projections of $X' \times Y$ to its factors. Then $p' : Y' \to X'$ is a covering: if U is a trivializing open subset of p, we verify that $f^{-1}(U)$ is a trivializing open subset of p'. As X' is simply connected, this covering is trivializable: the restriction of p' to each connected component of Y' is a diffeomorphism by Theorem 2.44. Let Y_0' be the connected component that contains (a', b), and q the inverse diffeomorphism of the restriction of p' to Y_0'. Then we can take $g = \mathrm{pr}_2 \circ q$.

We now verify uniqueness. Let g_1 and g_2 be two smooth maps such that $p \circ g_1 = p \circ g_2 = f$, with $g_1(a') = g_2(a') = b$. Now the set $x \in X'$ such that $g_1(x) = g_2(x)$ is nonempty, closed (this is a general property of continuous maps), and open (this uses the fact that p is a covering). As X' is connected, $g_1 = g_2$. $\qquad\square$

We may prove (see the same references above) that every manifold X is diffeomorphic to a quotient Y/Γ, where Y is a simply connected manifold, and Γ is a group acting freely and properly on Y. The group Γ and the manifold Y, which are unique up to isomorphism and diffeomorphism respectively, are called the *fundamental group* and *universal cover* of X. Thus, the fundamental group of T^n is $\mathbf{Z}^n$, and that of $P^n \mathbf{R}$ is the group with two elements, $\mathbf{Z}/2\mathbf{Z}$.

2.8. Countability at Infinity

We end this chapter by gathering several elementary but useful topological properties of manifolds.

By the definition, a smooth manifold (and even a topological manifold) is *locally compact*, and *locally connected* (this is to say every point admits a base of connected neighborhoods). In particular, every connected component of a manifold M is an open subset of M, and thus a submanifold of the same dimension. Therefore the study of a manifold reduces to that of its connected components. Moreover we have the following property.

Proposition 2.46. *A connected open subset of a (topological) manifold is path connected.*

PROOF. On such an open subset U, we define an equivalence relation by saying that two points x and y are equivalent if there exists a continuous map c from a closed interval $[a, b]$ in U such that $c(a) = x$ and $c(b) = y$. The equivalence classes are open, and therefore there is exactly one.   □

As in the Euclidean case, we would like a submanifold with strictly positive codimension to be of measure zero in the ambient manifold, and more generally to have a statement analogous to Sard's theorem. We first give a definition.

Definition 2.47. *Let X be a manifold. A subset $E \subset X$ is* negligible *if for every $x \in X$ there exists a chart (U, ϕ), with U containing x, such that $\phi(U \cap E)$ is of measure zero in $\phi(U)$.*

By Lemma 1.39, this property is independent of the chart chosen. Further, it is immediate that every countable union of negligible sets s negligible.

The proof of Sard's theorem seen in Chapter 1 (Theorem 1.41) then extends, but with one condition: as we applied arguments with charts in the domain space, we will need to assume the existence of a countable atlas. We are led to the following definition.

Definition 2.48. *A manifold X is* countable at infinity *if it is a union of a countable collection of compact subsets.*

Remark. This property is often called *the second axiom of countability*.

This reduces to saying that the point at infinity of the Alexandrov compactification $\widehat{X}$ of X has a countable neighborhood base, which explains the terminology. We recall (see [Dugundji 65, XI.8.4]) that $\widehat{X}$ is the disjoint union of X and a point ω, "the point at infinity". We define a topology on $\widehat{X}$ by taking a base of open subsets formed by the open subsets of X and the sets $\omega \cup (X \smallsetminus K)$, where K runs overs the set of compact subsets of X.

With this hypothesis satisfied, the manifold structure can be defined by an at most countable atlas. If $X = \bigcup_{n \in \mathbf{N}} K_n$, where the K_n are compact, and if $(U_i, \phi_i)_{i \in I}$ is an atlas that defines the differentiable structure on X, each K_n can be covered by a finite number of open subset U_i. Then the manifold X itself can be covered by a countable number of U_i, because every countable union of finite sets is countable. We have therefore proved the following result.

Theorem 2.49. *Let X and Y be two manifolds, and let $f : X \to Y$ be a smooth map. If X is countable at infinity, the set of critical values of f is a negligible subset of Y.*

We will use this result only in the case where $\dim X \leqslant \dim Y$, which we have proved. For the general case, see [Hirsch 76].

> **HENCEFORTH, WE ASSUME ALL MANIFOLDS
> ARE COUNTABLE AT INFINITY**

Manifolds that are not countable at infinity are relatively pathological. An amusing example, which uses the theory of ordinals, is that of the *long line*, or transfinite line obtained by gluing together an uncountable number of copies of $\mathbf{R}$. For an explicit description, see [Spivak 79, Appendix], or [Douady 05, pp. 14–15].

2.9. Comments

How to get manifolds

The real numbers are defined by Dedekind cuts or by equivalence classes of Cauchy sequences. These points of view are rarely used in everyday (mathematical) life, where we manipulate given real numbers defined for example as the solution of an equation or a sum of a series.

The same is true for manifolds. The simplest examples are submanifolds of Euclidean space. (This simplicity can be misleading: it is not always easy to extract precise topological information from a system of equations, just as it is not easy to know if a real number given as the sum of a series is rational, algebraic or transcendental.)

We have seen certain group actions on manifolds give manifolds by passing to the quotient, and we will see in Chapter 4 another type of quotient.

Finally, Exercise 28 suggests the possibility of obtaining manifolds by gluing two or more manifolds along diffeomorphic open subsets.

Topological manifolds and smooth manifolds

After we defined manifolds, we were very discreet about the question of existence and uniqueness. We will be very brief about this subject, as it leads to difficult questions. In two dimensions, one shows that every topological manifold has a unique smooth structure up to diffeomorphism. This result is false in higher dimension: there are topological manifolds that do not have even a C^1 structure, and others which have many non diffeomorphic smooth structures. The first example, that of the sphere S^7 was discovered by Kervaire and Milnor in the 1950s. This manifold admits simple descriptions: we can realize it (as the "standard" sphere S^7, see Exercise 16 in Chapter 7) as the total space of an appropriate fibration on S^4 with S^3 fibers, or as a submanifold of codimension 3 in $\mathbf{R}^{10}$ given by polynomial equations. Conversely it is difficult to show that the manifold then obtained is not diffeomorphic to S^7 with its standard structure. To get an idea of what happens, see [Dieudonné 88, VII.B].

If $n \neq 4$, the differentiable structure of $\mathbf{R}^n$ is unique (always up to diffeomorphism, see [Stalling 62]), but it has been known since the 1980s that there exists an infinite number of non diffeomorphic smooth structures on $\mathbf{R}^4$! For an idea of this construction see [Lawson 85].

To return to more accessible considerations, which are at the root of rich geometric theory, we can study atlases whose transitions functions preserve a local geometric property of $\mathbf{R}^n$ to some extent. One obtains richer structures than a simple manifold structure. We now give a few examples.

Foliations

A foliation of codimension q on an n-manifold M is a collection of charts on M with values in the open subsets of the form $U \times V$, where U and V are open in $\mathbf{R}^p$ and $\mathbf{R}^q$ respectively, and with transition functions of the form

$$(x, y) \longmapsto \big(f(x, y), g(y)\big) \quad \text{where } x \in U,\ y \in V.$$

The equations $y = constant$ then define *strictly immersed submanifolds* of codimension q in the ambient space X (these are not submanifolds in general) called the *leaves*, and which form a partition of X. The coordinates y show that locally we can parametrize the leaves by a "transverse" submanifold of dimension q. Even in the simple case where the leaves are defined by a global submersion, their topological type can change: the function $f(x, y, z) = (1 - x^2 - y^2)e^z$ defines a foliation of codimension 1 on $\mathbf{R}^3$ whose leaves $f^{-1}(c)$ are diffeomorphic to $\mathbf{R}^2$ if $c > 0$, and to $S^1 \times \mathbf{R}$ if $c \leqslant 0$. To learn more about this subject, see for example [Hector-Hirsch 81].

Flat structures

We can impose even stronger constraints on the transition functions. An extreme case consists of requiring that they be translations. This is actually too restrictive: one can show that a manifold of this type is a torus.

If the transition functions are affine isometries of $\mathbf{R}^n$, the situation is more interesting, but also restrictive. We obtain *flat Riemannian manifolds*. These are quotients of $\mathbf{R}^n$ by a discrete subgroup of isometries acting without fixed points. For each n, there are only finite many such topological types of such manifolds, and all are quotients of the torus (for more details, see [Wolf 84]).

Affine manifolds are obtained by requiring that the transition functions be affine transformations of $\mathbf{R}^n$. This situation is much richer, and is still mysterious even today: we must study not only quotients of $\mathbf{R}^n$ under the action of a discrete subgroup of the affine group, but also of quotients of certain open subsets of $\mathbf{R}^n$ by such actions. The simplest example of such a situation is that of the quotient of $\mathbf{R}^n \smallsetminus \{0\}$ by the group generated by the homothety $x \mapsto \lambda x$ ($\lambda \neq 1$).

In an analogous way, we can take charts with values in open subsets of $P^n\mathbf{R}$ or of S^n, and require that transition functions are given by elements of $PGl(n+1, \mathbf{R})$ – see Exercise 5 below – in the first case (projective manifolds), or of the Möbius group – see Exercise 12 – in the second (conformally flat manifolds). For such questions, see for example [Kulkarni-Pinkall 88].

2.10. Exercises

1. *A non Hausdorff space locally homeomorphic to* $\mathbf{R}$

Let X be the real line with two origins. In other words, $X = \mathbf{R} \coprod \{\alpha\}$, with open subsets of X being the unions of open subsets of $\mathbf{R}$ and sets of the form $U \smallsetminus \{0\} \cup \{\alpha\}$, where U is open in $\mathbf{R}$. Show that every point of X is contained in an open subset diffeomorphic to $\mathbf{R}$, but X is not Hausdorff.

2. Equip $\mathbf{R}^2$ with its canonical Euclidean structure. At every point a, associate the function f_a, defined on the manifold M of straight lines by the formula

$$f_a(d) = (\mathrm{dist}(a, d))^2.$$

Show that the function f_a is smooth.

3. Show that the set of points $(x, y, z, t) \in \mathbf{R}^4$ such that

$$x^2 + y^2 = z^2 + t^2 = \frac{1}{2}$$

is a submanifold of S^3, diffeomorphic to $S^1 \times S^1$. In the same way, give examples of submanifolds of S^{2n-1} diffeomorphic to $(S^1)^n$.

4. *The unitary and special unitary groups*

a) Show, by using an appropriate submersion, that the set $U(n)$ of unitary matrices (matrices of order n with complex coefficients such that ${}^t\overline{A}A = I$) is a submanifold of $\mathbf{R}^{2n^2}$ of dimension n^2. Use the exponential map to obtain a parametrization of $U(n)$.

b) In the same way show that the set $SU(n)$ of *special unitary* matrices (defined by the conditions $A \in U(n)$ and $\det A = 1$) is a submanifold of dimension $n^2 - 1$.

c) Show that $SU(2)$ is diffeomorphic to S^3.

5. *Projective group*

a) Show that every invertible linear map $A \in Gl(n+1, \mathbf{R})$ defined by passing to the quotient $P^n\mathbf{R}$ is a diffeomorphism, and that the group of diffeomorphisms so obtained is isomorphic to $Gl(n+1, \mathbf{R})/\mathbf{R}^*I$.

b*) Explicitly write the action of $Sl(2, \mathbf{R})$ on S^1 obtained.

Note. This group is denoted $PGl(n+1, \mathbf{R})$, and is called the projective group. Everything proceeds in the same way if we replace $\mathbf{R}$ by $\mathbf{C}$.

6. *Projective quadrics*

a) Let q be a quadratic form of maximum rank on $\mathbf{R}^4$, and let p be the canonical projection of $\mathbf{R}^4 \setminus \{0\}$ to $P^3\mathbf{R}$. Show that $p(q^{-1}(0))$ is a (possibly empty) submanifold of $P^3\mathbf{R}$.

b) Show that if q is of type $(1, 3)$ or $(3, 1)$ this submanifold is diffeomorphic to S^2.

c) Show that if q is of type $(2, 2)$ this submanifold is diffeomorphic to $P^1\mathbf{R} \times P^1\mathbf{R}$, which is to say $S^1 \times S^1$.

d*) More generally, given a non-degenerate quadratic from q on $\mathbf{R}^{n+1}$, study the topology of the manifold $p(q^{-1}(0))$.

7. *Generalities involving immersions and submersions*

a) Show that the composition of two immersions (resp. submersions) is an immersion (resp. a submersion).

b) Let X and Y be two manifolds, and let $f : X \to Y$ be a smooth map. Show that the graph of f is a closed submanifold of $X \times Y$, and that the map $g : x \mapsto (x, f(x))$ is an embedding.

8. Given an example of an embedding of T^3 into $\mathbf{R}^4$ and of $S^2 \times S^2$ into $\mathbf{R}^5$.

9. Let $v : I \to \mathbf{R}^{n+1} \setminus \{0\}$ be a smooth map ($I \subset \mathbf{R}$ denotes an open interval) and let $p : \mathbf{R}^{n+1} \setminus \{0\} \to P^n\mathbf{R}$ be the canonical projection. Show that $p \circ v$ is an immersion in t if and only if the vectors $v(t)$ and $v'(t)$ are independent.

10. *More examples of embeddings*

a) Let f be a T-periodic smooth map of $\mathbf{R}$ to a manifold X, which is injective on $[0, T)$. Show that $f(\mathbf{R})$ is a submanifold of X diffeomorphic to S^1. (We call such a submanifold a closed curve.)

b) Show that the map

$$(u, v) \longmapsto (u^n, \ldots, u^{n-k}, v^k, \ldots, v^n) \quad \text{from } \mathbf{R}^2 \setminus \{0\} \text{ to } \mathbf{R}^{n+1}$$

defines an immersion of $P^1\mathbf{R}$ to $P^n\mathbf{R}$. Is this immersion an embedding?

11. *A little more on submersions*

a) Let $f : X \to Y$ be a submersion of a manifold X to a manifold Y. Show that $f(X)$ is open in Y. Deduce that if X is compact and Y is connected, then f is surjective. Does this property persist if X is not compact?

b) Let Z be a submanifold of Y. Show (still supposing that f is a submersion) that $f^{-1}(Z)$ is a submanifold of X.

c*) Example: if $h : S^3 \to S^2$ is the Hopf fibration, show that the inverse image of a closed curve in S^2 is a submanifold of S^3 which is diffeomorphic to $S^1 \times S^1$.

12. Let M be a manifold. Show that the tangent space to the diagonal of $M \times M$ is the diagonal of $T_mM \times T_mM$.

13. *An embedding of the projective plane into $\mathbf{R}^4$: the Veronese surface*

a) Show that the map v from $\mathbf{R}^3$ to $\mathbf{R}^6$ given by

$$v(x, y, z) = \left(x^2, y^2, z^2, \sqrt{2}xy, \sqrt{2}yz, \sqrt{2}zx\right)$$

defines an immersion of S^2 into $\mathbf{R}^6$.

Hint. First show that v is an immersion of $\mathbf{R}^3 \setminus \{0\}$ into $\mathbf{R}^6$.

b) Is the map v injective? Show that it defines a homeomorphism V from $P^2\mathbf{R}$ to $v(S^2)$.

c) Show that $v(S^2)$ is a submanifold of $\mathbf{R}^6$ and that V is an embedding of $P^2\mathbf{R}$ into $\mathbf{R}^6$ (use the fact that the map $p : (x, y, z) \mapsto [(x, y, z)]$ from S^2 to $P^2\mathbf{R}$ is a local diffeomorphism).

d) Show that $v(S^2) = V(P^2\mathbf{R})$ is included in $H \cap S^5$, where H is an (affine) hyperplane of $\mathbf{R}^6$ and S^5 is the unit sphere. Deduce that there exists an embedding of $P^2\mathbf{R}$ into $\mathbf{R}^5$ and even into $\mathbf{R}^4$.

Conversely, there does not exist an embedding of $P^2\mathbf{R}$ into $\mathbf{R}^3$. Indeed, every compact hypersurface of $\mathbf{R}^n$ is orientable (see [Hirsch 76]), while $P^2\mathbf{R}$ is not orientable (see Chapter 6 for this notion and result). There are, on the other hand, many *immersions* of the projective space into $\mathbf{R}^3$. One can find explicit examples and beautiful pictures in [Apéry 87].

14. Define a map p from $SO(n+1)$ to S^n by

$$p(g) = g \cdot e_0,$$

where e_0 is the first vector in the canonical basis of $\mathbf{R}^{n+1}$.

a) Show that p is smooth, and that the inverse image of a point is a submanifold of $SO(n+1)$ diffeomorphic to $SO(n)$.

b*) Show that p is a fibration.

15. *A surjection from projective space to the sphere of the same dimension*

a) Show that the subset of points of $P^n\mathbf{R}$ where a homogeneous coordinate (the first for example) is zero forms a submanifold diffeomorphic to $P^{n-1}\mathbf{R}$.

b) Consider the map from $\mathbf{R}^{n+1} \smallsetminus \{0\}$ to $\mathbf{R}^{n+1}$ defined by

$$(t, x_1, \ldots, x_n) \longmapsto \left(\frac{2tx_1}{t^2 + \|x\|^2}, \ldots, \frac{2tx_n}{t^2 + \|x\|^2}, \frac{-t^2 + \|x\|^2}{t^2 + \|x\|^2} \right),$$

where

$$\|x\|^2 = \sum_{i=1}^n x_i^2.$$

Show that this map defines a map p from $P^n\mathbf{R}$ to S^n by passing to the quotient, and that p is smooth. What is the inverse image of the North pole $N = (0, \ldots, 0, 1)$? of the South pole $(0, \ldots, 0, -1)$?

c) Using stereographic projection from the North pole N, show that p induces a diffeomorphism from $P^n\mathbf{R} \smallsetminus p^{-1}(N)$ to $S^n \smallsetminus \{N\}$.

d) What can we say about p for $n = 1$?

e) Show that the set of points in $P^n\mathbf{R}$ where a homogeneous coordinate (the first for example) is nonzero is connected. Assuming the fact that the complement of a simple closed curve in S^2 has two connected components, deduce that $P^2\mathbf{R}$ is not homeomorphic to S^2.

f) Now consider the map from $\mathbf{C}^{n+1} \smallsetminus \{0\}$ to $\mathbf{C}^n \times \mathbf{R}$ given by

$$(\zeta, z_1, \ldots, z_n) \longmapsto \left(\frac{2\bar{\zeta} z_1}{|\zeta|^2 + \|z\|^2}, \ldots, \frac{2\bar{\zeta} z_n}{|\zeta|^2 + \|z\|^2}, \frac{\|z\|^2 - |\zeta|^2}{|\zeta|^2 + \|z\|^2} \right),$$

where

$$\|z\|^2 = \sum_{i=1}^{n} |z_i^2|.$$

Imitating the above, show that we can also define a smooth map q from $P^n \mathbf{C}$ to S^{2n}, which induces a diffeomorphism between $P^n \mathbf{C} \smallsetminus P^{n-1} \mathbf{C}$ and $S^{2n} \smallsetminus \{N\}$. What happens for $n = 1$?

16*. *Conformal compactification of $\mathbf{R}^n$; Möbius group* [4]

a) Equip the space $\mathbf{R}^n$ with its usual Euclidean norm, and define a map p from $\mathbf{R}^n$ to $P^{n+1}\mathbf{R}$ by the formula

$$p(x) = \left[\left(\frac{1}{2}, x, \frac{1}{2}\|x\|^2 \right) \right].$$

Show that p is a diffeomorphism from $\mathbf{R}^n$ to the "quadric" Q_n of $P^{n+1}\mathbf{R}$ define by the equation

$$4X_0 X_{n+1} - \sum_{i=1}^{n} X_i^2 = 0$$

with the point $[(1, 0, \ldots, 0)]$ excluded.

b) Show that Q_n is diffeomorphic to S^n.

c) We write $O(1, n + 1)$ for the subgroup of $Gl(n + 2, \mathbf{R})$ that leaves the quadratic form

$$4X_0 X_{n+1} - \sum_{i=1}^{n} X_i^2$$

(with signature $(1, n+1)$!) invariant, and $PO(1, n + 1)$ the corresponding subgroup of $PGl(n+2, \mathbf{R})$ (see Exercise 5). Show that $PO(1, n+1)$ is the subgroup of $PGl(n + 2, \mathbf{R})$ which leaves Q_n globally invariant. By using translation by p as in a), show that the following transformations extend in a unique way to transformations of $PO(1, n + 1)$:

1) linear isometries;

2) homotheties;

3) translations;

4) inversion $x \mapsto \frac{x}{\|x\|^2}$.

---

4. This exercise is addressed to readers having some notions of "elementary geometry", such as those expounded in [Berger 87] for example.

Conversely, let $r \in O(1, n+1)$ be a *reflection*. (Recall that this means that $r^2 = 1$, $r \neq I$, and that r leaves a hyperplane invariant point by point.) Show that the projective transformation associated to r is obtained by extending an inversion.

17*. *Blow up*

Let E be the subset of $P^1\mathbf{R} \times \mathbf{R}^2$ defined by the equation

$$xY - yX = 0$$

(here we write the coordinates of a point in $\mathbf{R}^2$ as (x, y), and the homogeneous coordinates of a point in $P^1\mathbf{R}$ as (X, Y)). In other words, E is the set of ordered pairs (p, D) formed by a point $p \in \mathbf{R}^2$ and a straight line D passing through the origin such that $p \in D$.

a) Show that E is a submanifold of dimension 2 in $P^1\mathbf{R} \times \mathbf{R}^2$ (E is called the blow up of $\mathbf{R}^2$ at 0; c) and d) give the reasons for this terminology).

b) Show that the restrictions to E of the projections from $P^1\mathbf{R} \times \mathbf{R}^2$ to the factors are smooth maps.

c) Let π be the restriction of E of the second projection. Show that $\pi^{-1}(0)$ is diffeomorphic to $P^1\mathbf{R}$. Show that π induces a diffeomorphism from $E \smallsetminus \pi^{-1}(0)$ to $\mathbf{R}^2 \smallsetminus \{0\}$.

d) Let r denote the inverse of the diffeomorphism in c). Then let c be a smooth map from $I = (-\epsilon, \epsilon)$ to $\mathbf{R}^2$ such that

$$c(t) \neq 0 \text{ for } t \neq 0, \quad c(0) = 0, \quad c'(0) \neq 0.$$

Show that the map
$$r \circ c: \ I \smallsetminus \{0\} \longrightarrow E$$
extends in a unique way to a continuous map $c: I \to E$. Show that c is smooth (use the Hadamard lemma, Lemma 3.12).

Application. If for example F is the "*folium* of Descartes" given by the equation $x^3 + y^3 - 3xy = 0$ in $\mathbf{R}^2$ (which we can draw), there exists a unique smooth submanifold $\widetilde{F}$ of E such that $\widetilde{F} \cap \pi^{-1}(0)$ consists of two points such that the restriction of π to $\widetilde{F} \smallsetminus \widetilde{F} \cap \pi^{-1}(0)$ is a diffeomorphism to its image.

e) Show that E is diffeomorphic to the manifold of straight lines M seen in subsection 2.2.2.

f) If ϕ is a diffeomorphism from an open subset U in $\mathbf{R}^2$ containing 0, such that $\phi(0) = 0$, show that there exists a unique diffeomorphism $\widehat{\phi}$ of $\pi^{-1}(U)$ such that $\pi \circ \widehat{\phi} = \phi \circ \pi$. Deduce a definition for the blowup at a point for any two dimensional manifold.

18. Let Γ be a group acting on a set X, and let Γ_o be the kernel of the group morphism $\gamma \mapsto (x \mapsto \gamma \cdot x)$ of Γ into the group of bijections of X. Show that this is a "natural" action of the quotient group Γ/Γ_o on X, and that this action is effective. Compare to Exercise 5.

Is the action of $SO(n+1)$ on $X = P^n\mathbf{R}$ obtained by passing the natural action on $\mathbf{R}^{n+1}$ to the quotient effective?

19. Show that the action of $\mathbf{Z}$ on $\mathbf{R}^2$ defined by

$$n \cdot (x, y) = (2^n x, 2^{-n} y)$$

is not proper, and neither is the induced action on $\mathbf{R}^2 \smallsetminus \{0\}$. Show that we obtain a proper action if we restrict to the half-plane $y > 0$.

20. Let $\Gamma = \mathbf{Z}/n\mathbf{Z}$ act on $\mathbf{R}^2 \simeq \mathbf{C}$ by rotations in the angle $2k\pi/n$ with respect to the origin. Show that the quotient space is a manifold (diffeomorphic to $\mathbf{R}^2$), and that the quotient map is smooth but not a submersion.

21. *The Möbius strip again*

Take $\Gamma = \mathbf{Z}$, $X = \mathbf{R} \times \mathbf{R}$, and

$$n \cdot (x, y) = (x + n, (-1)^n y).$$

Show that this is a free and proper action, and that $\mathbf{R}^2/\Gamma$ is diffeomorphic to the quotient of $S^1 \times \mathbf{R}$ by the action of the group with two elements given by the transformation

$$(u, y) \longmapsto (-u, -y).$$

Let M be the manifold obtained. Show that M is diffeomorphic to the manifold of straight lines of the plane seen in Section 2.2.2, as well as $P^2\mathbf{R}$ with a point removed.

Show that this manifold is also the total space of a vector bundle of dimension 1 on the circle, and that this vector bundle is nontrivial. (For the definition of a vector bundle, see Section 3.5.2.)

22. *Lens spaces*

a) Show that the only subgroup (non reducible to the identity element) of $O(2n+1)$ that acts freely on S^{2n} (the action induced by the linear action on $\mathbf{R}^{2n+1}$) is the group with two elements $\{Id, -Id\}$.

b) Consider S^3 as the set of points satisfying

$$\{(z, z') \in \mathbf{C} \times \mathbf{C} : |z|^2 + |z'|^2 = 1\}.$$

Let p be a positive integer, and u a p-th root of unity in $\mathbf{C}$. Show that

$$k \cdot (z, z') = (u^k z, u^k z'),$$

defines a free group action of the group $\Gamma = \mathbf{Z}/p\mathbf{Z}$ on S^3, and therefore that S^3/Γ is a manifold.

23. *Suspension of a diffeomorphism*

a) If X is a manifold and ϕ is a diffeomorphism of X, we can define an action of $\mathbf{Z}$ on $\mathbf{R} \times X$ by

$$n \cdot (t, x) = \big(t + n, \phi^n(x)\big).$$

Show that this action is free and proper.

b*) Show that the quotient manifold is a fibered space with base S^1 and typical fiber X.

24. *Coverings and local diffeomorphisms*

a) Let $f : X \to Y$ be a local diffeomorphism. Show that if X is compact and Y is connected, then f is a covering map.

b) Give an example of a surjective local diffeomorphism from a connected manifold to a compact manifold that is not a covering map.

25. Try to reprove the fundamental theorem of algebra for polynomials with *real* coefficients. What goes wrong?

26. Suppose the differentiable structures of M and N are given by *maximal* atlases $(U_i, \varphi_i)_{i \in I}$ and $(V_j, \psi_j)_{j \in J}$. Is the atlas

$$(U_i \times V_j, \varphi_i \times \psi_j)_{(i,j) \in I \times J}$$

on $M \times N$ maximal?

27. *Distinct diffeomorphic structures*

Show that the atlas $\big(\mathbf{R}, \sqrt[3]{t}\big)$ defines on $\mathbf{R}$ is a distinct differentiable manifold structure from the canonical structure given by $(\mathbf{R}, t)$. Now show these structures are diffeomorphic.

28**. *Connected sum*

Let M_1 and M_2 be two smooth manifolds of dimension n, and let (U_1, ϕ_1) (resp. (U_2, ϕ_2)) be a chart of M_1 (resp. M_2) such that ϕ_i is a diffeomorphism on U_i to the open ball $B(0, 2)$ (here $\mathbf{R}^n$ is equipped with its canonical Euclidean norm). Let C be the annulus $\{x \in \mathbf{R}^n : \frac{1}{2} < \|x\| < 2\}$.

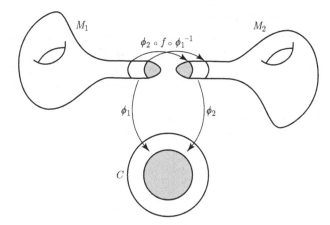

Figure 2.10: Connected sum

a) Show that $x \mapsto \frac{x}{\|x\|^2}$ is a diffeomorphism of C.

b) Consider the topological space X obtained by taking the disjoint union of

$$M_1 \smallsetminus \phi_1^{-1}\big(\overline{B(0,1/2)}\big) \coprod M_2 \smallsetminus \phi_2^{-1}\big(\overline{B(0,1/2)}\big)$$

and identifying $\phi_1^{-1}(C)$ and $\phi_2^{-1}(C)$ by means of the diffeomorphism $\phi_2 \circ f \circ \phi_1^{-1}$ (in other words, we quotient this sum by the equivalence relation: x and y are equivalent if they are equal, or if one of them (say x) belongs to $\phi_1^{-1}(C)$, and the other belongs to $\phi_2^{-1}(C)$ and if $y = (\phi_2 \circ f \circ \phi_1^{-1})(x)$.

Show that $M_i \smallsetminus \phi_i^{-1}\big(\overline{B(0,1/2)}\big)$ is homeomorphic to its image under the quotient map.

Let V_i be this image, equipped with the differentiable structure inherited by this homeomorphism (this is to say the unique differentiable structure such that this homeomorphism is a diffeomorphism).

Show that there exists a (unique) differentiable structure on X such that the inclusions $V_i \to X$ are diffeomorphic to their images in X. The space X, equipped with this differentiable structure, is the *connected sum* of M_1 and M_2, and is denoted $M_1 \sharp M_2$.

c) Show that $M \sharp S^n$ is diffeomorphic to M, and that $T^2 \sharp T^2$ is diffeomorphic to the "sphere with two holes" of Exercise 20 of Chapter 1.

d**) Show that $\mathbf{R}^2 \sharp P^2 \mathbf{R}$ is diffeomorphic to the blow up of $\mathbf{R}^2$ at 0 (cf. Exercise 17).

29*. *Weak transversality theorem*

Let X, T and Y be three manifold, and let $f : X \times T \to Y$ be a smooth map. Suppose that f is transverse to a submanifold M of Y. Show that for almost every $t \in T$, the map $f_t : X \to Y$ is transverse to M (use Sard's theorem in the case where the dimension of the domain is larger than that of the range).

30. Let M be a connected topological manifold of dimension at least two. Show that the complement of every finite set is connected.

Chapter 3

From Local to Global

3.1. Introduction

This chapter consists of variations on the following themes:

- on a smooth manifold, there are "many" smooth functions;
- there are also "many" diffeomorphisms.

We note that the first property is obvious for submanifolds of $\mathbf{R}^n$, but the second is not any easier for submanifolds than for the "abstract" manifolds seen in the previous chapter.

The existence of continuous functions of compact support (other than the zero function) on $\mathbf{R}^n$ is banal. The existence of *smooth* functions with compact is less so. This property, explained in detail in Section 3.2, is fundamental: if U is an open subset of a manifold that is diffeomorphic to $\mathbf{R}^n$, then we can transport every smooth function with compact support on $\mathbf{R}^n$ to this open subset. This function will again be of compact support within U, which allows us to extend the function to be zero outside of U. Thanks to this procedure, we find sufficiently many functions to embed every compact manifold into $\mathbf{R}^N$ (Theorem 3.7).

Warning. On a submanifold of $\mathbf{R}^N$, we get many smooth functions for free, by restricting smooth functions in $\mathbf{R}^N$ to this submanifold, but here things are inverted: it is the existence of smooth functions on a manifold that allows us to realize it as a submanifold.[1] This embedding theorem furnishes a convenient way to prove approximation results (see for example Theorem 7.20 or Lemma 8.19).

---

1. None of this is true for complex manifolds: on a compact complex manifold, the holomorphic functions are constant by the maximum principle.

The next part of the chapter is devoted to vector fields. These are used in modeling a lot of natural phenomena, the speed of points in a solid or a moving fluid, the gravitational vector field, the electric field, etc. For the mathematician, a vector field is above all a "infinitesimal transformation" (they were once called this) and even an infinitesimal diffeomorphism.

On an open subset $U \subset \mathbf{R}^n$, a vector field can be seen as a smooth map from U to $\mathbf{R}^n$. On a manifold, it is necessary to give for every $m \in M$ a vector v_m in the tangent space at m, and one may give a sense of smoothness with respect to m. The price we pay is a smooth manifold structure on the tangent bundle, defined as the disjoint union of tangent spaces. Things are facilitated by interchanging a geometric and algebraic point of view: vector fields on a smooth manifold are identified with derivations of $C^\infty(M)$, this is to say linear maps of $C^\infty(M)$ to itself that satisfy the Leibnitz rule (see Definition 3.14).

This point of view also permits us to define a fundamental notion, the bracket $[X, Y]$, which takes into account the non-commutativity of infinitesimal diffeomorphisms X and Y. This non-commutativity is expressed in Proposition 3.37 and Theorem 3.38.

We naturally associate to every vector field a differential equation, that of its trajectories. The term *flow* for the expression $\phi_t(x)$, which denotes the value of the solution at time t that starts at x when $t = 0$, evokes the example of a velocity vector field above. The setting of manifolds is then very convenient: unlike what can happen in $\mathbf{R}^n$, the solutions of this equation on a compact manifold extend to all of $\mathbf{R}$, and in this case the flow defines a one-parameter group of diffeomorphisms (Theorem 3.39).

An important consequence is the fact that, if f is a smooth function on a compact manifold, the sub-level sets $f \leqslant a$ and $f \leqslant b$ have the same topology between any two levels containing no critical points (Theorem 3.40).

Finally, we return to the embedding theorem to deduce that every manifold (which is countable at infinity) of dimension 1 is diffeomorphic to the circle or the line. The surprise: although this is a very natural result, it is relatively difficult to prove.

3.2. Bump Functions; Embedding Manifolds

As we explained in the introduction, we must start by proving the existence of smooth functions with compact support, first on $\mathbf{R}^n$, and then on any manifold.

Definition 3.1. *A* bump function *on a manifold M is a smooth function f with values in $[0,1]$ such that there exists two open subsets U and V with compact closure, and with $\overline{U} \subset V$, such that*

$$\operatorname{supp} f \subset V \quad \text{and} \quad f(x) = 1 \text{ for } x \in U.$$

We recall that the *support* of a continuous function is the closure of the set of points where it is nonzero.

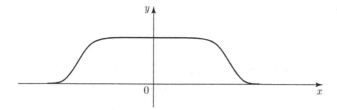

Figure 3.1: A bump function

Proposition 3.2. *If U and V are two open balls of $\mathbf{R}^n$ with the same center and with $\overline{U} \subset V$, there exists a smooth function equal to 1 on U and with support contained in V.*

PROOF. We begin by remarking that the function f_a of one real variable defined by

$$f_a(t) = \exp \frac{1}{t^2 - a^2} \quad \text{if } |t| < a \quad \text{and} \quad f_a(t) = 0 \text{ if } |t| \geqslant a$$

is smooth. The same is true for the function

$$g_a(t) = \frac{\int_{-\infty}^t f_a(u)\, du}{\int_{-\infty}^{+\infty} f_a(u)\, du}$$

which is zero for $t \leqslant -a$ and equal to 1 for $t \geqslant a$. Now if $b > a$, the function $g_a(b - t)$ equals 1 on $[0, b - a]$ and 0 if $t > a + b$. On $\mathbf{R}^n$, it suffices to take a function of the form $g_a(b - \|x\|^2)$. □

We immediately give a spectacular example using bump functions: this theorem, whose proof is due to Émile Borel, ensures that we can arbitrarily prescribe the sequence of derivatives at a point of a smooth function. As this result will not be used in the sequel, we will merely give a sketch of the proof.

Theorem 3.3. *Let $(a_n)_{n \geqslant 0}$ be a sequence of real numbers. Then there exists a smooth function $f \in C^\infty(\mathbf{R})$ such that*

$$f^{(n)}(0) = a_n \text{ for all } n.$$

PROOF. Let ρ be a bump function $\mathbf{R}$ such that

$$\begin{cases} \rho(x) = 1 & \text{if } |x| < \frac{1}{2} \\ \rho(x) = 0 & \text{if } |x| > 1. \end{cases}$$

Set

$$f(x) = \sum_{n=0}^{\infty} a_n \frac{x^n}{n!} \rho(b_n x), \quad \text{where } b_n = |a_n| + n.$$

One verifies that f is smooth, and that

$$\forall n, \quad f^{(n)}(0) = a_n. \qquad \square$$

To obtain the existence of smooth functions on any manifold, we repeatedly use the following elementary but fundamental property:

Proposition 3.4. *Let U and V be two open subsets of a manifold M, such that $M = U \cup V$, and let $f : U \to N$ and $g : V \to N$ be two smooth maps to a manifold N, whose restrictions to $U \cap V$ are equal. Then there exists a smooth function $h : M \to N$ such that*

$$h_{|U} = f \quad and \quad h_{|V} = g.$$

In particular, if U is an open subset of M, and if $N = \mathbf{R}$, every smooth function with compact support in U extends to M as a smooth function that is zero outside of U: it suffices to apply the above to U and $M \smallsetminus \operatorname{supp} f$.

Corollary 3.5

i) *Let U be an open subset of a manifold M. Then for every $a \in U$, there exists an relatively compact open subset V containing a such that $\overline{V} \subset U$, and a bump function equal to 1 on V and with support in U.*

ii) *If K is a compact subset of M, and $U \supset K$ is an open subset, then there exists a bump function supported in U and equal to 1 on K.*

PROOF

i) Let (U', φ) be a chart with domain included in U such that $\varphi(a) = 0$. We can then find two strictly positive real numbers r_1 and r_2 (with $r_1 < r_2$) such that

$$B(0, r_1) \subset B(0, r_2) \subset \varphi(U'),$$

and a bump function g on $\mathbf{R}^n$ equal to 1 on $B(0, r_1)$ and 0 outside of $B(0, r_2)$. A priori, the function $g \circ \varphi$ is not defined on U. However, since its support is included in $\varphi^{-1}\big(B(0, r_2)\big)$, it therefore extends by the above as a smooth function on M that is zero outside of U. By construction, it equals 1 on $\varphi^{-1}\big(B(0, r_1)\big)$.

ii) Every point $x \in K$ is contained in an open subset of a chart $V_x \subset U$ on which we can apply i). We obtain for each x an open subset W_x containing x such that $\overline{W_x} \subset V_x$, a bump function f_x equal to 1 on W_x, and with support contained in V_x. The open subsets W_x form an open cover of K, and we extract a finite subcover $(W_{x_i})_{1 \leqslant i \leqslant k}$. Then the function

$$g = 1 - \prod_{i=1}^{k} (1 - f_{x_i})$$

equals 1 on the union of the (W_{x_i}), and

$$\operatorname{supp} g \subset \bigcup_{i=1}^{k} V_{x_i}. \qquad \square$$

To fully use this result, we will need the following purely topological property.

Lemma 3.6. *Let $(U_i)_{i \in I}$ be a finite open cover of a compact manifold M. Then there exists an open cover $(V_i)_{i \in I}$ such that $\overline{V_i} \subset U_i$ for each i.*

PROOF. Every $x \in M$ is contained in an open subset $U_{i(x)}$ of the cover. There also exists an open subset W_x containing x such that

$$\overline{W_x} \subset U_{i(x)}$$

(this property, which is clear for $\mathbf{R}^n$, is true for topological manifolds by definition). Then the $(W_x)_{x \in M}$ from an open cover of M, from which we can extract a finite subcover $(W_{x_k})_{1 \leqslant k \leqslant p}$. The result follows, by taking

$$V_i = \left(\bigcup W_{x_k} \right)_{i(x_k)=i}. \qquad \square$$

Although these tools are quite rough, they allow us to show that every manifold is in fact a submanifold of $\mathbf{R}^n$. The conceptual importance of this result speaks for itself.

Theorem 3.7. *Every compact manifold admits an embedding into a space $\mathbf{R}^m$ for some m.*

PROOF. Let $(U_i, \varphi_i)_{1 \leqslant i \leqslant N}$ be a finite atlas of M. By Corollary 3.5 and Lemma 3.6 there exists an open cover $(V_i)_{1 \leqslant i \leqslant N}$ such that $\overline{V_i} \subset U_i$ for each i, and for each i a bump function f_i with support contained in U_i and equal to 1 on V_i. By Proposition 3.4, the function $f_i \varphi_i$, extended by 0 outside of U_i, gives a smooth map from M to $\mathbf{R}^n$, where $n = \dim M$. Let

$$F = (f_1 \varphi_1, \ldots, f_N \varphi_N, f_1, \ldots, f_N),$$

and observe we obtain a smooth map from M to $\mathbf{R}^{N(n+1)}$, which is an immersion. Indeed, each x belongs to an open subset V_i, and the i-th block of $T_x F$ is then equal to $T_x \varphi_i$, which is bijective, which shows that $T_x F$ is injective.

We now show that F is injective. Let x and y be two points of M such that $F(x) = F(y)$. In particular,

$$\forall i, \quad f_i(x) = f_i(y).$$

The V_i cover M, there exists i such that $f_i(x) \neq 0$. Then x and y belong to U_i, and for this i, the inequality

$$f_i(x)\varphi_i(x) = f_i(y)\varphi_i(y)$$

gives $\varphi_i(x) = \varphi_i(y)$, and $x = y$ because φ_i is bijective. By applying Proposition 2.28 we conclude an injective immersion into a compact manifold is an embedding. $\square$

By using completely different techniques, we can improve this result.

Corollary 3.8 (Whitney's "easy" embedding theorem). *Every compact manifold of dimension n embeds into $\mathbf{R}^{2n+1}$.*

PROOF. We start with an embedding of f of X into $\mathbf{R}^m$ for some m. We will see that by composing f with a well chosen projection, we can obtain an embedding into $\mathbf{R}^{m-1}$. To do this, we equip $\mathbf{R}^m$ with an inner product, and we introduce for every unit vector $v \in S^{m-1}$ the projection p_v onto the subspace orthogonal to v in $\mathbf{R}^m$.

Let $Y = f(X)$. For the restriction of p_v to Y to be injective, it is necessary and sufficient that for all distinct x and y in Y, the vector

$$\frac{x - y}{\|x - y\|}$$

is different from v, where again v does not belong to the image of the map

$$(x, y) \longmapsto \frac{x - y}{\|x - y\|} \quad \text{from } Y \times Y \setminus \Delta \text{ to } S^{m-1},$$

and where we have denoted the diagonal of Y by Δ.

By Sard's theorem (Theorem 2.49), there exists such v since $2n < m - 1$.

For $p_{v|Y}$ to be an immersion, it is necessary and sufficient – see Figure 3.2 – that v does not belong to any subspace tangent to Y (note that this is the infinitesimal version of the preceding condition). We introduce

$$Z = \big\{(x, v) \in X \times S^{N-1} : \ v \in T_{f(x)} Y\big\}.$$

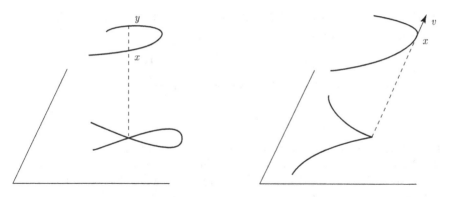

Figure 3.2: A projection can cause singularities to appear

We verify that Z is a submanifold of dimension $2n-1$ of $X \times S^{N-1}$ (see Exercise 15 of Chapter 1). In particular $\mathrm{pr}_2(Z)$ is of measure zero since $2n < m$.

By iterating the preceding argument, we see that X admits an immersion into $\mathbf{R}^{2n}$ and an embedding into $\mathbf{R}^{2n+1}$.   □

Remarks

a) The same property remains true for manifolds that are countable at infinity, and the proof is essentially the same (see [Hirsch 76]).

b) With much more work and completely different techniques, H. Whitney proved that every smooth compact manifold of dimension n is embedded in $\mathbf{R}^{2n}$ (see [Adachi 93]). This result is optimal, but we can do better in certain cases: for example $P^2\mathbf{R}$ is not embeddable in $\mathbf{R}^3$, but every orientable surface is, as we see from the classification of surfaces (see Section 7.10).

3.3. Derivations

3.3.1. Derivation at a Point

With now return to the tangent space with a different point of view, by exploiting the fact, already seen in Proposition 3.4, that smoothness is a local property.

Definition 3.9. *Let X be a topological space and $x \in X$. Two functions each defined on an open subset containing x are said to have the same germ at x if there exists a further open subset of x on which both are equal.*

In other words, we introduce the following equivalence relation on the set of functions defined on an open subset of x

$$(f : \ U \longrightarrow \mathbf{R}) \sim (g : \ V \longrightarrow \mathbf{R})$$

if and only if there exists an open subset W containing x, $W \subset U \cap V$ such that

$$f_{|W} = g_{|W}.$$

The set of germs is the quotient set of this equivalence relation. Continuity at x for example depends only on the germ. If f is a function defined on an open subset U of x, we denote its germ by $\dot{f}$. However we will often abuse notation and denote a germ by one of its representatives.

The set of germs is naturally equipped with a *ring structure*: we define addition and multiplication through the representatives.

We are interested in the case of manifolds, and denote by $\mathcal{F}_m M$ the germ of smooth functions at m on the manifold M. This is clearly a subring of the ring of germs.

Remarks

a) The notion of germs given here is also of interest for continuous, C^p or analytic functions. For example, the ring of germs of holomorphic functions of one complex variable at 0 is identified with the ring of infinite series with nonzero radius of convergence. For smooth functions, the situation is much more complicated: if $m \in \mathbf{R}^n$, the Taylor series at m of a smooth function in a neighborhood of m depends only the germ of f, but knowing this series does not determine the germ. See for example the function f_a of Proposition 3.2, where all of the derivatives at a are zero.

b) For every manifold M and every point $m \in M$, the space $\mathcal{F}_m M$ is isomorphic to $\mathcal{F}_0 \mathbf{R}^n$. Indeed we take a chart (U, φ) with $m \in M$ and $\varphi(m) = 0$, and we associate to every function f defined on an open subset V containing m the function $f \circ \varphi^{-1}$ defined on $\varphi(U \cap V)$. The map $f \mapsto f \circ \varphi^{-1}$ passes to the quotient and gives an isomorphism (amongst many others!) between $\mathcal{F}_m M$ and $\mathcal{F}_0 \mathbf{R}^n$.

Let v be a vector in $\mathbf{R}^n$. If f is a smooth function on an open subset containing 0, the directional derivative of f with respect to v, this is to say the number

$$T_0 f \cdot v = \sum_{j=1}^{n} \partial_j f(0) v^j$$

depends only on the germ of f at 0. We will use algebraic properties of the map $\dot{f} \mapsto T_0 f \cdot v$ to give a definition of the tangent space to a manifold which uses neither charts nor coordinates on $\mathbf{R}^n$.

Definition 3.10. *A derivation at a point m of M is a linear map $\delta : \mathcal{F}_m M \to \mathbf{R}$ such that for $\dot{f}, \dot{g} \in \mathcal{F}_m M$ we have*

$$\delta(\dot{f} \cdot \dot{g}) = f(m)\delta(\dot{g}) + g(m)\delta(\dot{f}).$$

The name derivation is justified by analogy with the product rule, and also by the following result:

Theorem 3.11. *There is a natural bijection between $\mathbf{R}^n$ and the derivations of $\mathcal{F}_m \mathbf{R}^n$. More precisely, for such a derivation, δ, there exists a unique vector $v \in \mathbf{R}^n$ such that*

$$\delta(f) = T_m f \cdot v = \sum_{j=1}^{n} \partial_j f(m) v^j.$$

We will use the following lemma, which is a fundamental technical tool.

Lemma 3.12 (Hadamard lemma). *For all $1 \leqslant p \leqslant \infty$, every C^p function defined on an open ball with center 0 in $\mathbf{R}^n$ can be written in the form*

$$f(x) = f(0) + \sum_{j=1}^{n} x^j h_j(x),$$

where the h_j are C^{p-1}. Moreover, $h_j(0) = \partial_i f(0)$.

PROOF. We have

$$f(x) - f(0) = \int_0^1 \frac{d}{dt} f(tx)\, dt = \sum_{j=1}^{n} x^j \int_0^1 \partial_j f(tx)\, dt,$$

and the existence of the h_j then results from the theorem of differentiation through an integral. A direct calculation shows that necessarily $h_j(0) = \partial_j f(0)$. (We note that the h_j are not unique, even if this proof gives possible explicit choices.) $\square$

PROOF OF THEOREM 3.11. Let $\dot{f} \in \mathcal{F}_m \mathbf{R}^n$, and f be a representative of $\dot{f}$. We have $\delta(f) = \delta(f - f(m))$, since by definition

$$\delta(1.1) = \delta(1) + \delta(1) \text{ and therefore } \delta(\text{constant}) = 0.$$

By the lemma,

$$\delta(f) = \sum_{j=1}^{n} \delta(x^j - m^j) h_j(m) \quad \text{and} \quad h_j(m) = \partial_j f(m).$$

It is clear that the map

$$f \longmapsto \partial_j f(m)$$

depends only on $\dot{f}$, and that it is itself a derivation. $\square$

This result extends directly to manifolds, and lets us give a new definition of the tangent space, which does not appeal to charts.

Theorem 3.13. *If M is a smooth manifold and $m \in M$, the set of derivations on $\mathcal{F}_m M$ is isomorphic to $T_m M$: for every derivation δ, there is a unique $\xi \in T_m M$ such that*

$$\delta(f) = df_m \cdot \xi.$$

PROOF. Let (U, φ) be a chart such that $m \in U$ and $\varphi(m) = 0$. If δ is a derivation in $\mathcal{F}_m M$, the map

$$g \longmapsto \delta(g \circ \varphi)$$

is a derivation in $\mathcal{F}_0 \mathbf{R}^n$. Now let v be the vector such that $\delta(g \circ \varphi) = g'(0) \cdot v$. Then, if $f = g \circ \varphi$, we have

$$\delta f = df_m \cdot \xi,$$

where $\xi = \theta_\varphi^{-1}(v)$ (see Definition 2.23).   □

3.3.2. Another Point of View on the Tangent Space

We will see another definition of the tangent space, which synthesizes between the velocity vector (Definition 2.21) and derivation point of view.[2] There is a price to pay: a little more algebraic formalism. This subsection can be omitted on a first reading, and will not be used in the sequel.

For a manifold M and a point $m \in M$, we consider the set $\mathcal{C}_m^M$ of curves $c : (-1, 1) \mapsto M$ such that $c(0) = m$, and we introduce the vector space $\mathcal{T}_m^M$ of (finite) real linear combinations of elements of $\mathcal{C}_m^M$, or in other words the $\mathbf{R}$-vector space whose basis consists of elements of $\mathcal{C}_m^M$.

For a germ $\dot{f} \in \mathcal{F}_m^M$ and $c \in \mathcal{C}_m^M$, we write

$$B(\dot{f}, c) = \frac{d}{dt} f\big(c(t)\big)\big|_{t=0}.$$

On the right hand side, f in fact denotes a function defined on an open subset containing m and representing $\dot{f}$, it is clear, as in the analogous situations studied in Section 3.3, that the result does not depend on the representative of $\dot{f}$ chosen.

We extend B to $\mathcal{F}_m^M \times \mathcal{T}_m^M$ by linearity, this is to say by setting

$$B\left(\dot{f}, \sum_{i=1}^k \lambda_i c_i\right) = \sum_{i=1}^k \lambda_i B(\dot{f}, c_i).$$

---

2. This enlightening point of view was communicated to me by Marc Troyanov.

By construction B is a bilinear form on $\mathcal{F}_m^M \times \mathcal{T}_m^M$. We also introduce the kernel of B, which by definition is

$$(\mathcal{F}_m^M)^0 = \{ \dot{f} \in \mathcal{F}_m^M : \ \forall u \in \mathcal{T}_m^M,\ B(\dot{f}, u) = 0 \}$$
$$(\mathcal{T}_m^M)^0 = \{ u \in \mathcal{T}_m^M : \ \forall \dot{f} \in \mathcal{F}_m^M,\ B(\dot{f}, u) = 0 \}.$$

Under these conditions, B passes to the quotient, and gives a *non-degenerate* bilinear form

$$b : \ \mathcal{F}_m^M / (\mathcal{F}_m^M)^0 \times \mathcal{T}_m^M / (\mathcal{T}_m^M)^0 \longrightarrow \mathbf{R}.$$

All of this input is "natural", which is to say equivariant with respect to diffeomorphisms. If ϕ is a diffeomorphism from an open subset U containing M to an open subset U' in a manifold M', then $f \mapsto f \circ \phi^{-1}$ defines a vector space isomorphism from $\mathcal{F}_m^M$ to $\mathcal{F}_{\phi(m)}^{M'}$, and $c \mapsto \phi \circ c$ defines, by extending linearly, an isomorphism of $\mathcal{T}_m^M$ to $\mathcal{T}_{\phi(m)}^{M'}$. We denote these isomorphisms by $\mathbb{T}'(\phi)$ and $\mathbb{T}(\phi)$. It is immediate that

$$B(\mathbb{T}'(\phi) \cdot f, \mathbb{T}(\phi) \cdot u) = B(f, u)$$
$$\mathbb{T}(\psi \circ \phi) = \mathbb{T}(\psi) \circ \mathbb{T}(\phi) \qquad (3.1)$$
$$\mathbb{T}'(\psi \circ \phi) = \mathbb{T}'(\psi) \circ \mathbb{T}'(\phi).$$

As a result of these remarks, $\mathcal{T}_m^M / (\mathcal{T}_m^M)^0$ is "naturally" identified with $T_m M$, this is to say that there exists an isomorphism $\mathrm{Is}_{M,m}$ such that, with the preceding notation,

$$T_m \phi \circ \mathrm{Is}_{M,m} = \mathrm{Is}_{M',\phi(m)} \circ \mathbb{T}(\phi).$$

Indeed, two curves c_1 and c_2 are tangent at m in the sense of definition 2.21 if and only if c_1 and c_2 are equivalent modulo $(\mathcal{T}_m{}^M)^0$: if $c_1 - c_2 \in (\mathcal{T}_m^M)^0$, testing the equality $B(\dot{f}, c_1) = B(\dot{f}, c_2)$ for the germs of the coordinate functions of a local chart ϕ, we see that $(\phi \circ c_1)'(0) = (\phi \circ c_2)'(0)$. Conversely, if this equality is true for a chart ϕ, this is to say if the curves $\phi \circ c_1$ and $\phi \circ c_2$ in $\mathbf{R}^n$ have the same velocity, then $\phi \circ c_1 - \phi \circ c_2 \in (\mathcal{T}_0{}^{\mathbf{R}^n})^0$ and by equations (3.1), $c_1 - c_2 \in (\mathcal{T}_m^M)^0$. In summary, we have a map from $\mathcal{C}_m^M$ to $T_m M$ which extends by linearity to $\mathcal{T}_m^M$, and passes to the quotient giving an injection $\mathcal{T}_m^M / (\mathcal{T}_m^M)^0$. The map is also surjective, again by equation (3.1), it suffices to verify this for $\mathbf{R}^n$, which is elementary: we take the curves $t \mapsto tv$.

Moreover, by the same definition of B, the map $\dot{f} \mapsto B(f, c)$ is a derivation at a point which depends only on the equivalence class of c. To see that every derivation at a point is of this form, it suffices, again by equation (3.1), to do this for $\mathbf{R}^n$ and to apply Theorem 3.11.

As a bonus, this point of view gives a natural realization of the cotangent vector space, because thanks to the bilinear form b the space $\mathcal{F}_m^M / (\mathcal{F}_m^M)^0$ appears as the dual of $T_m M$.

Considering again the formulas (3.1) for a not necessarily invertible map ϕ, we also see that $\mathbb{T}(\phi)$ gives a linear map from $\mathcal{T}_m^M/(\mathcal{T}_m^M)^0$ to $\mathcal{T}_{\phi(m)}^M/(\mathcal{T}_{\phi(m)}^M)^0$ by passing to the quotient which is identified with $T_m\phi$.

3.3.3. Global Derivations

We now "globalize" the preceding situation.

Definition 3.14. A derivation *(we will sometimes say global derivation if we want to insist on the difference with what preceded)* on a smooth manifold M is a linear map δ from $C^\infty(M)$ to itself such that

$$\text{for } f,g \in C^\infty(M), \quad \delta(f \cdot g) = f \cdot \delta(g) + g \cdot \delta(f).$$

This definition has a purely algebraic character, but allows us to "localize" derivations:

Theorem 3.15. *Let δ be a derivation on a manifold M. Then*

i) *If f and g are two smooth functions whose restrictions to an open subset $U \subset M$ are equal,*

$$(\delta f)_{|U} = (\delta g)_{|U}.$$

ii) *For every open subset $U \subset M$, there exists a unique derivation on U, denoted $\delta_{|U}$, such that*

$$\delta_{|U}\left(f_{|U}\right) = (\delta f)_{|U} \text{ for } f \in C^\infty(M).$$

PROOF

i) By linearity, it suffices to show that if f vanishes on U, then so does δf. For each $x \in U$, we take an open subset V containing x such that $\overline{V} \subset U$ and a bump function h with support in U, equal to 1 on V. Then $f = (1-h)f$ and therefore

$$\delta f = (1 - h)\delta f + f\delta(1 - h).$$

Therefore δf is zero for each such V, and so on U as well.

ii) There is a (small!) difficulty which comes from the fact that a smooth function on U does not in general extend to M. For each x in U, we take an open subset V and a function h as in i). Then, if $f \in C^\infty(U)$, by Proposition 3.4 the function fh is defined and smooth on all of M, and for $y \in V$ we write $(\delta_{|U}f)(y) = \left(\delta(fh)\right)(y)$. By i), the function $\delta_{|U}f$ is thus defined without ambiguity, and we obtain a derivation having the desired properties. $\qquad\square$

We come to the characterization of derivations on open subsets of $U \subset \mathbf{R}^n$. It is clear that for $i \in [1,n]$, the map $f \mapsto \partial_i f$ is a derivation, and more generally, if $X_1, \ldots, X_n$ are smooth functions,

$$f \longmapsto L_X f = \sum_{i=1}^{n} X_i \partial_i f$$

is a derivation. Better still:

Theorem 3.16. *The vector space of derivations on an open subset U of $\mathbf{R}^n$ is isomorphic to $C^\infty(U, \mathbf{R}^n)$.*

PROOF. Associate to X in $C^\infty(U, \mathbf{R}^n)$ the derivation L_X given by the formula above. The map $X \mapsto L_X$ is injective: if X is nonvanishing at $a \in U$, there exists a smooth function f such that $(L_X f)(a) = T_a f \cdot X(a) \neq 0$.

We now show the surjectivity in the case of $\mathbf{R}^n$. If δ is a derivation, we remark first that as in the case of derivations at a point, δ vanishes for constant functions. For $y \in \mathbf{R}^n$, we write the decomposition

$$f(x) - f(y) = \sum_{i=1}^{n} (x^i - y^i) h_{i,y}(x)$$

from the Hadamard lemma. Then

$$(\delta f)(y) = \left(\delta\left(f - f(y)\right)\right)(y) = \sum_{i=1}^{n} \delta(x^i - y^i)(y) h_{i,y}(y) = \sum_{i=1}^{n} \delta(x^i)(y) \partial_i f(y).$$

The same argument applies to an convex open subset: it is convexity that allows us to apply the Hadamard lemma to the function $x \mapsto f(x) - f(y)$. We pass to the general case by applying Theorem 3.15 to open balls in U. $\square$

Confronted with the analogous result for derivations at a point, this result leads us to adopt new terminology.

Definition 3.17. A vector field *on an open subset U of $\mathbf{R}^n$ is a smooth map from U to $\mathbf{R}^n$. The map L_X from $C^\infty(U, \mathbf{R})$ to itself, introduced above is called the* derivation associated to X.

We have just seen that if δ is a derivation on $U \subset \mathbf{R}^n$, the map $X : U \to \mathbf{R}^n$ which is associated to it by Theorem 3.16 can be interpreted as attaching to $x \in U$ a vector $X_x \in T_x U \simeq \mathbf{R}^n$ in a smooth fashion, which justifies the name vector field. We will see that derivations on any manifold admit an analogous characterization to that of Theorem 3.16. This requires certain results on derivations and vector fields on open subsets of $\mathbf{R}^n$, which extend easily to manifolds.

3.4. Image of a Vector Field; Bracket

If $X \in C^\infty(U, \mathbf{R}^n)$ is a vector field on an open subset U of $\mathbf{R}^n$, we hereafter denote its value at the point x by X_x. Then, the associated derivation is given by

$$(L_X f)(x) = T_x f \cdot X_x.$$

The derivation associated to the constant vector field equal to the i-th vector in the canonical basis of $\mathbf{R}^n$ is simply $f \mapsto \partial_i f$, and we denote this vector field by ∂_i. Then, X can be written

$$\sum_{i=1}^n X^i \partial_i, \quad \text{where the } X^i \text{ are smooth functions.}$$

If $\varphi : U \to V$ is a diffeomorphism, the composition $g \mapsto g \circ \varphi$ is a ring isomorphism between $C^\infty(V)$ and $C^\infty(U)$. Conjugation by this isomorphism allows us to move a derivation on U to a derivation on V.

Definitions 3.18. *Let $\varphi : U \to V$ be a diffeomorphism between open subsets of $\mathbf{R}^n$, and let δ be a derivation on U. The image of δ under φ is the derivation*

$$g \longmapsto \left(\delta(g \circ \varphi)\right) \circ \varphi^{-1}$$

on V. If X is the vector field associated to δ, we denote $\varphi_* X$ the vector field associated to the image of δ, and we also say that this vector field is the image of X under φ.

Then,

$$L_{\varphi_* X} f = \left(L_X(f \circ \varphi)\right) \circ \varphi^{-1}.$$

Proposition 3.19. *We have*

$$(\varphi_* X)_y = T_{\varphi^{-1}(y)} \varphi \cdot X_{\varphi^{-1}(y)}.$$

PROOF. By the chain rule,

$$\left(\delta(g \circ \varphi)\right)(x) = T_x(g \circ \varphi) \cdot X_x = T_{\varphi(x)} g(T_x \varphi \cdot X_x),$$

hence the result follows by replacing x by $\varphi^{-1}(y)$. $\qquad\qquad\square$

Examples

a) The image of the vector field ∂_i under a translation is the same vector field.

b) The image under the homothety $h_\lambda : x \mapsto \lambda x$ of the vector field

$$\sum_{i=1}^{n} X^i \partial_i$$

is the vector field

$$\sum_{i=1}^{n} \lambda (X^i \circ h_\lambda) \partial_i.$$

c) The image of the vector field $\frac{d}{dx}$ on $\mathbf{R}$ under the exponential map is the vector field $x \frac{d}{dx}$ on $\mathbf{R}^*_+$.

Remark. We can easily check for a smooth map $\varphi : U \mapsto V$ we can define the image of a *derivation at a point* δ_m by

$$\delta_{\varphi(m)} \cdot g = \delta_m \cdot (g \circ \varphi),$$

and that if δ_m is given by a tangent vector $v \in T_m U$, then $\delta_{\varphi(m)}$ is given by $T_m \varphi \cdot v$. Conversely, to define the image of a global derivation, we are obliged to "climb up" from V to U, thus to suppose that φ is invertible, and that φ^{-1} is smooth. We check for example that if φ is the map $t \mapsto t^3$ from $\mathbf{R}$ to $\mathbf{R}$, there does not exist a vector field X on $\mathbf{R}$ such that $X_{\varphi(t)} = T_t \varphi \cdot (d/dt)_t$.

If δ_1 and δ_2 are two derivations their composition is not a derivation, because

$$\delta_1(\delta_2(fg)) = \delta_1(\delta_2 f)g + (\delta_1 f)(\delta_2 g) + (\delta_1 g)(\delta_2 f) + \delta_1(\delta_2 g)f.$$

However this formula also lets us see that $\delta_1 \circ \delta_2 - \delta_2 \circ \delta_1$ (which is what the algebraists call the commutator of δ_1 and δ_2) is a derivation.

Definition 3.20. The Lie bracket *of two vector fields X and Y, denoted $[X, Y]$, is the vector field corresponding to the derivation $L_X L_Y - L_Y L_X$.*

Proposition 3.21. *If X and Y are given on an open subset U of $\mathbf{R}^n$ by*

$$X = \sum_{i=1}^{n} X^i \partial_i \quad and \quad Y = \sum_{i=1}^{n} Y^i \partial_i$$

then

$$[X, Y] = \sum_{i=1}^{n} Z^i \partial_i, \quad where \quad Z^i = \sum_{j=1}^{n} \left(X^j \partial_j Y^i - Y^j \partial_j X^i \right).$$

PROOF. If f is a smooth function,

$$L_X L_Y f = L_X \left(\sum_{i=1}^{n} Y^i \partial_i f \right)$$

$$= \sum_{i,j=1}^{n} \left(X^j \partial_j Y^i \partial_i f + X^j Y^i \partial_{ij}^2 f \right).$$

Reversing the roles of X and Y and applying the Clairaut-Schwarz theorem, we obtain the result. □

Remarks

a) It is often convenient, to avoid calculation errors, to use the definition of the bracket over the formula above. Thus we see much more rapidly, for example, that

$$[fX, gY] = f(L_X g)Y - g(L_Y f)X + fg[X, Y].$$

b) The bracket has a local character: if V is an open subset of U, and if X and Y are two vector fields on U, $[X_{|V}, Y_{|V}] = [X, Y]_{|V}$. This property can be seen using the explicit formula, but it is more instructive to note that it is a consequence of Theorem 3.15.

Example. Let $\overline{A}$ be the vector field $x \mapsto Ax$ on $\mathbf{R}^n$, where A is an $n \times n$ matrix. (Such a vector field is called *linear*.) In coordinates,

$$\overline{A}_x = \sum_{i=1}^{n} \left(\sum_{j=1}^{n} a_j^i x^j \right) \partial_i .$$

Then

$$\left[\overline{A}, \overline{B} \right]_x = (BA - AB) \cdot x.$$

(Watch the sign!)

Lemma 3.22 (Jacobi identity). *If X, Y, Z are three vector fields, then*

$$[X, [Y, Z]] + [Y, [Z, X]] + [Z, [X, Y]] = 0.$$

PROOF. On the corresponding derivations, the algebraic verification is immediate. □

The derivation point of view also allows us to reduce the proof the following result to an easy but tedious exercise.

Proposition 3.23. *If X and Y are two vector fields on U, and if φ is a diffeomorphism from U to V, then*

$$\varphi_*[X, Y] = [\varphi_* X, \varphi_* Y].$$

3.5. The Tangent Bundle

3.5.1. The Manifold of Tangent Vectors

On a manifold, as we have seen, the notion of derivation makes sense. Under these conditions, we would like to have analogous result to Theorem 3.11 for derivations at a point: a derivation on a manifold M should allow us to associate to each point m in M a tangent vector X_m of $T_m M$, with this correspondence being smooth in a sense that we will make precise. To do this, we will show that the set of tangent vectors is itself a manifold in a natural way. We first set

$$TM = \coprod_{m \in M} T_m M.$$

For the moment, TM is the disjoint union of different tangent vector spaces to M, without a topology. For each chart (U, φ), the map

$$\Phi : \quad (x, \xi) \longmapsto (\varphi(x), T_x \varphi \cdot \xi)$$

is a bijection from TU to $\varphi(U) \times \mathbf{R}^n$.

Given an atlas $(U_i, \varphi_i)_{i \in I}$ of M, we equip TM with a topology by imposing the following conditions:

1) the sets TU_i are open subsets of TM;

2) the maps Φ_i are homeomorphisms.

Then $\Omega \subset TM$ is open if and only if $\Phi_i(\Omega \cap TU_i)$ is an open subset of $\varphi(U_i) \times \mathbf{R}^n$ for every i. To see that these conditions are consistent, we remark that by the same definition of tangent space, if $U_i \cap U_j \neq \emptyset$, the map

$$\Phi_i \circ \Phi_j^{-1} : \quad \varphi_j(U_i \cap U_j) \times \mathbf{R}^n \longrightarrow \varphi_i(U_i \cap U_j) \times \mathbf{R}^n$$

given by

$$(y, v) \longmapsto \left((\varphi_i \circ \varphi_j^{-1})(y), T_y(\varphi_i \circ \varphi_j^{-1}) \cdot v \right)$$

is a homeomorphism and even a diffeomorphism.

We have therefore defined a topology on TM which makes it a topological manifold with the atlas $(TU_i, \Phi_i)_{i \in I}$. As this atlas is smooth, TM is a smooth manifold of dimension $2 \dim M$. At this stage, it is important to remark that if M is a C^p manifold (with $p > 0$), then TM is a C^{p-1} manifold. This manifold is called the *tangent bundle* to M. We justify this name.

Proposition 3.24. *The canonical projection p from TM to M is a fibration.*

PROOF. It suffices to introduce the map

$$\psi_i : \ p^{-1}(U_i) = TU_i \longrightarrow U_i \times \mathbf{R}^n \quad \text{given by} \quad \psi_i(\xi_x) = (x, T_x\varphi.\xi_x). \quad \square$$

We can say more: the restriction of ψ_i to the fiber T_xM is a vector space isomorphism from T_xM to $\mathbf{R}^n$. If M_1 and M_2 are two manifolds and $f : M_1 \to M_2$ is a smooth map, by considering all the $(T_xf)_{x \in M_1}$ simultaneously, we find a map that we denote Tf from TM_1 to TM_2, which is smooth (and C^{p-1} if f is C^p) and such that the restriction to each fiber T_xM_1 is the linear map T_xf. This translates into the commutative diagram

$$
\begin{array}{ccc}
TM_1 & \xrightarrow{\ Tf\ } & TM_2 \\
{\scriptstyle p_1}\downarrow & & \downarrow{\scriptstyle p_2} \\
M_1 & \xrightarrow{\ f\ } & M_2
\end{array}
$$

Moreover, if M_3 is a third manifold and $g : M_2 \to M_3$ a smooth map, by the "manifold" version of the chain rule (see Proposition 2.25), $T(g \circ f) = Tg \circ Tf$.

3.5.2. Vector Bundles

This situation is the prototype of a more general situation that merits description.

Definition 3.25. *A real (resp. complex) vector bundle of rank k on a manifold B is a fibered space (E, p, B) such that*

i) *the typical fiber F and the fibers $p^{-1}(b)$, $b \in B$ are real (resp. complex) vector spaces of dimension k;*

ii) *for every local trivialization φ, the restriction of φ to $p^{-1}(b)$ (which sends $p^{-1}(b)$ to $\{b\} \times F$) induces a vector space isomorphism on F.*

The fiber $p^{-1}(b)$ is denoted E_b.

Examples

a) The product bundle $M \times \mathbf{R}^k$, called *the trivial bundle of rank k* over M.

b) The tangent bundle to a manifold of dimension n is a real vector bundle of rank n.

c) If M is a submanifold embedded into an inner product space E, the *normal bundle* to M, denoted $N(M)$ is the set of ordered pairs (x, v) of $M \times E$ such that v is orthogonal to T_xM; the projection to M is the restriction to $M \times E$ of pr_1.

The rank of this bundle is the codimension of M in E. See Exercise 20.

The following definitions parallel those seen in Section 2.5 for bundles.

Definitions 3.26

a) *A morphism between two vector bundles (E_1, p_1, B_1) and (E_2, p_2, B_2) (not necessarily of the same rank) is a smooth map $f : E_1 \to E_2$ which maps fibers to fibers (this is to say that there exists a map $g : B_1 \to B_2$ that makes the following diagram*

$$
\begin{array}{ccc}
E_1 & \xrightarrow{\ f\ } & E_2 \\
{\scriptstyle p_1}\big\downarrow & & {\scriptstyle p_2}\big\downarrow \\
B_1 & \xrightarrow{\ g\ } & B_2
\end{array}
$$

commute), and such that the restriction to each fiber $(E_1)_b$ is a linear map from $(E_1)_b$ to $(E_2)_{g(b)}$.

b) *If f is a diffeomorphism (in which case the map from fiber to fiber is a vector space isomorphism) we say that f is a* vector bundle isomorphism.

c) *A bundle is* trivializable *if it is isomorphic to a trivial bundle. A manifold whose tangent bundle is trivializable is called* parallelizable.

Example. The circle S^1 is parallelizable, as seen in Figure 3.3.

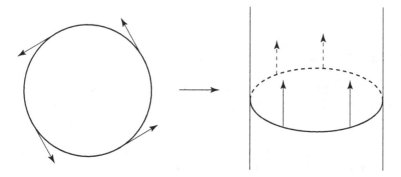

Figure 3.3: Tangent bundle of S^1

For more examples, see Exercise 12, Chapter 4, and the discussion which follows Proposition 3.28. These situations are something of an exception. The following notion is the key to better understand the tangent bundle, and vector bundles in general.

Definitions 3.27

a) *A* section *of a vector bundle E with base B is a smooth map s from B to E such that $p \circ s = Id_B$ (which is to say that $s(x) \in E_x$ for every x in B).*

b) *A* vector field *on a manifold is a section of the tangent bundle.*

We denote the set of (smooth) sections of E by $C^\infty(E)$, and therefore $C^\infty(TM)$ is the set of vector fields on M. There exists a natural vector space structure on $C^\infty(E)$ which is obtained by requiring that

$$(s+t)(x) = s(x) + t(x) \quad \text{(fiberwise addition in } E_x\text{)}.$$

The *zero section* defines an embedding of B into E.

Sections of a trivial bundle $B \times \mathbf{R}^k$ are of the form $x \mapsto (x, f(x))$, and are thus identified with functions defined on B with values in $\mathbf{R}^k$. In particular, if U is an open subset of $\mathbf{R}^n$, $C^\infty(TU)$ is identified with $C^\infty(U, \mathbf{R}^n)$. As in the case with open subsets of $\mathbf{R}^n$, we hereafter denote the value of a vector field X at x by X_x instead of $X(x)$.

If $(U_i, \varphi_i)_{i \in I}$ is an atlas of M, we see using the corresponding atlas of TM that a vector field may be given by a family of maps $X_i \in C^\infty(U_i, \mathbf{R}^n)$ such that

$$\forall x \in U_i \cap U_j, \quad X_i(x) = T_{\varphi_j(x)}\left(\varphi_i \circ \varphi_j^{-1}\right) \cdot X_j(x).$$

We can also give the functions $\overline{X}_i = X_i \circ \varphi_i^{-1}$. For each i, we thus have a vector field on $\varphi_i(U_i)$, and the condition above is equivalent to

$$\left(\varphi_j \circ \varphi_i^{-1}\right)_* \overline{X}_i = \overline{X}_j \quad \text{on} \quad \varphi_j(U_i \cap U_j).$$

This simply says that on U_i, $\varphi_{i*} X = \overline{X}_i$.

Proposition 3.28. *A vector bundle E of rank k over a manifold B is trivializable if and only if there exists k sections whose values at every point $x \in B$ form a basis for E_x.*

PROOF. If $(e_i)_{1 \leqslant i \leqslant k}$ is the canonical basis of $\mathbf{R}^k$ and if φ is a trivialization of E, we introduce k sections

$$x \longmapsto \varphi^{-1}(x, e_i).$$

Conversely, let $(s^\alpha)_{1 \leqslant \alpha \leqslant k}$ be k everywhere linearly independent sections. We decompose $e_x \in E$ relative the basis $s^\alpha(x)$, in other words write

$$\xi_x = \sum_{\alpha=1}^{n} \lambda_\alpha(\xi_x) X_x^\alpha.$$

We leave to the reader the task of verifying, using the property of local trivializations, that the functions λ_α are smooth on E. We find a trivialization φ by writing

$$\varphi(x) = (p(x), \lambda_1(x), \dots, \lambda_n(x)).$$

We have just seen that φ is smooth; as for its inverse, it is explicitly given by

$$\varphi^{-1}(x, v_1, \dots, v_k) = \sum_{\alpha=1}^{k} v_\alpha s^\alpha(x). \qquad \square$$

Examples

a) For a vector bundle of rank 1 to be trivializable, it is necessary and sufficient that it admit a nonvanishing section. An important example of a nontrivializable vector bundle of rank 1 is the infinite *Möbius strip*, seen as a bundle over S^1 (see Exercise 21).

b) We will see below (see Theorems 6.17 and 7.23) that for each vector field X on S^{2n}, there exists a point x such that $X_x = 0_x$ (such a point is called a *zero* of the vector field considered). It follows that spheres of even dimension are not parallelizable. On the other hand, by question c) of Exercise 4 in Chapter 2, the 3-dimensional sphere is parallelizable: the sphere is diffeomorphic to $SU(2)$, as the multiplicative group of quaternions of norm 1, and we will see in the next chapter that *every Lie group is parallelizable*.

c) **For analogous but more subtle algebraic reasons, S^7 is parallelizable (see [Steenrod 51] for example). Conversely, the only parallelizable spheres are S^1, S^3 and S^7. This is a profound result of A. Dold (see [Bott-Milnor 58]).**

3.5.3. Vector Fields on Manifolds; The Hessian

As in the case of open subsets of $\mathbf{R}^n$, we associate to every vector field X the map $f \mapsto L_X f$ from $C^\infty(M)$ to itself defined by

$$(L_X f)(x) = T_x f \cdot X_x$$

and we have the same characterization of derivations.

Theorem 3.29. *The map $L : X \mapsto L_X$ is a bijection from $C^\infty(TM)$ to the set of derivations on M.*

PROOF. We first note that L is injective. Indeed let $a \in M$ be such that $X_a \neq 0$. Then, if (U, φ) is a chart whose domain contains a, there exists a function $f \in C^\infty(U)$ such that $L_X f(a) \neq 0$: the vector $T_a \varphi \cdot X_a$ has at least one nonzero coordinate, which we suppose to be the k-th coordinate, and let f be the k-th component of the chart φ. If V is an open subset containing a such that $\overline{V} \subset U$, by Corollary 3.5 there exists a smooth function f equal to 1 on V and with support contained in U. Now $fg \in C^\infty(\mathbf{R}^n)$ and $L_X(fg)(a) = L_X(f)(a)$.

We now show that L is surjective. Let $(U_i, \varphi_i)_{i \in I}$ be an atlas of M, and let δ be a derivation. By Theorem 3.15, every open subset $U \subset M$, δ induces a derivation of $C^\infty(U)$. Then the map

$$h \longmapsto \varphi_i \circ \left(\delta(h \circ \varphi_i^{-1})\right)$$

is a derivation of $C^\infty(\varphi_i(U_i))$. We already know in this case by Theorem 3.16 that there exists a vector field Y^i on $\varphi_i(U_i)$ such that

$$\varphi_i \circ \left(\delta(h \circ \varphi_i^{-1})\right) = L_{Y^i} h.$$

Then, if $X^i = \varphi_{i\,*}^{-1} Y^i$, then clearly $\delta_{U_i} = L_{X^i}$. Therefore if $U_i \cap U_j \neq \emptyset$, the derivations L_{X^i} and L_{X^j} induce the same derivation on $U_i \cap U_j$, and by the first part the vector fields X^i and X^j have the same restriction to $U_i \cap U_j$. $\square$

AN IMMEDIATE CONSEQUENCE OF THIS THEOREM IS THAT THE DEFINITIONS AND PROPERTIES OF THE IMAGE AND LIE BRACKET OF VECTOR FIELDS ON OPEN SUBSETS OF $\mathbf{R}^n$ EXTEND TO MANIFOLDS

This can also be seen in a pedestrian yet instructive way by using the compatibility conditions given above.

Remark. The derivation point of view lets us justify a very convenient abuse of notation: if (U, φ) is a local chart of M, we will denote the image of the vector field ∂_i on $\varphi(U)$ under φ^{-1} again by ∂_i. We always write this with a choice of chart understood. A vector field on U can then be uniquely written in the form

$$\sum_{i=1}^n X^i \partial_i \quad \text{with } X^i \in C^\infty(U).$$

We finish this paragraph with an important application to critical points.

Theorem 3.30. *Let $f : M \to \mathbf{R}$ be a C^2 function on a manifold M, let a be a point where $df_a = 0$, and let u, v be two tangent vectors at a. Let X and Y be two vector fields defined on an open subset containing a, such that $X_a = u$, $Y_a = v$. Then*

$$X \cdot df(Y)(a) = Y \cdot df(X)(a) \tag{$*$}$$

depends only on u and v.

PROOF. By the definition of the bracket,

$$X \cdot df(Y) - Y \cdot df(X) = [X, Y] \cdot f = df([X, Y]),$$

which gives the first statement. The second statement follows from the fact that the left hand side of (*) shows that as a function of X, the quantity considered depends only on X_a, and the right hand side depends only on Y_a. $\qquad\square$

We now associate to every function having a critical point at a the symmetric bilinear form on $T_a M$ given above.

Definitions 3.31

a) *The symmetric bilinear form thus obtained is called the* Hessian *of f at a.*

b) *A critical point is called* non-degenerate *if this bilinear form is non-degenerate.*

Of course for two vector fields defined by local coordinates, with the notations of the remark above, we have

$$\text{Hess } f_a(\partial_i, \partial_j) = \partial_{ij}^2 f(\varphi(a)). \qquad (**)$$

We consider the associated quadratic form more often than this symmetric bilinear form, and this quadratic form is also called the Hessian, We see from (**) that if the Hessian is positive definite (resp. negative definite) at a, the critical point corresponds to a local minimum (resp. a local maximum).

Note. On $\mathbf{R}^n$, analysts commonly use the Hessian matrix $\partial_{ij}^2 f$ and the associated bilinear form. We then have

$$\text{Hess}(f \circ \varphi)(a)(u, v) = \text{Hess}(f)(T_a\varphi \cdot u, T_a\varphi \cdot v) + T_{\varphi(a)}f \cdot \text{Hess } \varphi(u, v).$$

This formula allows us to give an equivalent definition of Hessian at a critical point for C^2 functions on a manifold, using the definition of tangent space directly.

3.6. The Flow of a Vector Field

It is time to introduce a more geometric side of vector fields. For a vector field $x \mapsto V_x$ in the plane, simply tracing the vectors V_x seen as vectors with origin at x allows us to see curves to which these vectors are tangent.

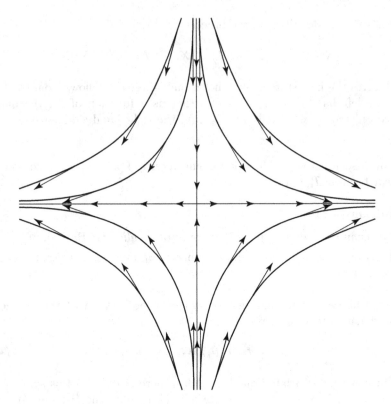

Figure 3.4: Trajectories of the vector field $(x, -y)$ in the plane

Definition 3.32. *A* trajectory *or* integral curve *of a vector field X on a manifold M, is any C^1 curve $t \mapsto c(t)$, defined on an open interval $I \subset \mathbf{R}$ with values in M such that*

$$\forall t \in I, \quad c'(t) = X_{c(t)}.$$

For an open subset U of $\mathbf{R}^n$, if $X = \sum_{i=1}^{n} X^i \partial_i$, this amounts to saying that the components of c are solutions of the system of first order differential equations

$$\frac{dc^i}{dt} = X^i(c^1, \ldots, c^n) \quad (1 \leqslant i \leqslant n).$$

As the functions X^i are smooth, we can apply the classical result concerning the existence and uniqueness of systems of differential equations seen in Section 1.9.

Theorem 3.33. *Let X be a vector field on an open subset U of $\mathbf{R}^n$ and x a point in U. Then there exists an open interval I containing 0 such that $c(0) = x$; if $c_1 : I_1 \to U$ is another trajectory with the same property, c and c_1 coincide on $I \cap I_1$.*

Using the terminology of Section 3.3, this means that the germ of c is unique. We may then show that there exists a unique *maximal* interval of definition of c. We denote the corresponding trajectory by c_x. We can now give a more precise version of the preceding statement ("dependence on initial conditions"):

If I_x is the (maximal) interval of definition of c_x, the union of the $I_x \times \{x\}$, as x varies over U, is an open subset Ω of $\mathbf{R} \times U$, containing $\{0\} \times U$, for which the map

$$(x, t) \longmapsto c_x(t)$$

is smooth.

Examples. If $U = \mathbf{R}$ and $X_x = x$, we have $c_x(t) = xe^t$, and $\Omega = \mathbf{R} \times \mathbf{R}$. However if $X(x) = x^2$, then

$$c_x(t) = \frac{x}{1 - tx} \quad \text{and} \quad \Omega = \left\{(x, t) \in \mathbf{R}^2 : \ xt < 1\right\}.$$

Thus the trajectories are not necessarily defined on all of $\mathbf{R}$.

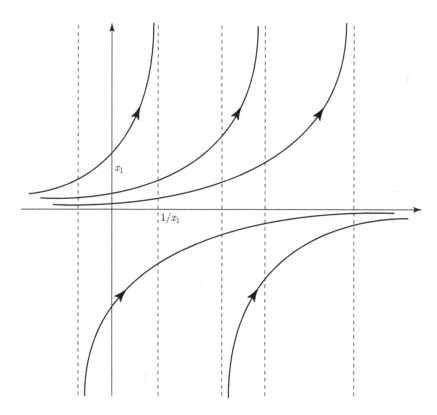

Figure 3.5: Blow up in finite time

We return to the general case. The system of differential equations given in Definition 3.32 is *autonomous* (this is to say that the variable t does not explicitly enter into the right hand side of the system). As a consequence, if $t \mapsto c(t)$ is a trajectory, the same is true for $t \mapsto c(t+a)$ for any real number a. Taking the uniqueness into account, we obtain the identity

$$c_x(t + a) = c_{c_x(a)}(t).$$

We will rewrite this relation (valid under the condition that both sides make sense) by emphasizing the "spatial" variable over the "temporal". In other words, we write

$$\varphi_t^X(x) = c_x(t).$$

In particular $\varphi_0^X(x) = x$, and the identity above gives

$$\varphi_{t+t'}^X(x) = \varphi_t^X\left(\varphi_{t'}^X(x)\right).$$

This abuse of notation allows us to write the following assertion: if the right hand side is defined (this is to say if (t', x) and $\left(t, \varphi_{t'}^X(x)\right)$ are in Ω) then the left hand side is also defined, and they are equal. We then write, with the obvious abuse of notation

$$\varphi_{t+t'}^X = \varphi_t^X \circ \varphi_{t'}^X = \varphi_{t'}^X \circ \varphi_t^X.$$

Definition 3.34. *The map* $\varphi^X : \Omega \to U$ *is called the* flow *of the vector field X.*

The φ_t^X are *diffeomorphisms* onto their image. (But note: their domain of definition can be different from U if $t \neq 0$. See the example which follows Theorem 3.33.) Because of the identity above, we also call φ^X the *local one-parameter group of diffeomorphisms* associated to the vector field X. This is not always a group! The word local is there to emphasize the possible restrictions on the domains of definition of φ^X and the φ_t^X. We omit the X in this notation when there is no ambiguity. Before we proceed further, we recall a few typical examples in $\mathbf{R}^n$.

1. If $X = \sum_{i=1}^n a^i \partial_i$ (constant vector field), $\varphi_t(x) = x + ta$.

2. If $X = \sum_{i=1}^n x^i \partial_i$ (*radial* vector field), $\varphi_t(x) = e^t x$.

3. If $\overline{A}$ is the linear vector field associated to the endomorphism A, defined by $\overline{A}_x = Ax$, we have $\varphi_t(x) = (\exp tA) \cdot x$.

We note that as a bonus, the preceding discussion gives a proof of the fact that

$$\exp(t + t')A = (\exp tA)(\exp t'A).$$

The following property, which is very useful in practise, allows us to pass from the flow to the vector field.

Lemma 3.35. *Let h be a smooth map defined on an open subset $I \times U$ containing $\{0\} \times U$ with values in U such that*

 i) $h(t, h(t', x)) = h(t+t', x)$ *as soon as both sides of this equation are defined;*

 ii) $h(0, x) = x$;

iii) $\frac{d}{dt} h(t, x)_{|t=0} = X_x$.

Then $h(t, x) = \varphi_t^X(x)$ everywhere where h is defined.

PROOF. It suffices to remark that

$$\frac{d}{dt} h(t, x)_{|t=t_0} = \frac{d}{dt} h(t + t_0, x)_{|t=0} = \frac{d}{dt} h(t, h(t_0, x))_{|t=0} = X_{h(t_0, x)}. \qquad \square$$

Definitions 3.36. *A map h satisfying the conditions of Lemma 3.35 is a* local one-parameter group, *and the vector field*

$$X_x = \frac{d}{dt} h(t, x)_{|t=0}$$

is the infinitesimal generator *of h.*

We now study the flow of the image of a vector field.

Proposition 3.37. *Let X be a vector field on an open subset $U \subset \mathbf{R}^n$ and let $\psi : U \mapsto V$ be a diffeomorphism. If φ_t is the flow of X, then the flow of the image $\psi_* X$ is*

$$\psi \circ \varphi_t \circ \psi^{-1}.$$

PROOF. By Lemma 3.35 applied to $h(t, x) = \psi \circ \varphi_t \circ \psi^{-1}(x)$, it suffices to remark that

$$\frac{d}{dt} \left(\psi \circ \varphi_t \circ \psi^{-1} \right)(x)_{|t=0} = T_{\psi^{-1}(x)} \psi \cdot X_{\psi^{-1}(x)} = \psi_* X_x. \qquad \square$$

Example. Recall the linear vector field $\overline{A}$ associated to $A \in End(\mathbf{R}^n)$, and consider the diffeomorphism given by an invertible matrix P. Then

$$P_* \overline{A}_x = PAP^{-1} \cdot x,$$

and we recover the fact that $\exp PAP^{-1} = P(\exp A)P^{-1}$.

Moreover, if the matrix P is itself of the form $\exp tB$, then

$$(\exp tB)(\exp A)(\exp -tB) = \left(1 + tB + O(t^2)\right) A \left(1 - tB + O(t^2)\right)$$
$$= A + t(BA - AB) + O(t^2)$$

and it follows

$$\frac{d}{dt} (\exp tB)(\exp A)(\exp -tB) \cdot x = (BA - AB) \cdot x.$$

But this can also be written

$$\frac{d}{dt}\left((\exp tB_*\overline{A})_x\right)_{t=0} = \left[\overline{A}, \overline{B}\right]_x.$$

This final property is true in a much more general context.

Theorem 3.38. *Let X and Y be two vector fields on an open subset U of $\mathbf{R}^n$, and let φ_t^Y be the flow of Y. Then*

$$\left(\frac{d}{dt}\right)(\varphi_{t*}^Y X)_{|t=0} = [X, Y].$$

PROOF. We use the following property, which is a version of the Hadamard lemma with parameters.

Let $f : [-\epsilon, \epsilon] \times U \to \mathbf{R}$ be a smooth map such that $f(0, x) = 0$. Then f can be written

$$f(t, x) = tg(t, x)$$

where g is smooth and

$$\frac{\partial f}{\partial t}(0, x) = g(0, x).$$

We now reason using derivations, and consider the effect of $\varphi_{t*}X$ on a function f. So let $f(t, x) = f(\varphi_t(x)) - f(x)$, and set $g_t = g(t, .)$ with $g_0 = L_Y f$. Then

$$(L_{\varphi_{t*}X})f = \left(L_X(f \circ \varphi_t)\right) \circ \varphi_{-t} \quad \text{(by Theorem 3.16)}$$
$$= \left(L_X(f + tg_t)\right) \circ \varphi_{-t}$$
$$= (L_X f) \circ \varphi_{-t} + t(L_X g_t) \circ \varphi_{-t}.$$

The derivative with respect to t of the first term, evaluated at $t = 0$ is $-L_Y(L_X f,)$ and that of the second is $L_X g_0 = L_X L_Y f$. □

Remark. Let X, Y and Z be three vector fields, and φ_t the flow of Z. Differentiating the identity

$$\varphi_{t*}^Z[X, Y] = [\varphi_{t*}^Z X, \varphi_{t*}^Z Y]$$

with respect to t, we obtain

$$[[X, Y], Z] = [[X, Z], Y] + [X, [Z, Y]],$$

again recovering Jacobi's identity.

EVERYTHING WE HAVE SEEN IN THIS PARAGRAPH APPLIES TO MANIFOLDS

The only place where we used a result specific (at least superficially) to open subsets of $\mathbf{R}^n$ is Theorem 3.33: this result appeals to the classical theory of differential equations. But if X is a vector field on a manifold M equipped with an atlas $(U_i, \varphi_i)_{i \in I}$, and if $\theta_{i,t}$ is the flow of the vector field $\varphi_{i*} X$ on $\varphi_i(U_i)$, then

$$\varphi_i^{-1} \circ \theta_{i,t} \circ \varphi_i$$

is a one-parameter group of diffeomorphisms on U_i, which by construction has infinitesimal generator given by X. We deduce that

$$\varphi_i^{-1} \circ \theta_{i,t} \circ \varphi_i = \varphi_j^{-1} \circ \theta_{j,t} \circ \varphi_j$$

on $U_i \cap U_j$, because a one-parameter group is determined by its infinitesimal generator.

Moreover, to study flows of vector fields, compact manifolds are actually a more natural and agreeable setting than open subsets of $\mathbf{R}^n$ for the following reason:

Theorem 3.39. *If X is a vector field on a compact manifold M, then the flow of X is defined on $\mathbf{R} \times M$. In particular, for every t, φ_t is a a diffeomorphism of M.*

PROOF. At the outset, the domain of definition of the flow φ is an open subset of $\Omega \subset \mathbf{R} \times M$, containing $\{0\} \times M$, and the domain of definition of the trajectory c_x is the interval $I_x = \Omega \cap \mathbf{R} \times \{x\}$. Set $I_x = (\alpha, \beta)$, and suppose for example that $\beta < +\infty$. If (t_n) is a sequence of real numbers which increases to β, then by compactness of M, the sequence $(c_x(t_n))$ has a limit point y (after taking a subsequence if necessary). Then there exists an $\epsilon > 0$ and an open subset U containing y such that $(\epsilon, \epsilon) \times U \subset \Omega$.

Then let t_n be such that $|t_n - \beta| < \epsilon$ and $c_x(t_n) \in U$. The trajectory $s \mapsto d(s)$ with initial condition $d(0) = c_x(t_n)$ is defined on an interval (ϵ, ϵ), and by Theorem 3.33, $d(s) = c_x(t_n + s)$ if $t_n + s \in I_x$. Let

$$\widetilde{c}_x(t) = \begin{cases} c_x(t) & \text{if } t \in I_x \\ d(t - t_n) & \text{if } t_n \leqslant t < t_n + \epsilon \end{cases}$$

and we obtain a trajectory with extends c_x beyond the limit β of I_x, contradicting the maximality hypothesis of I_x. The relation

$$\varphi_{t+t'}^X = \varphi_t^X \circ \varphi_{t'}^X = \varphi_{t'}^X \circ \varphi_t^X$$

is now true for any t, t' and x. In particular, φ_t is a diffeomorphism with inverse φ_{-t}. $\qquad\square$

Remarks

a) This result is still true for vector fields with compact support on noncompact manifolds: it suffices to remark that nonconstant trajectories of a vector field are contained in the support of the vector field.

b) We also deduce that a smooth manifold, whether compact or not, admits "many" diffeomorphisms. It suffices to see, by using a chart and a bump function with support contained in the domain of the chart, that a manifold always admits nontrivial vector fields with compact support. We can deduce (see Exercise 15) that the group of diffeomorphism is k-transitive for all k.

The next result shows why it is useful to be able to construct such diffeomorphisms.

Theorem 3.40. *Let f be a smooth real-valued function on an compact manifold M. For every real number a, set*

$$M^a = \{x \in M : \ f(x) \leqslant a\}.$$

If $a < b$ and $f^{-1}([a,b])$ does not contain a critical point, then there exists a diffeomorphism of M from M^a to M^b.

PROOF. Suppose M is embedded in Euclidean space $\mathbf{R}^N$, which is possible by Theorem 3.7. Then we send M^a to M^b by "pulling" along the orthogonal trajectories of the submanifolds $f = $ constant (see Figure 3.6).

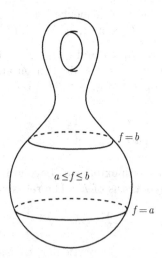

Figure 3.6: From one level-set to another

More precisely, given f we define a vector field ∇f by requiring

$$\langle \nabla_m f, u \rangle = df_m(u), \quad \forall m \in M,\ u \in T_m M,$$

where we denoted the Euclidean inner of $\mathbf{R}^N$ product by $\langle , \rangle$. The set of critical points of f is closed, thus $f^{-1}([a,b])$ is contained in an open subset of U without critical points. Now let g be a bump function, equal to 1 on $f^{-1}([a,b])$ and supported in U. We define a vector field on M by

$$X_m = \frac{g(m)}{\langle \nabla f_m, \nabla f_m \rangle} \nabla f_m.$$

If φ_t is the corresponding flow, we have

$$\frac{d}{dt} f\big(\varphi_t(m)\big) = df_{\varphi_t(m)} \cdot X_{\varphi_t(m)} = g(m) \geqslant 0,$$

and

$$\frac{d}{dt} f\big(\varphi_t(m)\big) = 1 \quad \text{if } m \in f^{-1}([a,b]).$$

As a result, $\varphi_{b-a}(M^a) = M^b$ and $\varphi_{a-b}(M^b) = M^a$. $\qquad\qquad \square$

Remarks

a) **The embedding of M into $\mathbf{R}^N$ is not essential. We took the *gradient* of f with respect to the Riemannian metric $\langle , \rangle_{|M}$, but any other Riemannian metric would have also worked.**

b) If we use the notion of manifold with boundary, we can say that M^a and M^b are diffeomorphic.

c) The same result is true if M is not compact, provided that $f^{-1}([a,b])$ is compact. On the other hand this last hypothesis is essential, as we can see by removing a point.

3.7. Time-Dependent Vector Fields

To find the trajectories of a vector field on an open subset of $\mathbf{R}^n$ or even a manifold reduces to solving a system of autonomous differential equations. On the other hand, a non-autonomous system of differential equations on an open subset of $\mathbf{R}^n$, that is to say a system of the form

$$\frac{dc^i}{dt} = X^i(t, c^1(t), \ldots, c^n(t)) \quad (1 \leqslant i \leqslant n)$$

can be reduced to an autonomous system on $\mathbf{R} \times U$ given by

$$\frac{d\tau}{dt} = 1$$

$$\frac{dc^i}{dt} = X^i(\tau(t), c^1(t), \ldots, c^n(t)) \quad (1 \leqslant i \leqslant n).$$

This extends without difficulty to manifolds.

Definition 3.41. *If M is a manifold and $I \subset \mathbf{R}$ is an interval, we call a smooth map X from $I \times M$ to TM such that $p(X(t,x)) = x$ (here we denoted the projection of TM to M by p) a time-dependent vector field.*

We hasten to associate to X the vector field $\widetilde{X}$ on $I \times M$ defined by

$$\widetilde{X}_{(t,x)} = (1, X(t,x)).$$

Adapting Definition 3.32, we call a *trajectory* of X any curve $c : J \to M$, where $J \subset I$, such that $c'(t) = X_{c(t)}(t)$. The trajectories of X are then projections to M of trajectories of $\widetilde{X}$ on $I \times M$. The theorem of local existence and uniqueness of trajectories immediately extends: it is no more difficult to pass to systems of autonomous differential equations than to the non-autonomous ones. From the flow point of view, we note that if c is the trajectory such that

$$c'(s) = X(s, c(s)) \quad \text{and} \quad c(t) = x,$$

then we have

$$\varphi_s^{\widetilde{X}}(t,x) = (t+s, c(t+s)).$$

If the manifold M is compact, we have a weaker but still useful version (we no longer have a one-parameter group of diffeomorphisms) of Theorem 3.39.

Theorem 3.42. *Let X be a time dependent vector field on a compact manifold M. Suppose that X is defined on $I = (\alpha, \beta) \times M$ (we assume $-\infty \leqslant \alpha < \beta \leqslant +\infty$). Then:*

i) *The trajectories of X are defined for all $t \in I$.*

ii) *If $s \mapsto F_s(t,x)$ is the position at time $t+s$ of the trajectory that passes through x at time t, then $x \mapsto F_s(t,x)$ is a diffeomorphism of M for all $s \in (\alpha - t, \beta - t)$.*

iii) *Conversely, let $s \mapsto G_s$ be a path of diffeomorphisms of M (in other words, $(s,x) \mapsto G_s(x)$ is a smooth map of $J \times M$ to M and $x \mapsto G_s(x)$ is a diffeomorphism for all $s \in J$). Suppose that $0 \in J$ and further $G_0 = Id_M$. Then*

$$Y(t,x) = \frac{d}{ds}\left(G_s(G_t^{-1}(x))\right)_{s=t}$$

is a time-dependent vector field on M, and $s \mapsto G_s(x)$ is the trajectory which passes through x at time 0.

PROOF

i) It suffices to consider the argument of Theorem 3.39, which did not use the fact that the differential equation under consideration was autonomous.

ii) Now let φ_s be the flow of $(1, X)$ on $I \times M$. We have

$$\varphi_s(t, x) = \big(s + t, F_s(t, x)\big).$$

Applying the property of being a one-parameter group to the component in M, we obtain

$$F_{s_1 + s_2}(t, x) = F_{s_1}\big(t + s_2, F_{s_2}(t, x)\big).$$

We see in particular that for fixed $t \in (\alpha, \beta)$ and $s \in (\alpha - t, \beta - t)$, $x \mapsto F_s(t, x)$ is a diffeomorphism with inverse $x \mapsto F_{-s}(s + t, x)$.

iii) Now let $s \mapsto G_s$ be a smooth path of diffeomorphisms, and

$$Y(t, x) = \frac{d}{ds}\big(G_s\big(G_t^{-1}(x)\big)\big)_{|s=t}$$

be the associated time-dependent vector field. Then

$$Y\big(t, G_t(x)\big) = \frac{d}{ds} G_s(x)_{|s=t},$$

which shows that the curve $t \mapsto G_t(x)$ is a trajectory of the vector field Y. This is by hypothesis the trajectory which passes through x at time 0. $\quad\square$

In the same way that a vector field gives rise to a one-parameter group of diffeomorphisms, a time-dependent vector field gives rise to a family of one-parameter diffeomorphisms. This justifies the following definition:

Definition 3.43. *The time dependent vector field defined in 3.42 is called the* infinitesimal generator *of the family of diffeomorphisms $s \mapsto G_s$.*

The construction of families of diffeomorphisms from their infinitesimal generator is a very useful tool, invented by Jürgen Moser (1928–1999), and called *Moser's trick*. As an example we use it for the following result (proved also in Exercise 11 of Chapter 1). For other examples of Moser's trick, see Exercise 17 of Chapter 5 and Exercise 19 of Chapter 7.

Theorem 3.44 (Morse lemma). *Let $f : U \to \mathbf{R}$ be a smooth function defined on an open subset of $\mathbf{R}^n$ and let $a \in U$ be a non-degenerate critical point of f (see Definition 3.31). Then there exists open subsets V and W*

containing a and contained in U and a diffeomorphism φ from W to V such that $\varphi(a) = a$ and

$$f(\varphi(x)) - f(a) = \sum_{i,j} \partial^2_{ij} f(a)(x^i - a^i)(x^j - a^j).$$

PROOF. We may always suppose that $a = 0$ and $f(a) = 0$. Two applications of Hadamard's lemma allows us to write f in the form

$$f(x) = \sum_{1 \leqslant i,\, j \leqslant n} h_{ij}(x) x^i x^j,$$

with $h_{ij} = h_{ji}$ and $h_{ij}(0) = \partial^2_{ij} f(0)$, for some smooth functions h_{ij}. In a ball with center 0 contained in U we write, for $t \in [0,1]$,

$$f_t(x) = \sum_{1 \leqslant i,\, j \leqslant n} h_{ij}(tx) x^i x^j.$$

Thus $f_1 = f$, while f_0 is the desired local model. We will show that there exists a path of diffeomorphisms $t \mapsto \varphi_t$, such that

$$f_t \circ \varphi_t = f_0 \quad \text{and} \quad \varphi_0 = \mathrm{Id}.$$

Let $f'_t(x)$ be the partial derivative with respect to t of the function $(t,x) \mapsto f_t(x)$. In differentiating the preceding relation with respect to t, we see that the infinitesimal generator X of φ_t satisfied

$$f'_t(\varphi_t(x)) + (df_t)_{\varphi_t(x)} \cdot X_{\varphi_t(x)}(t) = 0.$$

We must therefore find a time-dependent vector field such that

$$f'_t(y) + (df_t)_y \cdot X_y(t) = f'_t(y) + \sum_k \partial_k f_t(y) X^k(t,y) = 0.$$

The difficulty comes from $(df_t)_0 = 0$. However, knowing that

$$f'_t(y) = \sum_{k,i,j} \partial_k h_{ij}(ty) y^k y^i y^j,$$

and

$$\partial_k f_t(y) = \sum_{i,j} t h_{ij}(ty) y^i y^j + 2 \sum_i h_{ki}(ty) y^i,$$

we are assured of finding such a vector field if the coordinates X^k are solutions of the linear system

$$\sum_k \left(2 h_{ki}(ty) + t \sum_j \partial_k h_{ij}(ty) y^j \right) X^k(t,y) + \sum_{k,j} \partial_k h_{ij}(ty) y^k y^j = 0$$

$(1 \leqslant i \leqslant n)$. As 0 is a non-degenerate critical point, the determinant of the matrix $(h_{ki}(0))_{1 \leqslant k, \, i \leqslant n}$ is nonzero. The same will be true of the determinant of a system if y is sufficiently close to 0, say for $\|y\| < r$. To be sure that the time-dependent vector field thus found is the infinitesimal generator of a path of diffeomorphisms defined up to time 1, we employ a method proved many times in this chapter: we multiply by a bump function with support in $B(0, r)$ equal to 1 on $B(0, r/2)$.   □

Remark. In particular, the reduced form obtained shows that non-degenerate critical points are isolated.

3.8. One-Dimensional Manifolds

We will use the embedding theorem from earlier in this Chapter to prove the following result.

Theorem 3.45. *A manifold which is connected and countable at infinity of dimension 1 is diffeomorphic to S^1 if it is compact, and to $\mathbf{R}^n$ if it is not compact.*

The proof of this "intuitive" result is delicate. By Theorem 3.7, such a manifold C admits an embedding in $\mathbf{R}^n$ (we can even say in $\mathbf{R}^3$ by using Theorem 3.8, but this will not be necessary). We will start with such an embedding, but first a few preliminaries on curve parametrizations.

Definition 3.46. *Let $\varphi \, : \, (a, b) \, \rightarrow \, C$ be a parametrization of a one-dimensional submanifold of Euclidean space. We say φ is an* arc-length parametrization *if for every t in (a, b) the velocity vector $\varphi'(t)$ has norm 1.*

Such parametrizations always exist, and we have uniqueness up to a change of orientation and starting point. More precisely:

Lemma 3.47. *Let $\psi \, : \, J \, \rightarrow \, C$ be a parametrization of C. Then there exists an arc-length parametrization with the same image $\psi(J)$. Moreover, if $\varphi_1 : I_1 \rightarrow C$ and $\varphi_2 : I_2 \rightarrow C$ are two such parametrizations, there exists a c such that $\varphi_2(t) = \varphi_1(c + t)$ where $\varphi_2(t) = \varphi_1(c - t)$.*

PROOF. Choose $t_o \in J$ and set

$$S(t) = \int_{t_o}^{t} \|\psi'(u)\| \, du.$$

The function S is smooth and its derivative is everywhere nonzero. If $I = S(J)$, the function $\psi \circ S^{-1} : I \rightarrow C$ gives an arc-length parametrization.

Now let φ_1 and φ_2 be two parametrizations with image $\psi(J)$, defined on open intervals I_1 and I_2. Then function $\varphi_1^{-1} \circ \varphi_2$ is a diffeomorphism from I_2 to I_1. Moreover, by differentiating

$$\varphi_2 = \varphi_1 \circ \left(\varphi_1^{-1} \circ \varphi_2 \right),$$

we see that $|(\varphi_1^{-1} \circ \varphi_2)'(t)| = 1$, and thus that the derivative $(\varphi_1^{-1} \circ \varphi_2)'(t)$ is constant and equal to 1 or -1. $\square$

PROOF OF THEOREM 3.45. Let $\varphi : (a, b) \mapsto C$ be an arc-length parametrization. We may suppose this parametrization to be maximal (possibly with $a = -\infty$ or $b = -\infty$). We will see that φ is surjective by showing that its image $\varphi((a, b))$ is open and closed in C. First $\varphi((a, b))$ is open as $\varphi : (a, b) \to C$ is a local diffeomorphism. Now let

$$x \in \overline{\varphi((a, b))} \smallsetminus \varphi((a, b)).$$

We have $x = \lim_{n \mapsto +\infty} \varphi(t_n)$, where (t_n) is a sequence of points in (a, b). Passing to a subsequence, we may suppose that (t_n) converges. The limit is either a point of (a, b) or a or b. The first case is impossible as we supposed that x is not in the image of φ. Suppose for example that the limit is b, and let $\psi : (-\epsilon, \epsilon) \to C$ be an arc-length parametrization of a neighborhood of x in C, such that $\psi(0) = x$. Then $\varphi(t_n) \in \psi(J)$ for n sufficiently large. More precisely, we have $\varphi(t_n) = \psi(\eta_n)$, for a unique η_n which tends to 0 as n tends to infinity. By the lemma,

$$\varphi(t) = \psi(\eta_n \pm (t - t_n)).$$

For n sufficiently large, the right hand side furnishes an extension of φ past b, contrary to the maximality hypothesis.

If φ is injective, we have a injective immersion from an interval (a, b) to E, thus the image is a submanifold. Such an immersion is an embedding (we leave this last point which is not immediately obvious to the reader), and is a diffeomorphism between (a, b) and C, which proves that C is diffeomorphic to $\mathbf{R}$.

Now suppose that φ is not injective, and let t_1 and t_2 be such that $\varphi(t_1) = \varphi(t_2)$. By translating the parameter t, we may suppose that $t_1 = 0$ and $t_2 = c > 0$. The vectors $\varphi'(0)$ and $\varphi'(c)$ are two vectors with norm 1 tangent to C at the same point, thus $\varphi'(0) = \pm\varphi'(c)$. The second case is impossible: the lemma applied to the parametrizations $\varphi(t)$ and $\varphi(c - t)$ on $[0, c]$ gives $\varphi(t) = \varphi(c - t)$, and by differentiating $\varphi'(t) = -\varphi'(c - t)$, and $\varphi'(c/2) = 0$, which is impossible.

So $\varphi'(0) = \varphi'(c)$. The lemma then gives $\varphi(t) = \varphi(c + t)$ until $c + t \in (a, b)$; however if $(a, b) \neq \mathbf{R}$, this relation also gives an extension of φ past (a, b). Thus $(a, b) = \mathbf{R}$ and φ is periodic. Let T be its smallest period.

If $\varphi(t_1) = \varphi(t_2)$, then by the reason above $\varphi(t_2 - t_1 + t) = \varphi(t)$ for any t, and $t_2 - t_1$ is an integer multiple of T. Then, by passing to the quotient, φ gives an *injective* immersion of $\mathbf{R}/\mathbf{Z}$ to C, thus an embedding. $\square$

Remark. We have given very simple "intrinsic characterization" of one-dimensional manifolds. On the other hand, their embeddings in $\mathbf{R}^3$ can be much more complicated that the "standard" embedding $t \mapsto (\cos t, \sin t, 0)$. This embedding extends to an embedding of the closed disk. The same is not true for the *trefoil knot*

$$t \longmapsto \big((2 + \cos 3t)\cos 2t, (2 + \cos 3t)\sin 2t, \sin 3t\big)$$

(this is not an obvious assertion, despite Figure 3.7).

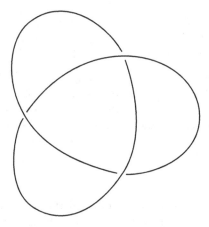

Figure 3.7: Trefoil knot

For a taste of Knot Theory, which is now experiencing an explosion of interest, see [Adams 94].

3.9. Comments

More about immersions and embeddings

The proof of Whitney's theorem foreshadows the fact that once we know that immersions and embeddings of a given manifold exist, they are numerous. For example one shows (cf. [Hirsch 76, Chapter 2]) that if M is compact, the set of immersions of M to $\mathbf{R}^{2 \dim M}$ is open and dense in $C^\infty(M, \mathbf{R}^{2 \dim M})$ **for the compact-open topology**.

Hence the question of the classification of these immersions. To answer we introduce the notion of *isotopy*. Two immersions f and g from M to N are isotopic if there exists a smooth map $H : [0,1] \times M \to N$ such that $H(0,x) = f(x)$, $H(1,x) = g(x)$, and such that the map $x \mapsto H(t,x)$ is an immersion for all t. It is easy to see for example that two immersions sufficiently close in C^1 on a compact manifold are isotopic. The classification of immersions up to isotopy is started for example in [Adachi 93].

We explain what happens for $M = S^1$. This is easier than in higher dimensions, but still interesting.

1. Two immersions of S^1 to $\mathbf{R}^n$ are always isotopic if $n \geqslant 3$ (we invite the reader to prove this by imitating the proof of Whitney's theorem).

2. Two immersions f and g of S^1 to $\mathbf{R}^2$ are isotopic if and only if their "hodographs" $t \mapsto f'(t)/\|f'(t)\|$ and $t \mapsto g'(t)/\|g'(t)\|$ are homotopic (in other words, if their oriented tangent vectors make the "same number of turns"). This is the Whitney-Gauenstein theorem, proved for example in [Adachi 93], [Berger-Gostiaux 88, 9.4.8] or [Do Carmo 76, 1.7].

If we now replace immersion by embedding in the definition of isotopy, we may show (*ibidem*) that there are two embedding isotopy classes of S^1 in $\mathbf{R}^2$ (the trigonometric sense and the needles of a watch sense). If we replace $\mathbf{R}^2$ by $\mathbf{R}^3$, the problem of classifying embedding isotopies of S^1 is similar to the problem of classifying knots mentioned in the previous paragraph.

Morse theory

We saw in this Chapter that the topological type of a "sub level" $\{f \leqslant a\}$ of a smooth function on a manifold does not change so long as we do not cross a critical level (which is to say a level containing a critical point). What happens when we meet a critical point? The simplest case is that where there are only two critical points (this is the minimum number on a compact manifold, a constant function does not have a distinct maximum and minimum value). The answer is given by Reeb's theorem:

If M is a compact manifold that admits a smooth function with exactly two critical points then M is homeomorphic to a sphere.

We cannot conclude that M is diffeomorphic to S^n (we will see why in Exercise 25, where we prove Reeb's theorem in the easier case where the critical points are non-degenerate: it is clear that a homeomorphism of S^n can be extended to a homeomorphism of the closed ball, but the corresponding property for diffeomorphisms is false).

In the general case, if $p \in M$ is a *non-degenerate* critical point (and therefore isolated by Theorem 3.44), the Morse lemma allows us to control the change of topology as we pass from the sublevel $f \leqslant f(p) - \epsilon$ to $f \leqslant f(p) + \epsilon$, as a function of the *index* of the critical point p, which is defined as the number of negative squares in the Morse lemma (see [Milnor 63] and [Hirsch 76] for more detailed statements and suggestive pictures). From this we deduce for example, the weak Morse inequality (*ibidem*):

If f is a function on M having only non-degenerate critical points, and if $c_k(f)$ is the number of critical points of index k, then

$$\dim H^k(M, \mathbf{R}) \leqslant c_k(f)$$

and

$$\sum_k (-1)^k c_k(f) = \sum_k (-1)^k \dim H^k(M, \mathbf{R}).$$

The cohomology spaces $\dim H^k(M, \mathbf{R})$ will be defined in Chapter 7. **However, the differential form point of view adopted in this book makes the proof of the Morse inequalities inconvenient. On the other hand they are easy to prove with a little algebraic topology.**

Dynamical systems

By this we mean the study of the global trajectories of vector fields (and more generally the study of iterates of a diffeomorphism whether obtained by a flow or not).

Exercises 6, 7 and 10 give some of the few cases where the structure of the trajectories is well understood. Another "trivial" example is that of the vector field $\partial_x + \alpha \partial_y$ on $T^2 = \mathbf{R}^2/\mathbf{Z}^2$: if α is rational, then the trajectories are all closed curves, if α is irrational, the trajectories are all strictly immersed submanifolds which are everywhere dense.

In the general case, even for vector fields on $\mathbf{R}^2$ with polynomial coefficients[3] it is out of the question to hope for an explicit expression for the trajectories.

Whence the importance of a *qualitative* study: the behavior of the trajectories when t tends to infinity, stabilities, existence of dense orbits, or periodic ones, etc. The first to recognize this was without doubt Henri Poincaré, see for example the book [Charpentier-Ghys-Lesne 10].

For these questions, see also [Demazure 00], [Arnold 78; 92], [Hirsch-Smale-Devaney 03] and above all [Katok-Hasselblatt 95].

---

3. This example is not a random example: the fact that there exists only a finite number of closed trajectories is a difficult result ($\simeq$ 1987).

Why vector bundles?

The justification for most the notions introduced in this book is the usage made of them in the sequel. This is not true for vector bundles which will be seldom used here.

The fact that all of the algebraic operations on vector spaces have analogues in the vector bundle setting is useful to obtain a convenient description of the tangent bundle, see for example Exercises 20 to 23.

Another justification comes from the study of complex manifolds. The fact that there are no nonconstant holomorphic functions on a compact complex manifold could render a mathematician very unhappy. However there exists certain holomorphic vector bundles with numerous holomorphic sections that can take the place of the smooth functions in this book. Exercise 27 gives an example of this situation. Complex geometry uses holomorphic bundles intensively, see for example [Debarre 05], [Griffiths-Harris 94] or [Demailly 12].

Connections

We give a few words on this notion that will barely be touched in this book.

We can see connections as a convenient way to do higher-order differential calculus. The differential of a vector field or the second differential of a function require the introduction of an iterated tangent bundle $T(TM)$, where manipulation is difficult. To avoid this, we introduce the following definition, which mimics the properties of the directional derivative in $\mathbf{R}^n$.

Definition 3.48. *A connection on a manifold M is a bilinear map ∇ from $C^\infty(TM) \times C^\infty(TM)$ to $C^\infty(TM)$ with the following properties.*

$$\nabla_{fX} Y = f \nabla_X Y$$
$$\nabla_X fY = df(X)Y + f \nabla_X Y$$

(X and Y are vector fields, and f is a smooth function).

The same argument using bump functions as in Theorem 3.15 shows that, like derivations, connections localize. Namely

$$(\nabla_X Y)_{|U} = \nabla_{X_{|U}} Y_{|U}$$

for every open subset of U. The first property then implies that the value at m of $\nabla_X Y$, for a given vector field Y, depends only on X_m. The second is essentially the Leibnitz rule. Localization allows us to work in the open set of chart. If in such an open subset $X = \sum_{i=1}^n X^i \partial_i$ and $Y = \sum_{i=1}^n Y^i \partial_i$, we define the Christoffel symbols by the formula

$$\nabla_{\partial_j} \partial_k = \sum_{i=1}^n \Gamma_{jk}^i \partial_i.$$

A direct calculation then shows

$$\nabla_X Y = \sum_{i=1}^{n} Z^i \partial_i$$

with

$$Z^i = \sum_{k=1}^{n} \partial_k Y^i X^k + \sum_{j,k=1}^{n} \Gamma^i_{jk} X^j X^k.$$

Connections always exist. For example, if M is embedding in $\mathbf{R}^n$, we can define a connection $\nabla_X Y$ as follows: if $T_m Y$ is the linear tangent map to Y at m *seen as a map of M to $\mathbf{R}^N$*, we may take $\nabla_X Y$ to be the orthogonal projection of $T_m Y \cdot X_m$ onto the tangent space to M at m.

The word connection is derived from connect: a connection permits us to associate to each piecewise C^1 path $c : [a, b] \to M$ from x to y a linear map from $T_x M$ to $T_y M$. See any book on Riemannian Geometry for details. **The example immediately above is that of the Levi-Civita connection of the Riemannian metric on M induced from the Euclidean metric on $\mathbf{R}^N$. This Euclidean metric was implicitly introduced when we spoke of the orthogonal projection.**

3.10. Exercises

1. Show that if M is a compact smooth manifold, the set of smooth functions is dense in $C^0(M)$ in the topology of uniform convergence (use the Stone-Weierstrass theorem).

2*. *Another theorem of Whitney*

a) Show that every open subset of $\mathbf{R}^n$ is a countable union of open balls.

b) Now let F be a closed subset of $\mathbf{R}^n$. We can then write for points x_p and certain positive numbers r_p,

$$\mathbf{R}^n \setminus F = \bigcup_{p \in \mathbf{N}} B(x_p, r_p).$$

Let f_p be a smooth nonnegative function which is strictly positive on $B(x_p, r_p)$ and with support equal to $\overline{B(x_p, r_p)}$ (we will give an explicit example of such a function). Let

$$M_p = \sup_{x \in \mathbf{R}^n, |\alpha| \leqslant p} \left| \frac{\partial^\alpha f_p}{\partial x_1^{\alpha_1} \dots \partial x_n^{\alpha_n}}(x) \right|, \quad \text{where } \alpha = \sum_{i=1}^{n} \alpha_i.$$

Show that the function

$$f(x) = \sum_{p \in \mathbf{N}} \frac{1}{p! M_p} f_p(x)$$

is smooth, and that $f^{-1}(0) = F$. (In other words, every closed subset of $\mathbf{R}^n$ is the zero set of a smooth function.)

3. Let f be a smooth function on $\mathbf{R}$ such that $f(\frac{1}{n}) = 0$ for each integer n (by the previous exercise, there exists such functions whose restriction to every neighborhood of zero is not identically zero). Show that every derivative of f is zero at the origin.

4. Show that the ring of germs of *continuous* functions at 0 on $\mathbf{R}^n$ does not admit a nonzero derivation.

5. Show that the ring of germs of smooth functions at 0 is not an integral domain. What happens for the ring of germs of analytic functions?

6. *The cross product seen as a bracket*

Consider $\mathbf{R}^3$ with the vector fields

$$X = z\partial_y - y\partial_z; \quad Y = x\partial_z - z\partial_x; \quad Z = y\partial_x - x\partial_y.$$

a) Show these vector fields are linearly independent over $\mathbf{R}$. What is their flow? Let E be the vector space on $\mathbf{R}$ which they span. Show that E is stable under bracket.

b) Let $\varphi : E \mapsto \mathbf{R}$ be given by

$$\varphi(aX + bY + cZ) = (a, b, c).$$

Show that φ is an isomorphism, and that

$$\varphi([V, W]) = \varphi(V) \wedge \varphi(W),$$

where $\wedge$ denotes the cross product.

c) What is the flow of $aX + bY + cZ$?

7. *North-South and North-North dynamics on the sphere*

Consider the sphere S^2 embedded in $\mathbf{R}^3$ in the usual way. Let i_N denote stereographic projection from the north pole, and let h_t denote the homothety with center $(0, 0, 0)$ and factor e^t of the plane $z = 0$.

a) Show that $i_N^{-1} \circ h_t \circ i_N$ extends in a unique way to a diffeomorphism g_t of S^2, and that $g_t \circ g_{t'} = g_{t+t'}$.

b) Show that the only fixed points of g_t are the north and south poles. Verify that for all $x \in S^2$ we have

$$\lim_{t \to +\infty} g_t(x) = N \quad \text{and} \quad \lim_{t \to -\infty} g_t(x) = S.$$

c) Verify that the infinitesimal generator of the one-parameter group g_t is given for all $x \in S^2$ by the orthogonal projection of $(0, 0, 1)$ onto $T_x S^2$.

d*) Taking inspiration from the construction done in a), give an example of a vector field on S^2 having a single zero, and sketch its trajectories.

8. Let X and Y be two vector fields on a manifold, with φ and ψ the corresponding flows. Show that φ_s and ψ_t commute for all s and t sufficiently small if and only if $[X, Y] = 0$.

9. Let X be a vector field on a manifold such that $[X, Y] = 0$ for any vector field Y. Show that $X = 0$.

10. Give an example of a vector field on S^{2n-1} which is nowhere vanishing. (Note that

$$S^{2n-1} = \left\{ (z_1, \ldots, z_n) \in \mathbf{C}^n : \sum_{i=1}^{n} |z_i|^2 = 1 \right\},$$

and use the one-parameter group $v_t(z) = (e^{it} z_1, \ldots, e^{it} z_n).$)

Note. We will see later that on even dimensional spheres, every vector field has at least one zero.

11. Show that TS^{n-1} is diffeomorphic to the submanifold of $\mathbf{C}^n$ defined by the equation

$$\sum_{i=1}^{n} z_i^2 = 1.$$

12*. *Examples of parallelizable manifolds*

a) Give three vector fields on S^3 that are everywhere independent.

b**) Show that the manifold $S^1 \times S^n$ is parallelizable.

13. Let $\varphi : M \to N$ and $\psi : N \to P$ be diffeomorphisms. Show that

$$\psi_* \circ \varphi_* = (\psi \circ \varphi)_* .$$

14*. Prove the inverse function theorem by Moser's trick. What can we say of this proof?

15*. *Transitivity of the group of diffeomorphisms*

a) Show that for all positive real numbers r and r' (with $r' > r$), and points x and y in the open ball $B(0, r)$, there exists a diffeomorphism v of $\mathbf{R}^n$ such that $v(x) = y$ and $v(z) = z$ if $\|z\| > r'$, using the flow of an appropriate vector field.

b) Let M be a manifold. Show that every point $x \in M$ has a neighborhood V with the following property:

   For every $y \in V$, there exists a diffeomorphism j of M such that $j(x) = y$ (we can reduce to a) by using suitable charts).

c) Show that if M is connected, the group of diffeomorphisms acts transitively on M.

d) Show that if $\dim M > 1$, this group is k-*transitive* for all k, in other words if $(x_1, \ldots, x_k)$ and $(y_1, \ldots, y_k)$ are two k-tuples of mutually distinct points, there exists a diffeomorphism φ of M such that $\varphi(x_i) = y_i$ for all $i \in [1, k]$.

16*. *Normal form of a nonvanishing vector field*

a) Let $X = \sum_{i=1}^{n} X^i \partial_i$ be a vector field on an open subsets U of $\mathbf{R}^n$ containing the origin. Suppose that $X^1(0) \neq 0$. Let φ_t be the local flow of X. Show that the map

$$F : (x^1, \ldots, x^n) \longmapsto \varphi_{x^1}(0, x^2, \ldots, x^n)$$

is a local diffeomorphism in a neighborhood of 0.

b) Show that $F_*^{-1} X = \partial_1$.

   Hint. Use Proposition 3.37.

c) Let M be a smooth manifold, and X a vector field on M. Show that for all a such that $X_a \neq 0$, there exists a chart (U, φ), where U contains a, such that $X_{|U} = \partial_1$.

The local classification of vector fields having an isolated zero is far from being this easy. It involves arithmetic conditions on the differential of X at the zero. See for example [Demazure 00].

17*. *Simultaneous reduction of a commuting system of vector fields*

Let $L_1 \ldots, L_p$ be p everywhere independent vector fields on a manifold M that satisfy

$$\forall i, j \quad [L_i, L_j] = 0.$$

We will prove that for all $m \in M$, there exists an open subset U containing m and a system of local coordinates on U such that $L_{i|U} = \frac{\partial}{\partial x^i}$ if $1 \leqslant i \leqslant p$.

The preceding exercise shows that this property is true for $p = 1$, and we proceed by induction.

a) Suppose the property is true for p, and let $L_1, \ldots, L_{p+1}$ be $p+1$ mutually commuting independent vector fields. Show that there exists local coordinates (x, y), where $x = (x^1, \ldots, x^p)$ and $y = (y^1 \ldots y^{n-p})$ in a neighborhood of every point m such that

$$L_i = \frac{\partial}{\partial x^i} \quad \text{if} \ \ 1 \leqslant i \leqslant p;$$

$$L_{p+1} = \sum_{i=1}^{p} a^i(y) \frac{\partial}{\partial x^i} + \sum_{j=1}^{n-p} b^j(y) \frac{\partial}{\partial y^j}.$$

b) Using the previous exercise, show that we can arrange this system (x, y) so that

$$L_{p+1} = \sum_{i=1}^{p} a_i(y) \frac{\partial}{\partial x^i} + \frac{\partial}{\partial y^1}.$$

c) Conclude the result with the help of a change of coordinates of the form

$$\xi^i = x^i + \varphi^i(y) \ \ (1 \leqslant i \leqslant p), \quad \eta^j = y^j \ \ (1 \leqslant j \leqslant n - p).$$

18*. *Frobenius theorem*

A system of p everywhere linearly independent vector fields $L_1, \ldots, L_p$ on a manifold M is said to be *completely integrable* if for every $m \in M$ there exists an open subset U containing m and a p-dimensional submanifold Y (called an *integral submanifold*), contained in U and containing m such that $L_i(y) \in T_y Y$ for all $y \in Y$ and for all $i \in [1, p]$. (Here the adverb "completely" indicates that we can find submanifolds tangent to the "field of p-planes" generated by the L_i whose dimension is as large as possible.)

We will show that a system of p vectors is completely integrable if and only if there exists p^3 smooth functions c^i_{jk} such that

$$[L_j, L_k] = \sum_{i=1}^{p} c^i_{jk} L_i. \tag{F}$$

(This condition is uninteresting when $p = 1$: Theorem 3.33 says precisely that a nonzero vector field is a completely integrable "system".)

a) Let $L'_1, \ldots, L'_p$ be a second system of vector fields such that for every $m \in M$ the vectors $L'_i(m)$ generate the same vector subspace as the $L_i(m)$ (so that this system is completely integrable if and only if the first system is). Show that

$$L'_i(m) = \sum_{j=1}^{p} a^j_i(m) L_i(m),$$

where the a^j_i are smooth functions. Deduce the systems $L_1, \ldots, L_p$ and $L'_1, \ldots, L'_p$ simultaneously satisfy condition (F).

b) Show this condition is necessary.

c) To show that this condition is sufficient, we reduce to the previous exercise, by showing that if the system $L_1, \ldots, L_p$ satisfies (F), every point of M is contained in an open subset U on which there exists a system $L'_1, \ldots, L'_p$ such that:

1) the $L'_i(x)$ are linearly independent for all $x \in U$;

2) $L'_i = \sum_{k=1}^p a_i^k L_k$, where $a_i^k \in C^\infty(U)$;

3) for all i and j in $[1, p]$, $[L'_i, L'_j] = 0$.

Being a local question, it suffices to work on an open subset of $\mathbf{R}^n$. From a linear change of variables, we may suppose that $L_i(0) = \partial_i$ for all $i \in [1, p]$. If

$$L_i(x) = \sum_{i=1}^n a_i^j(x)\partial_j,$$

then there exists an open subset containing 0 where the matrix

$$\left(a_i^j(x)\right)_{1 \leqslant i,\ j \leqslant p}$$

is invertible. Let (b_i^j) be the inverse matrix, and let

$$L'_i = \sum_{j=1}^p b_i^j L_i.$$

c1) Show that

$$L'_i(x) = \partial_i + \sum_{k>p} c_i^k(x)\partial_k.$$

c2) By using (F), show that $[L'_i, L'_j] = 0$.

19*. *Geometry of completely integrable systems*

Let M be a manifold equipped with a completely integrable system. We say that two points $x, y \in M$ are equivalent if there exists a finite sequence $Y_1, \ldots, Y_k$ of integral submanifolds such that $x \in Y_1$, $y \in Y_k$ and $Y_i \cap Y_{i+1} \neq \emptyset$ for $1 \leqslant i \leqslant k-1$. Show that equivalence classes are strictly immersed submanifolds.

Note. This equivalence classes are called *maximal integrals* of the system.

20. *Normal bundle*

a) Let f be an immersion, not necessarily injective, of a manifold M into Euclidean space E. Show that the set of pairs (x, v) of $M \times E$ such that v is orthogonal to $T_x f(T_x M)$ is a vector bundle over M of rank $\dim E - \dim M$ (the projection is of course the restriction of pr_1). This bundle is called the normal bundle, as in the case of submanifolds.

b) Show that the normal bundle to S^n seen as the set of vectors of norm 1 in Euclidean space of dimension $n + 1$ is trivializable.

c) Repeat the same question for a submanifold X of $\mathbf{R}^n$ of the form $f^{-1}(0)$, where f is a submersion of $\mathbf{R}^n$ to $\mathbf{R}^p$.

21*. *Tautological bundle*

a) We define a vector bundle of rank 1 on $P^n\mathbf{R}$, called the *tautological bundle*, and denoted γ_n, as follows. The total space ξ is the subset of $P^n\mathbf{R} \times \mathbf{R}^{n+1}$ of pairs $([x], v)$ such that v belongs to the line defined by $[x]$, with the projection being the restriction of pr_1. It is the same to say that if $x = (x^0, \ldots, x^n)$ is a system of homogeneous coordinates of $[x] \in P^n\mathbf{R}$, the total space is characterized by the equations

$$x^i v^j - x^j v^i = 0 \quad , (0 \leqslant i < j \leqslant n)$$

(compare with the first question of Exercise 17 in Chapter 2).

b) Show that for $n = 1$ the total space is diffeomorphic to the manifold of affine lines in the plane, and that the projection associates to each line its (non oriented!) direction.

c**) Show that γ_n is not trivializable.

22*. *Construction of some vector bundles*

a) Direct product.

If $\xi_1 = (E_1, p_1, B_1)$ and $\xi_2 = (E_2, p_2, B_2)$ are two vector bundles, then so is

$$(E_2 \times E_2, p_1 \times p_2, B_1 \times B_2),$$

with each fiber

$$(p_1 \times p_2)^{-1}(b_1, b_2) = (E_1)_{b_1} \times (E_2)_{b_2}$$

equipped with the product vector space structure. This bundle is denoted $\xi_1 \times \xi_2$.

b) Pullback bundle.

If $\xi = (E, p, B)$ is a vector bundle and $f : B' \to B$ a smooth map, then

$$E' = \{(b', e) \in B' \times E : \ f(b') = p(e)\},$$

equipped with the restriction of the first projection is the total space of a vector bundle on B', denoted $f^*\xi$. The restriction of the second projection gives a bundle morphism. A silly but important example: if M is a submanifold of $\mathbf{R}^n$, or more generally if $f : M \to \mathbf{R}^n$ is an immersion, the pullback of $T\mathbf{R}^n$ by f is the trivial bundle $M \times \mathbf{R}^n$.

c) Direct sum, or Whitney sum.

If $\xi_1 = (E_1, p_1, B)$ and $\xi_2 = (E_2, p_2, B)$ are two vector bundles with the same base B, the Whitney sum is the bundle $\Delta^*(\xi_1 \times \xi_2)$, where $\Delta : B \to B \times B$ is the diagonal map. It is denoted $\xi_1 \oplus \xi_2$.

Example. The Whitney sum of the tangent bundle and the normal bundle to a submanifold of $\mathbf{R}^n$ is the trivial bundle.

23**. *Tangent bundle to* $P^n\mathbf{R}$

a) Identify TS^n with the set of pairs $(x, v) \in \mathbf{R}^{n+1} \times \mathbf{R}^{n+1}$ such that $\|x\| = 1$ and $\langle x, v \rangle = 0$. Show that $TP^n\mathbf{R}$ is identified with the quotient of TS^n obtained by identifying (x, v) and $(-x, -v)$, and that the quotient map is a morphism of vector bundles.

b) Define a bundle of rank n on $P^n\mathbf{R}$, denoted $\gamma_n^\perp$, whose total space is

$$\{([x], v) \in P^n\mathbf{R} \times \mathbf{R}^{n+1} : \langle x, v \rangle = 0\}.$$

Thus the Whitney sum $\gamma_n \oplus \gamma_n^\perp$ (see Exercise 21) is the trivial bundle of rank $(n + 1)$ on $P^n\mathbf{R}$.

Show that $TP^n\mathbf{R}$ is isomorphic to $Hom(\gamma_n, \gamma_n^\perp)$.

24*. *Tubular neighborhood of a submanifold*

Let M be a compact submanifold of Euclidean space, and $N(M)$ the normal bundle to M. We define a map $f : N(M) \to E$ by

$$f(x, v) = x + v \quad \text{(for } (x, v) \in N(M) \subset M \times E, \text{ see Exercise 20).}$$

a) Show that f is a local diffeomorphism at $(x, 0)$.

b) For $r > 0$, write

$$N_r(M) = \{(x, v) \in N(M) : \|v\| < r\}.$$

Show that there exists a $\epsilon > 0$ such that f is a diffeomorphism from $N_r(M)$ to its image if $r < \epsilon$ (take inspiration from the proof of Lemma 6.18).

The image $f(N_r(M))$ is called a tubular neighborhood of M, and is denoted $V_r(M)$.

We mention in passing that if y is in a tubular neighborhood, and if $y = f(x, v)$, then $\text{dist}(y, M) = \|v\|$.

25*. *Reeb's theorem: the easy case*

Let M be a compact n-dimensional manifold and f a smooth real-valued function on M admitting exactly two *non-degenerate* critical points. Show that M is homeomorphic to S^n.

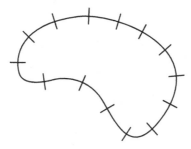

Figure 3.8: Tubular neighborhood

Hint. Using Theorems 3.40 and 3.44 we decompose M into three pieces, two of which are homeomorphic to the closed disk D^n, and the third to a product $I \times S^{n-1}$. We can glue these homeomorphisms and obtain an homeomorphism of M to S^n by remarking that every homeomorphism of S^{n-1} extends to a homeomorphism of D^n.

26. *A useful technical lemma*

a) Show that in Euclidean space, the open subset $B(0,1) \smallsetminus \{0\}$ is diffeomorphic to $B(0,1) \smallsetminus \overline{B}(0,r)$ (with $r < 1$ of course!).

b) Show that we can require such a diffeomorphism to equal the identity on the annulus $X = \{x : r' < \|x\| < 1, \text{ for any } r' \in (0,r)\}$.

c) Let M be a manifold and $p_1, \ldots, p_r$ distinct points. Show that there exists disjoint open subsets $U_1, \ldots, U_r$ such that $p_i \in U_i$ for all i and $M \smallsetminus \{p_1, \ldots, p_r\}$ are diffeomorphic to $M \smallsetminus \bigcup_{1 \leqslant i \leqslant r} \overline{U}_i$.

27*. *A modest invitation to complex geometry*

a) Show that the tautological bundle of $P^n\mathbf{C}$, this is to say the bundle of complex lines defined in the same way as the tautological bundle of $P^n\mathbf{R}$, is a *holomorphic* vector bundle (in other words that the local trivializations are holomorphic).

b) Show that this bundle does not admit nonzero holomorphic sections.

c) Show that its complex dual admits $n+1$ linearly independent holomorphic sections.

28*. *A supplement to Section 3.3.2*

In the ring $\mathcal{F}_x^M$ of germs of smooth functions at x, introduce the ideal $\mathfrak{m}_x$ of germs of functions vanishing at x and the ideal $\mathfrak{m}_x^2$ generated by finite sums $\sum_{i=1}^{k} g_i h_i$, where the g_i and the h_i are in $\mathfrak{m}_x$. Show that the cotangent bundle $T_x M^*$ is naturally (in the sense of 3.3.2) isomorphic to $\mathfrak{m}_x / \mathfrak{m}_x^2$ (use Hadamard's lemma).

Chapter 4

Lie Groups

4.1. Introduction

The notion of group was singled out around 1830 by Évariste Galois in his work on algebraic equations. This initial work was with finite groups.

Forty years later, the work of Galois inspired the Norwegian mathematician Sophus Lie, who rather than studying invariance of algebraic equations was studying the invariance properties of ordinary and partial differential equations and put the need for other types of groups into focus. These were formerly called "finite and continuous groups", which in today's language conveys groups of finite topological dimension. In fact many of the examples discovered were smooth manifolds, with smooth group operations. Today we call such groups Lie groups.

It was rapidly conjectured, but proved only in 1950 (the proof occupies an entire book, [Montgomery-Zippin 55]) that every topological manifold equipped with a continuous group operation is a Lie group. It is not surprising that many of the groups that arise in geometry and physics are Lie groups. The examples that we will see are all subgroups of the linear group $Gl(n, \mathbf{R})$. Conversely, it can be proved that every Lie group is locally isomorphic to a subgroup of a linear group. Nonetheless, the general theory loses none of its relevance: it is needed for the proof of this result which is far from obvious. Even at the elementary level where we begin, the theory sheds light on essential phenomena, where we study the link between group properties and those of a particular algebraic structure on the tangent space to the identity element. This structure is naturally identified with left invariant vector fields, and is stable under the bracket seen in the preceding chapter. This resulting algebraic structure, called the *Lie algebra*, reflects the properties of the group.

In the group to algebra direction, everything passes remarkably. The Lie algebra of a commutative group is commutative, and that of a subgroup is a subalgebra of the Lie algebra of the original group. The linear tangent map to a group morphism gives a morphism of Lie algebras.

In the direction from algebra to group, things are not as nice. The Lie algebra of a Lie group depends only on the properties of the group in a neighborhood of the identity element. Two groups having the same identity component have the same Lie algebra, as well as two groups where one is a covering of the other. There are examples of this situation in dimension 1: the group $(\mathbf{R}, +)$ (simply connected), $SO(2) \simeq S^1$ (connected but not simply connected), $O(2)$ (whose identity component is $SO(2)$) have the same Lie algebra. Here is a second example. The groups $SU(2)$ (simply connected as it is homeomorphic to S^3), $SO(3)$ (connected but not simply connected as it is a quotient of order 2 of $SU(2)$, see Exercise 5) and $O(3)$ (whose identity component is $SO(3)$) have the same Lie algebra, which turns out to be $\mathbf{R}^3$ equipped with the cross product.

This chapter gives the rudiments of this theory, with numerous examples. The existence and properties of the Lie algebra of a Lie group comes from the properties of the flow of a vector field and the bracket seen in the preceding chapter. We also give several topological results that allow one to understand passing in the direction from algebra-group.

We finish with the important notion of *homogeneous space*. At the algebraic level, this is a set E on which a group G acts transitively (to avoid ambiguity we should speak of a G-homogeneous space). There is then a bijection from E to the quotient of G by the isotropy subgroup G_x of a point x, this is to say to the subgroup of g such that $g \cdot x = x$; if we change the point, this subgroup is replaced by a conjugate. If G is a Lie group and if G_x is closed in G, we have a manifold structure on which G acts by diffeomorphisms. This gives a convenient way to see many geometric objects as manifolds. An important and typical example is that of the set of vector subspaces of a fixed dimension of a finite dimensional vector space (the Grassmannians).

For a historical and epistemological study of Lie's original point of view, see [Merker 10]. The elementary but instructive example of the Riccati equation is treated in [Berger 87, 6.8.7].

4.2. Left Invariant Vector Fields

Definition 4.1. *A Lie group is a group G equipped with a smooth manifold structure such that the maps*

$$(g, h) \longmapsto gh \quad from \ G \times G \ to \ G$$

and

$$g \longmapsto g^{-1} \quad \text{from } G \text{ to } G$$

are smooth.

It is enough to assume, of course, that the map

$$(g, h) \longmapsto gh^{-1} \quad \text{from } G \times G \text{ to } G$$

is smooth.

Examples

a) The additive group $\mathbf{R}^n$ is a Lie group.

b) The *linear group* $Gl(n, \mathbf{R})$ of automorphisms of the vector space $\mathbf{R}^n$ is a Lie group: this is an open subset of $\text{End}(\mathbf{R}^n)$, and therefore a manifold of dimension n^2, and the smoothness of the group operations can be seen from explicit formulas (this is one of the very rare occasions where the explicit formulate for the inverse of a matrix gives something!).

c) If G and H are two Lie groups, $G \times H$, equipped with the product group structure and the product manifold structure, is a Lie group.

d) The circle S^1, seen as the multiplicative group of complex numbers of unit modulus is a Lie group. The product Lie group $(S^1)^n$ is called the *n-dimensional torus* (by analogy with the case $n = 2$). We will see that this is, up to isomorphism, the only compact connected Abelian Lie group of dimension n.

e) By Section 1.5 and Proposition 2.8 the orthogonal group $O(n)$ is a Lie group. The same result and Exercise 4 of Chapter 2 shows that $U(n)$ and $SU(n)$ are Lie groups.

f) The group $A(n, \mathbf{R})$ of *affine* automorphisms $\mathbf{R}^n$, which is to say the transformations

$$x \longmapsto Ax + b, \quad \text{where } A \in Gl(n, \mathbf{R})$$

is a Lie group. Indeed, $A(n)$ is identified with $Gl(n, \mathbf{R}) \times \mathbf{R}^n$ equipped with the law of composition

$$(A, b) \times (A', b') = (AA', Ab' + b).$$

In particular,

$$(A, b)^{-1} = (A^{-1}, -A^{-1}b).$$

It is convenient to see $A(n, \mathbf{R})$ as the subgroup of $Gl(n+1, \mathbf{R})$ formed by the matrices

$$\begin{pmatrix} A & b \\ 0 \ldots 0 & 1 \end{pmatrix}.$$

g) We again mention the *Heisenberg group* $\mathcal{H}$ (see Exercise 11 for a justification of this name) of matrices of the form

$$\begin{pmatrix} 1 & x & z \\ 0 & 1 & y \\ 0 & 0 & 1 \end{pmatrix}$$

and the *Galilean group* $\mathcal{G}$ of classical mechanics. The action of an element of $\mathcal{G}$ on space-time parametrized by $(x,t) \in \mathbf{R}^3 \times \mathbf{R}$ is defined by

$$(A, v, b, \tau){\cdot}(x, t) = (Ax+vt+b, t+\tau), \quad \text{where } A \in SO(3),\ v, b \in \mathbf{R}^3,\ \tau \in \mathbf{R}$$

(v as in velocity!). In an analogous way to f), $\mathcal{G}$ can be considered as the subgroup of $Gl(5, \mathbf{R})$ of matrices of the form

$$\begin{pmatrix} A & & b & c \\ 0 & 0 & 0 & 1 & \tau \\ 0 & 0 & 0 & 0 & 1 \end{pmatrix}.$$

h) Finally we give the example of the *Lorentz group* $O(3, 1)$ of automorphisms of the quadratic form

$$x^2 + y^2 + z^2 - t^2$$

(see Exercise 19 of Chapter 1 for a justification), which comes up in special relativity.

Definitions 4.2

a) *A morphism between two Lie groups G and H is a map $f : G \to H$ which is both a group homomorphism and a smooth map.*

b) *Lie groups G and H are said to be* isomorphic *if f is both a group isomorphism and diffeomorphism.*

c) *A Lie subgroup of G is a submanifold which is also a subgroup.*

Examples

a) Of course $O(n)$ is a Lie subgroup of $Gl(n, \mathbf{R})$. The group $SO(n)$ (called the *special orthogonal group*) of orthogonal transformations of determinant 1 is a Lie subgroup of $O(n)$, because it is a subgroup and open subset of $SO(n)$: we have a partition

$$O(n) = SO(n) \cup O^-(n) \quad \text{where } O^-(n) = \big\{ A \in O(n) : \ \det(A) = -1 \big\}.$$

b) The connected component of the identity element of a Lie group is a Lie subgroup of G (of the same dimension since it is open in G; it is a subgroup by Proposition 4.30 below). We call it *the identity component* of G and we denote it by G_o. We will see for example that $O(n)_o = SO(n)$.

To each element g of a Lie group G, we associate *right and left translations*, denoted by R_g and L_g respectively, and defined by

$$R_g \cdot x = xg \quad \text{and} \quad L_g \cdot x = gx.$$

By virtue of the associativity of the group operation, we have

$$L_g \circ L_h = L_{gh} \quad \text{and} \quad R_g \circ R_h = R_{hg}.$$

In particular, R_g and L_g are diffeomorphisms of G. Moreover, again associatively implies that R_g and L_h *commute*, as $h(xg) = (hx)g$!

By Section 3.4, if X is a vector field on G, we can define for every $g \in G$ the vector field $L_{g*}X$ and $R_{g*}X$.

Definition 4.3. *A vector field X on a Lie group G is* left *invariant (resp.* right *invariant) if*

$$\forall g \in G, \quad L_{g*}X = X \quad (resp. \ R_{g*}X = X).$$

We denote the diffeomorphism $x \mapsto x^{-1}$ of G by $\mathcal{I}$.

Proposition 4.4. *The (left or right) invariant vector fields have the following properties:*

  i) *The sum and the Lie bracket of two left invariant (resp. right invariant) vector fields are left invariant (resp. right invariant).*

 ii) *The image of a left invariant (resp. right invariant) vector field under $\mathcal{I}$ is right invariant (resp. left invariant).*

iii) *If X is left invariant (resp. right invariant), the same is true for $R_{g*}X$ (resp. $L_{g*}X$).*

PROOF. The first property is a direct consequence of Proposition 3.23. For the second, it suffices to note that the relation

$$(gx)^{-1} = x^{-1}g^{-1}$$

can be written

$$\mathcal{I} \circ L_g = R_{g^{-1}} \circ \mathcal{I}.$$

Finally for the third property, if X is left invariant, then using the fact that, right translations commute with left translations we have

$$L_{h*}(R_{g*}X) = R_{g*}(L_{h*}X) = R_{g*}X. \qquad \square$$

However, we need to know that left invariant vector fields exist. The following result does the job.

Proposition 4.5. *If G is a Lie group, the map $X \mapsto X_e$ is an isomorphism between the vector space of left invariant vector fields on G and the tangent space T_eG.*

PROOF. If a vector field X on G is left invariant, then necessarily

$$X_g = T_eL_g \cdot X_e,$$

which shows injectivity. Conversely, given a vector $v \in T_eG$, it is necessary that the map of G to TG given by

$$g \longmapsto T_eL_g \cdot v$$

is a vector field (which will then be left invariant by construction). It suffices to show that this map defines a derivation. For $f \in C^\infty(G)$, we denote by L_vf the function

$$g \longmapsto T_gf \cdot (T_eL_g \cdot v).$$

It is clear that $L_v(fg) = (L_vf)g + f(L_vg)$, and it remains to show that the function L_vf is smooth.

Let $c : (-\epsilon, \epsilon) \mapsto G$ be a smooth curve such that $c(0) = e$ and $c'(0) = v$. Now, by the chain rule,

$$L_vf(g) = \frac{d}{dt}f\big(gc(t)\big)_{|t=0}.$$

However the function $(t, g) \mapsto f\big(gc(t)\big)$ is smooth on $(-\epsilon, \epsilon) \times G$, and the same is true of its partial derivative with respect to t.   □

Corollary 4.6. *Every Lie group is a parallelizable manifold.*

PROOF. It suffices to take any basis of T_eG and then form the corresponding left invariant vector fields from the preceding proposition. Now apply Proposition 3.28.   □

In particular, every vector field on G can be written as a linear combination (with coefficients in $C^\infty(G)$) of left invariant (or right invariant) vector fields.

We now examine the trajectories of a left invariant vector field.

Definition 4.7. *A one-parameter subgroup of a Lie group G is a morphism from $(\mathbf{R}, +)$ to G, in other words, a smooth map h from $\mathbf{R}$ to G such that*

$$h(t + s) = h(t)h(s)$$

for all $t, s \in \mathbf{R}$.

Examples

a) If $A \in M_n(\mathbf{R})$, then $t \mapsto \exp tA$ is a one-parameter subgroup of $Gl(n, \mathbf{R})$ by Lemma 1.27. If A is antisymmetric (resp. is traceless) we obtain one-parameter subgroups of $O(n)$ (resp. $Sl(n, \mathbf{R})$).

b) In fact, if G is a Lie subgroup of $Gl(n, \mathbf{R})$ and ϕ is a *continuous* morphism from $(\mathbf{R}, +)$ to G, then ϕ is a one-parameter subgroup of G, and there exists a unique $A \in T_I G \subset \mathrm{End}(\mathbf{R}^n)$ such that $\phi(t) = \exp(tA)$. Indeed we know by Theorem 1.30 that $\phi(t) = \exp(tA)$ for some A in $\mathrm{End}(\mathbf{R}^n)$. Now $t \mapsto \phi(t)$ is a smooth curve contained in G, and $A = \phi'(0) \in T_I G$.

Given a one-parameter subgroup of G, we can associate to it a one-parameter group of diffeomorphisms of G defined by

$$\phi(t, x) = xh(t).$$

In other words, ϕ_t is the *right* translation $R_{h(t)}$.

Theorem 4.8. *If h is a one-parameter subgroup of a Lie group G, the infinitesimal generator of*

$$\phi(t, x) = xh(t)$$

is a left *invariant vector field. Conversely, the flow ϕ of a left invariant vector field X on G is defined on $\mathbf{R} \times G$, and there exists a unique one-parameter subgroup h_X such that*

$$\phi_t(g) = gh_X(t)$$

for all $t \in \mathbf{R}$ and $g \in G$. Further, $h_X(t) = \phi_t(e)$.

PROOF. Let X be the infinitesimal generator of ϕ_t. By Proposition 3.37, the flow of $L_{g*}X$ is $L_g \circ \phi_t \circ (L_g)^{-1}$. However

$$L_g \circ \phi_t \circ (L_g)^{-1}(x) = g\phi_t(g^{-1}x) = gg^{-1}xh(t) = \phi_t(x),$$

which shows that $L_{g*}X = X$.

Conversely, if X is left invariant and if ϕ_t is its flow, the same argument shows that

$$\phi_t(g) = g\phi_t(e).$$

The domain of definition of ϕ therefore contains an open subset of the form $(-\epsilon, \epsilon) \times G$. From the relation

$$\phi_{t+t'} = \phi_t \circ \phi_{t'} \quad \text{for } |t|, |t'| < \frac{\epsilon}{2},$$

we deduce the conditions

$$\phi_t(e)\phi_{t'}(e) = \phi_{t+t'}(e).$$

Set $h_X(t) = \phi_t(e)$, the same argument as that of Theorem 3.39 allows us to extend h_X as a morphism from $\mathbf{R}$ to G, and therefore ϕ to the whole of $\mathbf{R} \times G$.   □

4.3. The Lie Algebra of a Lie Group

4.3.1. Basic Properties; The Adjoint Representation

For a Lie group, we have just seen that it is the same thing to have
- a left invariant vector field;
- a vector in the tangent space to the identity;
- a one-parameter subgroup.

In particular, every algebraic operation defined on one of these objects, such as the *bracket* for left invariant vector fields, can be transported to the others.

Definition 4.9. *A Lie algebra over a field K is a vector space L over K, equipped with a bilinear map from $L \times L$ to L, called the* bracket, *denoted $[\,,\,]$, such that*

i) $\forall X \in L$, $[X, X] = 0$.

ii) $\forall X, Y, Z \in L$, $[[X, [Y, Z]] + [Y, [Z, X]] + [Z, [X, Y]] = 0$ *(Jacobi identity)*.

Examples

a) Any vector space equipped with the zero bracket is a Lie algebra. This is the only case (at least in characteristic not equal to 2) where a Lie algebra is commutative, as calculating $[X+Y, X+Y]$ shows that $[X, Y]+[Y, X] = 0$.

b) The results of Section 3.6 may be reformulated by saying that for every smooth manifold M, the vector space $C^\infty(TM)$ equipped with the Lie bracket is a Lie algebra (of infinite dimension, since $C^\infty(M)$ is already infinite dimensional).

c) By Proposition 4.4, the left invariant (or right invariant) vector fields on a Lie group form a finite-dimensional Lie algebra.

Definition 4.10. *A morphism of Lie algebras L and L' over the same field K is a linear map f from L to L' such that*

$$\forall X, Y \in L, \quad f([X, Y]) = [f(X), f(Y)].$$

If f is invertible, it is clear that f^{-1} is also a morphism. We then say that f is a Lie algebra isomorphism.

Example. Again by Proposition 4.4, $\mathcal{I}_*$ is an isomorphism between the algebras of left and right invariant vector fields on G.

By transporting the structure, the tangent space to the identity of a Lie group is equipped with a Lie algebra structure.

Definition 4.11. The Lie algebra *of a Lie group G, denoted* $\mathfrak{G}$*, is* T_eG *equipped with the bracket above.*

In the sequel, we will frequently use the identification between T_eG and the left invariant vector fields.

Remark. A Lie group and its identity component have the same Lie algebra, since $T_eG = T_eG_o$.

Definition 4.12. *Given a Lie group G with Lie algebra* $\mathfrak{G}$*, the map which associates to* $X \in \mathfrak{G}$ *the value of the one-parameter group associated to X at time* 1 *is called the* exponential map, *and is denoted* exp.

If for example $G = Gl(n, \mathbf{R})$, $T_eGl(n, \mathbf{R}) = \text{End}(\mathbf{R}^n)$, and $X \in \text{End}(\mathbf{R}^n) = \mathfrak{G}$ give rise to the vector field whose value at A is AX (with the identifications used when we work with open subsets of $\mathbf{R}^n$). The exponential map is obtained by integrating the differential equation

$$A'(t) = A(t)X \quad \text{with initial condition } A(0) = I.$$

We recover the exponential of matrices seen in Chapter 1. It is immediate, but a bit pedestrian to show that in this case $[A, B] = AB - BA$. We will soon see a more conceptual proof. Returning to the general case, we easily verify that the principal properties of the exponential map subsist.

Proposition 4.13. *The exponential map is a smooth map from* $\mathfrak{G}$ *to* G*, and a local diffeomorphism from a neighborhood of* $0 \in \mathfrak{G}$ *to a neighborhood of* $e \in G$*.*

PROOF. The fact that exp is smooth comes from the smoothness of solutions of a family of differential equations depending smoothly on a parameter (here the space of parameters is $\mathfrak{G}$!). By the same definition of exp, we have

$$\frac{d}{dt} \exp tX_{|t=0} = X.$$

The differential of exp at 0 is therefore the identity, and we apply the inverse function theorem.   □

We have seen that if $X \in C^\infty(TG)$ is left invariant, then the same is true for $R_{g*}X$. A question arises to interpret this property in terms of tangent vectors at e and in terms of one-parameter subgroups.

Proposition 4.14. *We have*

$$(R_{g*}X)_e = \frac{d}{dt}\left(g^{-1}\exp tXg\right)_{|t=0}.$$

PROOF. The flow of X is given by $\phi_t(x) = x \exp tX$, and by Proposition 3.37 that of $R_{g*}X$ is given by

$$\left(R_g \circ \phi_t \circ R_{g^{-1}}\right)(x) = xg^{-1} \exp tX g.$$

The one-parameter subgroup of G associated to $R_{g*}X$ is then

$$t \longmapsto g^{-1} \exp tX g. \qquad \square$$

We define

$$\operatorname{Ad} g \cdot X = \left(R_{g^{-1}*}X\right)_e = \frac{d}{dt} g \exp tX g^{-1}_{|t=0}.$$

Now, for $X, Y \in \mathfrak{G}$ we have

$$\operatorname{Ad} g \cdot [X, Y] = [\operatorname{Ad} g \cdot X, \operatorname{Ad} g \cdot Y].$$

Further

$$\operatorname{Ad} g_1 g_2 = \operatorname{Ad} g_1 \circ \operatorname{Ad} g_2.$$

(It is this relation that explains the choice of $R_{g^{-1}}$ over R_g.) In other words, Ad is a morphism of G to the group of automorphisms of the Lie algebra $\mathfrak{G}$.

Definition 4.15. *The map* Ad *is called the* adjoint representation *of G.*

When we study differential calculus on a Lie group, it is very useful to write everything at the identity element, and the adjoint representation is made for this.

Proposition 4.16. *For $X, Y \in \mathfrak{G}$, we have*

$$\frac{d}{dt} \operatorname{Ad}(\exp tX) \cdot Y_{|t=0} = [X, Y].$$

In particular, if G is commutative, the bracket is identically zero.

PROOF. This result is a particular case of Theorem 3.38. $\qquad \square$

Example. We have seen that $T_e G$ is identified with $\operatorname{End}(\mathbf{R}^n)$, and exp with the exponential map of endomorphisms. Now

$$\operatorname{Ad} g \cdot A = \frac{d}{dt} \left(g \exp tA g^{-1}\right)_{|t=0} = gAg^{-1},$$

and

$$[B, A] = \frac{d}{dt}\left((\exp tB)A(\exp -tB)\right)_{|t=0} = BA - AB.$$

4.3.2. From Lie Groups to Lie Algebras

We are now ready to state one of the basic results of the theory, which makes one hope for the existence of a dictionary between properties of Lie groups and those of Lie algebras.

Theorem 4.17. *Let G and H be two Lie groups, and $f : G \to H$ a morphism. Then $T_e f : \mathfrak{G} \to \mathfrak{H}$ is a Lie algebra morphism. Further, if f is an isomorphism, $T_e f$ is an isomorphism.*

PROOF. The starting point is to note – clearly – that the image under f of the one-parameter subgroup $t \mapsto h(t)$ of G is a one-parameter subgroup of H. So if $Y \in T_e G$, there exists a unique $Z \in T_e H$ such that

$$f(\exp tY) = \exp tZ$$

and taking the derivative with respect to t of both sides of this equation at $t = 0$, we see that $Z = T_e f \cdot Y$. Now, for fixed $g \in G$,

$$f(g \exp tY g^{-1}) = f(g)(\exp tZ) f(g^{-1}).$$

By the definition of Ad, we also have

$$f(g \exp tY g^{-1}) = f\big(\exp t(\operatorname{Ad} g \cdot Y) \big).$$

Taking the derivative with respect to t at $t = 0$ of both sides of the equality

$$f\big(\exp t(\operatorname{Ad} g \cdot Y) \big) = f(g)(\exp tZ) f(g^{-1})$$

we find

$$T_e f(\operatorname{Ad} g \cdot Y) = \operatorname{Ad} f(g) \cdot (T_e f \cdot Y).$$

It suffices now to write this equality for $g = \exp tY$, to take the derivative at $t = 0$, and then apply the preceding proposition. Indeed, if f is an isomorphism, $T_e f$ is bijective. $\qquad\square$

Remark. If $G = H$, and if f is conjugation by an element of g of G, then $T_e f = \operatorname{Ad} g$.

Example: determinant and trace

We consider the homomorphism $\det : Gl(n, \mathbf{R}) \mapsto \mathbf{R}^*$. We have $T_e \det = \operatorname{tr}$ (see Section 1.2.2). On the other hand, it follows from the proof of Theorem 4.17 that the bracket of the Lie algebra of a commutative group ($\mathbf{R}^*$ here) is the zero bracket. We deduce from this theorem that for $A, B \in \operatorname{End}(\mathbf{R}^n)$

$$\operatorname{tr}[A, B] = 0 \quad \text{whence} \quad \operatorname{tr}(AB) = \operatorname{tr}(BA).$$

This result is well known, but the argument above gives a conceptual reason.

Remark. The example of $Gl(n, \mathbf{R})$ reminds us the exponential map has no reason to be a group homomorphism. Moreover, it has no reason to be either injective or surjective. It is indeed surjective for compact connected Lie groups. The simplest general proof uses Riemannian geometry (cf. [Spivak 79] or [Gallot-Hulin-Lafontaine 05, p. 100]), and for this reason we will not give it here. See Exercise 7 for different examples and counterexamples.

Another application of Theorem 4.17 is the following result:

Corollary 4.18. *If H is a Lie subgroup of G, the Lie algebra of H is a Lie subalgebra of the Lie algebra of G.*

PROOF. This is immediate: if j is the natural injection of $T_e H = \mathfrak{H}$ into $T_e G = \mathfrak{G}$, then again by Theorem 4.17 $j[X, Y] = [jX, jY]$. $\qquad\square$

4.3.3. From Lie Algebras to Lie Groups

Using Lie algebras allows us to prove the following result, which is a beautiful generalization of Theorem 1.30:

Theorem 4.19. *Every continuous group homomorphism from one Lie group to another is necessarily smooth. In particular, a topological group has at most one Lie group structure.*

PROOF. This is a consequence of the following three results, which are important in their own right.

Proposition 4.20. *If h is a continuous group homomorphism from $\mathbf{R}$ to a Lie group G (in other words, a one-parameter subgroup of G), there exists a unique $X \in \mathfrak{G}$ such that $h(t) = \exp tX$.*

PROOF. The fact that exp is a local diffeomorphism at 0 allows us to repeat the arguments of 1.30. $\qquad\square$

Lemma 4.21. *Let $(X_1, \ldots, X_n)$ be a basis of $\mathfrak{G}$. Then the map from $\mathbf{R}^n$ to G given by*

$$(t_1, \ldots, t_n) \longmapsto (\exp t_1 X_1) \ldots (\exp t_n X_n)$$

is a local diffeomorphism in a neighborhood of 0.

PROOF. The map above is smooth, and the image under the differential of 0 of the i-th canonical coordinate vector of $\mathbf{R}^n$ is clearly X_i. $\qquad\square$

Lemma 4.22. *Let $f : G \to H$ be a homomorphism of groups. Then*

i) *For f to be smooth, it suffices that f be smooth in a neighborhood of the identity element of the domain.*

ii) *Suppose that f is smooth. To see that f is a local diffeomorphism (respectively an immersion or a submersion), it suffices that the differential at the identity element be an isomorphism (resp. an injection or surjection).*

PROOF. We will only need the first of these two properties for the proof of the theorem at hand, but we have grouped them together since their proofs are similar. It suffices to observe that as f is a homomorphism, we may write it in the form

$$f(gh) = L_{f(g)} f(h).$$

This shows that if f is smooth on an open subset U containing e, it is also smooth on gU for any g, and therefore on the whole group.

The second property is obtained in the same fashion, as under the hypotheses f is already a local diffeomorphism (resp. immersion or submersion) in a neighborhood of $e \in G$. $\qquad\square$

REMAINDER OF THE PROOF OF THEOREM 4.19. Let $X_1, \ldots, X_n$ be a basis of $\mathfrak{G}$. For each $i \in [1, n]$, the map

$$t \longmapsto f(\exp t X_i)$$

is a one-parameter subgroup of H, and there exists a $Y_i \in \mathfrak{H}$ such that

$$f(\exp t X_i) = \exp t Y_i$$

for any t. Consequently,

$$
\begin{aligned}
f\big((\exp t_1 X_1) \ldots (\exp t_n X_n)\big) &= f(\exp t_1 X_1) \ldots f(\exp t_n X_n) \\
&= (\exp t_1 Y_1) \ldots (\exp t_n Y_n).
\end{aligned}
$$

Write

$$\phi(t_1, \ldots, t_n) = (\exp t_1 X_1) \ldots (\exp t_n X_n),$$

this shows that $f \circ \phi$ is smooth. By the first lemma, f is therefore smooth in a neighborhood of e, and therefore everywhere by the second lemma. $\qquad\square$

The essential question now is to place Lie algebras on the same level as Lie groups. More precisely:

a) Is a Lie algebra $\mathfrak{G}$ the Lie algebra of a Lie group?

b) Suppose $\mathfrak{G}$ is the Lie algebra of a Lie group G, and suppose $\mathfrak{H}$ is a Lie subalgebra of $\mathfrak{G}$. Does there exist a Lie subgroup H of G whose Lie algebra is $\mathfrak{H}$?

c) Suppose two Lie groups G_1 and G_2 are given, as well as a Lie algebra morphism $\phi : \mathfrak{G}_1 \mapsto \mathfrak{G}_2$. Does there exists a morphism of groups $F : G_1 \mapsto G_2$ where $T_e F = \phi$?

The answer to question a) is yes. It is known as *Lie's third theorem*. For a modern proof, see [Duistermaat-Kolk 99, 1.8 and 1.14].

Question c) can in principle be reduced to b): if F exists, its graph will be a Lie subgroup of $G_1 \times G_2$ whose Lie algebra will be the graph of ϕ. Further, since F is continuous, its graph is *closed*. Here is where things can go wrong: there are Lie subalgebras that cannot be the Lie algebra of a closed Lie group. The answer to questions b) and c) is in general *no*.

Counterexamples. For b), it suffices to take $G_1 = G_2 = S^1$, and the morphism (of commutative Lie algebras!) from $\mathbf{R}$ to $\mathbf{R}$ given by $x \mapsto \alpha x$, where $\alpha \notin \mathbf{Z}$. The morphisms of S^1 are of the form $t \mapsto t^k$, where $k \in \mathbf{Z}$ (see Exercise 8), and so it is impossible to lift ϕ to a morphism of S^1.

For a), we take $G = S^1 \times S^1$ and for $\mathfrak{H}$ a line with irrational slope α in $\mathbf{R}^2$. As the one-parameter subgroup

$$x \longmapsto (e^{ix}, e^{i\alpha x})$$

is tangent to $\mathfrak{H}$ at the identity, the only possible candidate for H is the image of this one-parameter subgroup, which is not a submanifold as we saw in Section 1.5, but merely a *strictly immersed* submanifold.

This example is typical of what happens in general.

We finish this section with results that will not be used in the sequel.

Theorem 4.23. *Let G be a Lie group and let $\mathfrak{H}$ be a subalgebra of the Lie algebra $\mathfrak{G}$ of G. There exists a Lie group H and a strict injective immersion of H into G whose image is the subgroup of G given by $\exp \mathfrak{H}$.*

For a proof, see for example [Godement 05, 6.49] or [Duistermaat-Kolk 99, 1.10].

Remark. This property justifies the terminology employed in the case where the image of H is not a Lie subgroup. A given injective immersion of H into G is called an *immersed Lie subgroup*.

Warning. Certain authors, for example Duistermaat-Kolk still call these Lie subgroups.

We can deduce from this result a partial (and optimal in view of the counterexamples above) answer to the question of whether we can lift Lie algebra morphisms to Lie groups.

Corollary 4.24. *Let G and H be two Lie groups, and let $f : \mathfrak{G} \to \mathfrak{H}$ be a morphism of Lie algebras. Then there exists neighborhoods U and V of the identity elements of G and H, and a smooth map F from U to V such that*

i) $T_e F = f$;

ii) if x_1, x_2 and $x_1 x_2$ lie in U, then $F(x_1) F(x_2) = F(x_1 x_2)$.

Moreover, the germ of f at e is determined uniquely by these properties.

PROOF. The graph of F is a Lie subalgebra of $\mathfrak{G} \times \mathfrak{H}$, where we apply the first part of Theorem 4.23. Let Y be a integral submanifold passing through the identity element of $G \times H$. By restricting to Y we may suppose this is a graph: indeed, the differential of the projection to G at e, restricted to Y is the projection of the graph of f onto $\mathfrak{G}$, which is an isomorphism. This projection is therefore locally invertible, and the inverse is of the form $x \mapsto (x, F(x))$, where F is a smooth map from an open subset U of G containing e to an analogous open subset V of H. Again by Theorem 4.23, if x_1 and x_2 lie in U, the product $(x_1, F(x_1))(x_2, F(x_2)) = (x_1 x_2, F(x_1) F(x_2))$ lies in Y, and as Y is the graph of F, we have $F(x_1 x_2) = F(x_1) F(x_2)$. The uniqueness of the germ of F comes from the uniqueness of the germ of an integral manifold of the graph of f.   □

Remark. If G is *simply connected*, every Lie algebra morphism of $\mathfrak{G}$ to the Lie algebra $\mathfrak{H}$ of a Lie group H lifts to a morphism of G to H (see [Stillwell 08, 9.6]; note that the author supposes that H is also simply connected, but this is not used).

The difficulties we have just seen justify a topological detour.

4.4. A Digression on Topological Groups

A *topological group* is a group equipped with a topology such that the group operation and its inverse are continuous. In particular, right and left translations are homeomorphims, as is the inverse map.

For two open subsets A and B of the group, we write

$$AB = \{ab, \ a \in A, \ b \in B\}$$
$$A^n = \{a_1 a_2 \ldots a_n, \ a_i \in A\}$$
$$A^{-1} = \{a^{-1}, \ a \in A\}.$$

Now, if $U \subset G$ is open, AU and UA are both open: there are clearly unions of open subsets. Further, continuity of multiplication and of the inverse can be explained in the following way.

Lemma 4.25. *For every open subset U containing e, there exists an open subset V containing e such that $V^2 \subset U$, and an open subset W containing e such that $W^{-1} \subset U$. In particular, the symmetric open subsets (those such that $U = U^{-1}$) form a neighborhood basis at e.*

Consequently:

Theorem 4.26. *Let G be a topological group. The following are equivalent:*

i) *G is Hausdorff.*

ii) *The set $\{e\}$ is closed.*

iii) *For every $g \in G$, $\{g\}$ is closed.*

PROOF. It is clear that i) implies ii). Further, since translations are homeomorphisms, properties ii) and iii) are equivalent, and it suffices to show that the identity element e and any $g \neq e$ may be separated in order to show ii) implies i).

So let V be an open subset containing e but not g. By virtue of continuity of multiplication, there exists an open subset W containing e such that $W \cdot W \subset V$, is open and we may suppose it is symmetric by using the continuity of $x \mapsto x^{-1}$. Then W and $g \cdot W$ are two disjoint open subsets containing e and g respectively.   □

To study subgroups of a topological group, it is useful to take the same approach as in algebra, this is to say to allow the subgroup to act on the group. Recall that G acts on itself on the left and right (by means of left and right translations respectively), and that these actions are transitive. The actions of a subgroup $H \subset G$ however, are not transitive as soon as $H \neq G$. We therefore introduce the equivalence relation

$$x \mathcal{R} y \iff \exists h \in H \text{ such that } y = R_h x$$
$$\iff x^{-1} y \in H.$$

The equivalence classes are of the form gH; these are the *orbits* of the right action of H on G. We also call these the *left equivalence classes of G modulo H*, which is natural as each of them is of the form $L_g H$. The quotient set is denoted G/H (G on the left and H on the right, to remind us of the situation just described).

The left translation of two equivalent elements are equivalent, therefore left translations pass to the quotient. If we continue to denote the translations just obtain on G/H by L_g, we again have

$$L_{g_1} \circ L_{g_2} = L_{g_1 g_2}.$$

In other words, there is a *left action* of G on G/H. This action is *transitive*. Indeed, if $\overline{x}, \overline{y} \in G/H$ are represented by $x, y \in G$, we have

$$y = yx^{-1}x \quad \text{and consequently} \quad \overline{y} = L_{yx^{-1}}\overline{x}.$$

Finally we note that L_g fixes the equivalence class of the identity (which is equivalent to saying that H is globally invariant at the group level) if and only if $g \in H$.

Example. Let G be the group $SO(n+1)$ and let H be the subgroup formed by the matrices of the form

$$\begin{pmatrix} 1 & 0 \cdots 0 \\ 0 & \\ \vdots & A \\ 0 & \end{pmatrix}$$

where $A \in SO(n)$. Then g and g' are equivalent modulo H if and only if the matrices which represent them have the same first column. G/H is in bijection with S^n, and the action of G on G/H is simply the usual action of $SO(n+1)$ on S^n (compare with Exercise 14 in Chapter 2).

Thus, S^n is identified with $SO(n+1)/SO(n)$ (with the understanding that the realization of $SO(n)$ as a subgroup of $SO(n+1)$ is the one written above).

If we now study the situation from the topological point of view, we equip G/H with the quotient topology: if $p : G \to G/H$ is the canonical map, $U \subset G/H$ is open if and only if $p^{-1}(U)$ is open in G. Then every continuous map of G to a topological space X which is constant on equivalence classes passes to the quotient as a continuous map of G/H to X. In particular we have a continuous bijection of $SO(n+1)/SO(n)$ to S^n, which is a homeomorphism as these two spaces are compact. Once we know (see the end of the chapter) that $SO(n+1)/SO(n)$ is a smooth manifold, we will show that this bijection is a diffeomorphism.

Theorem 4.27. *Let G be a topological group with subgroup H. Then*

i) *the map $p : G \to G/H$ is open;*

ii) *G/H is Hausdorff if and only if H is a closed subgroup of G;*

iii) *if H is open, G/H is endowed with the discrete topology. In particular, every open subgroup of a topological group is closed.*

PROOF

i) If U is open, then so is UH. However

$$p^{-1}(p(U)) = UH.$$

ii) If G/H is Hausdorff, every point is closed, therefore equivalence classes are closed in G. Conversely suppose H is closed. Let $\overline{x}, \overline{y}$ be distinct points of G/H, and $x, y \in G$ such that $p(x) = \overline{x}$ and $p(y) = \overline{y}$. Then x and y are not equivalent: x is not a member of the closed set yH. There exists an open subset U containing E such that

$$Ux \cap yH = \emptyset,$$

and an open subset V containing e such that $V^2 \subset U$. We have

$$V^2 xH \cap yH = \emptyset \quad \text{and therefore} \quad VxH \cap VyH = \emptyset.$$

Then $p(VxH)$ and $p(VyH)$ contain $\overline{x}$ and $\overline{y}$ respectively, and are open by i) and disjoint by construction.

iii) If H is open, every equivalence class is open, and the points of G/H are open by i).   □

Example. As we saw, the space $SO(n+1)/SO(n)$ is homeomorphic to S^n, the homeomorphism given by the map of $SO(n+1)$ to S^n which has an orthogonal matrix associated to its first column. We may then deduce the result below. We can show this in an elementary manner by using the reduced form of orthogonal matrices. However the method used, which can be generalized to other situations, merits attention.

Theorem 4.28. *For each n, the group $SO(n)$ is connected.*

We proceed by induction on n. First, $SO(1)$ is the trivial group, and $SO(2) \simeq S^1$. The induction step comes from the following lemma, which is important in its own right.

Lemma 4.29. *Let H be a subgroup of a topological group G. If G/H and H are connected, then G is also connected.*

PROOF. It suffices to verify that every continuous map f from G to a discrete set with two elements D is constant. First, as H is connected, every equivalence class of G modulo H is connected, therefore f is constant on equivalence classes, and passes to the quotient as a continuous map $\overline{f} : G/H \to D$. However since G/H is connected, $\overline{f}$ is constant, as $f = \overline{f} \circ p$.   □

Remark. We can easily show using the same idea that the total space of a fibration is connected if the base and fiber are connected. In fact, this lemma is merely a particular case of this result: we will see that $p : G \to G/H$ is a fibration. The statement below however, is specific to groups.

Proposition 4.30. *Let V be a connected symmetric $(V^{-1} = V)$ open subset containing e and let G_0 be the connected component of the identity of G.*

Then

$$G_0 = \bigcup_{n=1}^{\infty} V^n,$$

and G_0 is a normal subgroup of G.

PROOF. To begin, $\bigcup_{n=1}^{\infty} V^n$ is a subgroup of G which is open because it is union of open subsets, and closed by Theorem 4.27. This subset is also a connected subset of G: indeed, V^n is the image of the connected set

$$\overbrace{V \times \cdots \times V}^{n \text{ times}} \quad \text{under the continuous map} \quad (g_1, g_2, \ldots, g_n) \longmapsto g_1 g_2 \cdots g_n$$

from G^n to G. We therefore have a union of connected subsets whose intersection contains e. Finally, $\bigcup_{n=1}^{\infty} V^n$, is open, closed and connected, thus it must be the connected component containing the identity in G.

Further, for any $g \in G$, the conjugate $g G_0 g^{-1}$ of G_0 is connected and contains e (it is a subgroup!), thus is included inside G_0. □

Corollary 4.31. *If G is a Lie group, G_0 is given by $\exp \mathfrak{G}$. In particular, G_0 is countable at infinity.*

PROOF. We apply the preceding results to $V = \exp U$, where $U \subset \mathfrak{G}$ is a symmetric open subset containing 0 on which $\exp$ is a diffeomorphism. □

We now give some results on discrete subgroups of topological and Lie groups.

Definition 4.32. *A subgroup Γ of a topological group G is discrete if the topology induced on Γ by G is the discrete topology.*

We begin with an elementary property.

Proposition 4.33. *In a Hausdorff topological group, every discrete subgroup is closed.*

PROOF. Let Γ be a discrete subgroup of G, and $x \in \overline{\Gamma} \smallsetminus \Gamma$. For every symmetric open subset containing e, the open subset xV meets Γ. Let $\gamma \in xV \cap \Gamma$. By the Hausdorff hypothesis, there exists a symmetric open subset W containing e such that $\gamma \notin xW$. We can take $W \subset V$.

However xW also contains a point γ' in Γ. Thus $\gamma^{-1}\gamma' \in V^2$. However $\gamma \neq \gamma'$, and Γ is discrete: this is impossible for the good choice of V. □

To say a little more, it is natural to assume the ambient group is connected: if not, by taking the product of the group with another equipped with the discrete topology, we can realize any group as a discrete subgroup of a Lie group.

Proposition 4.34. *Let G be a connected group, and Γ a discrete normal subgroup. Then Γ is contained in the center of G.*

PROOF. For fixed γ in Γ, the map $x \mapsto x\gamma x^{-1}$ sends G to Γ as Γ is normal. As this is a continuous map, its image is connected. As the image is discrete, the image must be a point. Taking $x = e$, we see this point is γ. $\qquad\square$

Such groups play an important role when we consider covering spaces of topological groups. We will only brief consider coverings of Lie groups, and leave the (easy) generalizations to the reader.

Definition 4.35. *A covering map of a Lie group is a group homomorphism $p : G \to H$ which is a covering map of the underlying manifolds.*

The characterization of these maps is very simple.

Theorem 4.36. *Let Γ be a discrete normal subgroup of a Lie group G. Then there exists a unique Lie group structure on the quotient G/Γ such that the canonical map $\pi : G \to G/\Gamma$ is a covering map of Lie groups.*

Conversely, let $p : G \to H$ be a Lie group homomorphism such that $T_e p$ is an isomorphism. Then p is a covering map, the kernel $p^{-1}(e)$ is a discrete normal subgroup, and p passes to the quotient as a Lie group isomorphism between $G/p^{-1}(e)$ and H.

PROOF. The action of $\Gamma = p^{-1}(e)$ by translations (say on the right) is clearly free, and also proper: if A and B are two compact subsets of G, the set of $\gamma \in \Gamma$ such that $\gamma(A)$ meets B is equal to $\Gamma \cap AB^{-1}$. This is the intersection of a compact subset with a discrete subset, and is therefore a finite set. By Theorem 2.38, the quotient G/Γ admits a unique manifold structure such that $\pi : G \to G/\Gamma$ is a covering map.

Moreover, π is a group homomorphism as Γ is normal. It remains to show that G/Γ is a Lie group. It suffices to note that the action of $\Gamma \times \Gamma$ on $G \times G$ obtained by $(\gamma, \gamma')(x, y) = \big(\gamma(x), \gamma'(y)\big)$ is once again free and proper. The map $(x, y) \mapsto \pi(xy)$ passes to the quotient as a smooth map and is the multiplication in G/Γ.

Conversely, if $p : G \to H$ is a homomorphism such that $T_e p$ is invertible, then by Lemma 4.22 p is a local diffeomorphism. It follows that $p^{-1}(e)$ is discrete, and is closed because it is a subgroup, by Proposition 4.33. This subgroup is normal as p is a group homomorphism.

Let U be an open subset containing the identity of G, such that $p_{|U}$ is diffeomorphic to its image V and such that $U \cdot U \cap p^{-1}(e) = \{e\}$. Then $p^{-1}(V)$ is a disjoint union of $\gamma \cdot U$ for γ in $p^{-1}(e)$, and the restriction of p to each of these is a diffeomorphism. For $h \in H$, $p^{-1}hV$ has an analogous property, which shows that p is a covering map.

Let $\pi : G \to G/p^{-1}(e)$ be the covering obtained in the first part of the theorem. Then there exists a map $\bar{p} : G/p^{-1}(e) \to H$ such that $p = \bar{p} \circ \pi$. As π and p are local diffeomorphisms, so is $\bar{p}$. Moreover, for algebraic reasons, $\bar{p}$ is a group isomorphism. It is therefore a Lie group isomorphism.   □

Remarks

a) The fact that in this situation $p^{-1}(e)$ is Abelian and is even contained in the center was not used. But this property is important, and it is tied to the fact that the fundamental group of a Lie group is Abelian (see Exercise 18).

b) The start of the proof uses only that Γ is normal. From this we deduce that if Γ is a discrete subgroup of a Lie group G, there exists a unique manifold structure on the quotient G/Γ such that $\pi : G \to G/\Gamma$ is a covering map. Thus in principle we can construct many manifolds. However it is very difficult in general to construct "sufficiently large" discrete subgroups (this is to say such that G/Γ is compact) of a given Lie group. For an example of such a construction with $G = Sl(2, \mathbf{R})$, see [Godement 05, 1.5].

c) Examples of covering maps are given in Exercises 5 and 9.

This theorem dashes the hope of characterizing a Lie group, even a connected one, by its Lie algebra. We may however state "Lie's third theorem" (also called Cartan's theorem) in the following way.

Theorem 4.37. *Let $\mathfrak{G}$ be a finite dimensional Lie algebra. There exists a unique simply connected Lie group G, up to isomorphism, whose Lie algebra is $\mathfrak{G}$. Further, every connected Lie group with Lie algebra $\mathfrak{G}$ is isomorphic to the quotient of G by a discrete normal subgroup.*

One finds in [Duistermaat-Kolk 99, 1.14] an interesting proof that uses Banach Lie groups.

4.5. Commutative Lie Groups

4.5.1. A Structure Theorem

We will now see a simple but important example, where we can use the techniques above to describe all of connected Lie groups of a given Lie algebra.

Theorem 4.38. *The Lie algebra $\mathfrak{G}$ of a commutative Lie group is Abelian. Conversely, a connected Lie group whose Lie algebra is Abelian is commutative. More precisely, such a group is isomorphic to the quotient of its Lie algebra (seen as an additive group) by a discrete subgroup.*

PROOF. If G is commutative, then in particular $g \exp tX g^{-1} = \exp tX$ for any $g \in G$ and $X \in \mathfrak{G}$, and differentiating with respect to t at $t = 0$:

$$\mathrm{Ad}\, g \cdot X = X.$$

Writing this relation for $g = \exp tY$ and differentiating again with respect to t we find that $[X, Y] = 0$. Conversely, we have the

Lemma 4.39. *If two elements X and Y of a Lie algebra of a group G commute, then $\exp X$ and $\exp Y$ commute, and $\exp(X + Y) = (\exp X)(\exp Y)$.*

PROOF. It suffices to "lift" the preceding reasoning. Write $f(t) = \mathrm{Ad} \exp tX$, and we first see that

$$\frac{d}{dt} f(t) \cdot Y_{|t=0} = 0,$$

and then

$$\frac{d}{dt} f(t) \cdot Y_{|t=u} = \frac{d}{dt} f(t+u) \cdot Y_{|u=0}$$

$$= f(u) \cdot \frac{d}{dt} f(t) \cdot Y_{|t=0} = 0.$$

Thus, $\mathrm{Ad}(\exp tX) \cdot Y = Y$ for every t, which says that the one-parameter subgroups

$$s \longmapsto \exp sY \quad \text{and} \quad s \longmapsto \exp tX \exp sY \exp -tX$$

have the same infinitesimal generator, and are therefore equal. To show the second assertion, it suffices to note that under these conditions, $t \mapsto (\exp tX)(\exp tY)$ is again a one-parameter subgroup. Then we have

$$(\exp tX)(\exp tY) = \exp tZ, \text{ where } Z = \frac{d}{dt}(\exp tX)(\exp tY)_{t=0} = X + Y. \;\; \square$$

END OF THE PROOF OF THEOREM 4.38. We have just seen that $\exp$ is a morphism of the additive group $\mathfrak{G}$ to G. This is a local diffeomorphism at 0, and therefore everywhere by Lemma 4.22. Consequently, $\exp \mathfrak{G}$ is an open subgroup of G_0, and is therefore closed by Theorem 4.27, and equal to G everywhere since G is connected. Moreover, $\exp$ is a covering map by Exercise 4. By Theorem 4.36, G is therefore Lie group isomorphic to the quotient of $\mathfrak{G}$ by the discrete subgroup $\mathrm{Ker}(\exp)$. $\;\;\square$

It remains to determine the structure of discrete subgroups of a real vector space.

Theorem 4.40. *The discrete subgroups of a finite-dimensional real vector space are the subgroups generated* as additive groups *by k independent vectors.*

PROOF. We first show that if $v_1, \ldots, v_k$ are k independent vectors, the set Γ of linear combinations with integer coefficients of the v_i is discrete in the vector space E. It suffices to check that for each compact subset K in E, $K \cap \Gamma$ is finite. We complete the v_i to a basis of E. Let N be the sup norm associated to this basis. As all norms on a finite dimensional vector space are equivalent, there exists a $C > 0$ such that $K \subset B_N(0, C)$. Then, if $\gamma = \sum_{i=1}^{k} p_i v_i$ is in Γ, we have $|p_i| \leqslant C$ for all i, and therefore $\operatorname{card}(\Gamma \cap K) \leqslant (2C + 1)^n$, if n is the dimension of E.

The converse is proved by induction on n. Let Γ be a discrete subgroup. Each compact neighborhood of 0 contains only a finite number of elements of Γ. If $n = 1$ and $\Gamma \neq \{0\}$, there then exists a nonzero $v \in \Gamma$ with absolute minimum value, which we may suppose to be positive. For $x \in \Gamma$, we write

$$x = nv + r \quad \text{with } n \in \mathbf{Z}, \, 0 \leqslant r < v.$$

Then $r = x - nv \in \Gamma$, which is only possible if $r = 0$.

Now suppose the result is true for n, and we show it is true for $n + 1$. As Γ is discrete, if it does not reduce to $\{0\}$ it contains a nonzero element v_0 with minimal norm. Then the distance from $\Gamma \setminus \mathbf{Z}v_0$ to the line $\mathbf{R}v_0$ is strictly positive. If not, there exists a sequence (w_k) of points of $\Gamma \setminus \mathbf{Z}v_0$ and a sequence of real numbers λ_k such that

$$\lim_{k \to \infty} \| w_k - \lambda_k v_0 \| = 0.$$

We write $\lambda_k = [\lambda_k] + \mu_k$, with $[\lambda_k] \in \mathbf{Z}$ and $|\mu_k| < 1/2$, and we deduce that

$$\lim_{k \to \infty} \left\| (w_k - [\lambda_k]v_0) - \mu_k v_0 \right\| = 0.$$

Now for k sufficiently large, the element $w_k - [\lambda_k]v_0$ of $\Gamma \setminus \mathbf{Z}v_0$ has norm strictly less than that of v_0, which contradicts the hypothesis.

If p denotes the quotient map from E to $E/\mathbf{R}v_0$, the subgroup $p(\Gamma)$ of $E/\mathbf{R}v_0$ is then discrete. By the induction hypothesis there exists vectors $f_1, \ldots, f_k$ of $E/\mathbf{R}v_0$ such that

$$p(\Gamma) = \bigoplus_{i=1}^{k} \mathbf{Z} f_i.$$

Let $v_1, \ldots, v_k \in \Gamma$ be such that $p(v_i) = f_i$. Then, if $x \in \Gamma$, we have

$$p(x) = \sum_{i=1}^{k} n_i f_i, \quad \text{with } n_i \in \mathbf{Z}$$

and therefore

$$x - \sum_{i=1}^{k} n_i v_i \in \mathbf{Re} \cap \Gamma = \mathbf{Z} v_0. \qquad \square$$

Definitions 4.41

a) *The integer k, which is independent of the choice of v_i since it is the dimension of the vector space generated by Γ, is called the* rank *of Γ.*

b) *A discrete subgroup with maximal rank of a vector space is called a* lattice.

Simultaneous application of the two theorems above gives the following result:

Corollary 4.42. *Every connected commutative Lie group is isomorphic to either $\mathbf{R}^p \times (S^1)^q$, or $(S^1)^q$ if it is compact.*

Recalling two dimensions, it is common to call compact commutative Lie groups *tori*, and to *denote* the torus of dimension q by T^q.

Remark. The *metric* study of lattices is a profound problem in geometry of numbers. See [Martinet 03].

4.5.2. Towards Elliptic Curves

Let $P \in \mathbf{C}[X, Y, T]$ be a homogeneous polynomial of degree 3, and $E \subset P^2\mathbf{C}$ the set of points of the complex projective plane whose projective coordinates satisfy $P(x, y, t) = 0$. By Section 2.6.2, if the partial derivatives of P are not simultaneously vanishing away from the origin, E is a submanifold. We say that X is a smooth cubic curve. Using algebraic techniques which we omit, we can reduce to the case where

$$P(X, Y, T) = Y^2 T - X^3 - pXT^2 - qT^3 \quad \text{(Weierstrass curve)}.$$

It turns out E is a manifold if and only if the three roots of the polynomial $X^3 + pX + q$ are distinct; furthermore we see that E is path connected and so connected.

Moreover, we can equip E with a group structure in the following way. We choose a point O once and for all. For two points A and B of E, if R is the third point of intersection of the projective line AB with E (if $A = B$ we take the tangent at A), $A + B$ is the third point of intersection of the line OR with E. It is clear that this is a commutative operation with identity O, and the group operation is smooth. It is more delicate (in fact it is a pretty argument in classical algebraic geometry explained in [Hellegouarch 01, 4.4]) to see that the operation is associative. Then, by Corollary 4.42, E is diffeomorphic as a real manifold to a torus of dimension 2 (another proof of this result is given in Exercise 8 of Chapter 8).

In fact, by using the remark that follows the definition of complex projective space (see Section 2.5), we see that E is a complex manifold of dimension 1 and that the composition law that we have just defined is holomorphic. Adapting the argument of this section to the complex case, we see that E is **C-diffeomorphic to C/Γ, where Γ is a lattice of dimension 2.

Conversely, using the function $\wp_\Gamma$ associated to the lattice Γ (Weierstrass's elliptic function), we construct a **C**-diffeomorphism of C/Γ to a Weierstrass cubic curve (see [Hellegouarch 01, 2.5]). Such a curve is called an *elliptic curve* as it may be parameterized by elliptic functions.**

We also note, even though we will not need this property, that a compact complex connected Lie group is commutative (see Exercise 18). This opens the door to complex tori, this is to say *complex* manifolds $\mathbf{C}^n/\Gamma$, where Γ is a lattice of dimension $2n$. This is also a rich theory (see for example [Debarre 05]).

4.6. Homogeneous Spaces

We have seen that if G is a topological group and $H \subset G$ is a closed subgroup, the quotient space G/H of left equivalence classes of G modulo H is a Hausdorff topological space, and left translations pass to the quotient and give a left transitive action of G on G/H. If we suppose further that G is a Lie group, we can say much more. By virtue of the following result, whose proof we omit, H is automatically a Lie subgroup. (See for example [Duistermaat-Kolk 99, 1.10.7].)

Theorem 4.43 (Cartan-von Neumann).[1] *Every closed subgroup of a Lie group is Lie subgroup.*

Remark. Conversely, every Lie subgroup H of G is closed in G: as a submanifold, it is open in its closure $\overline{H}$, therefore closed in $\overline{H}$ by 4.27.

Example. The group of automorphisms of a Lie algebra $\mathfrak{G}$ of finite dimension is a Lie group: it is clearly a closed subgroup of $Gl(\mathfrak{G})$.

Remark. This result allows us to dispense with proof that $SO(n)$, $U(n)$, $SU(n)$, are Lie groups, since they are closed subgroups of the real or complex linear group. However, if we want further information (think of the dimension for example) we must determine the Lie algebra of the group anyhow.

---

1. Proof omitted.

Theorem 4.44. *If H is a closed subgroup of a Lie group G, there exists a unique smooth manifold structure on the quotient G/H such that the quotient map $p : G \to G/H$ is a submersion.*

As always with manifolds, the idea is to be guided by linear algebra, and more precisely in this case, by the fact that the quotient vector space $\mathfrak{G}/\mathfrak{H}$ is isomorphic to a complement of $\mathfrak{H}$ in $\mathfrak{G}$.

Lemma 4.45. *Let $\mathfrak{M} \subset \mathfrak{G}$ be a complement of $\mathfrak{H}$. Then there exists open subsets $U \subset \mathfrak{M}$ and $V \subset \mathfrak{H}$, containing 0, such that the map f defined by*

$$f(X, Y) = (\exp X)(\exp Y)$$

is a diffeomorphism from $U \times V$ to its image in G.

PROOF. This is immediate by the inverse function theorem, the differential of f at 0 is the map

$$(X, Y) \longmapsto X + Y \quad \text{from} \quad \mathfrak{G} \quad \text{to} \quad \mathfrak{M} \bigoplus \mathfrak{H}. \qquad \square$$

Lemma 4.46. *With the same hypotheses, there exists an open subset U containing 0 in $\mathfrak{M}$ such that for all $g \in G$, the map ϕ_g given by*

$$X \longmapsto p(g \exp X)$$

is a homeomorphism of U to its image.

PROOF. It suffices to study the case where $g = e$. Choose U, V as in the preceding lemma, with the additional condition that if $W = f(U \times V)$, W^3 is again contained in an open subset for which the lemma applies. Then if X, X' have the same image under ϕ, we have

$$\exp X' = \exp X h \quad \text{with} \quad h \in H \cap W^2,$$

and applying Lemma 4.45 to W^3, we see that $X = X'$ and $h = e$. Further, ϕ is open, as if U' is an open subset of U, then $f(U' \times V)$ is open in G, and therefore so is its image under p by Theorem 4.27. However

$$\phi(U') = p(f(U' \times V)). \qquad \square$$

PROOF OF THEOREM 4.44. Choose an open subset U containing 0 in $\mathfrak{M}$ as in the preceding lemma. This lemma says precisely that G/H is a C^0 manifold with charts

$$(p(g \exp U), \phi_g^{-1}).$$

If we want that p is a submersion for a C^∞ structure, we need that the ϕ_g are diffeomorphisms. This will yield the uniqueness of the smooth structure, and to show its existence, it suffices to check the compatibility condition.

Let g and g' be such that

$$T = p(g \exp U) \cap p(g' \exp U) \neq \emptyset.$$

Any $x \in T$ may be written in a unique way as $p(g \exp X)$ or $p(g' \exp X')$, and we must show the map $X \mapsto X'$ from $\phi_g^{-1}(T)$ to $\phi_{g'}^{-1}(T)$ is smooth. For $X_0 \in \phi_g^{-1}(T)$, there exists a (unique) $h \in H$ such that

$$g \exp X_0 h = g' \exp X_0'.$$

Right translation R_h is a diffeomorphism, and we can thus find open subsets W_1 and W_1' of $g \exp W$ and $g' \exp W$ respectively, containing $g \exp X_0$ and $g' \exp X_0'$ respectively, such that $R_h W_1 = W_1'$. However if X is such that $g \exp X \in W_1$, Lemma 4.45 and the chain rule imply that $g \exp X h$ may be written $g' \exp X' \exp Y$, and the map $X \mapsto X'$ is smooth. On the other hand

$$p(g \exp X) = p(g \exp X h) = p(g' \exp X' \exp Y) = p(g' \exp X'). \qquad \square$$

Corollary 4.47. *Under the same hypotheses:*

i) *p admits local sections, in other words, for every $x \in G/H$ there exists an open subset U containing x and a smooth map*

$$\sigma : U \longrightarrow G \quad \text{such that} \quad p \circ \sigma = Id_{|U};$$

ii) *in particular, p is a fibration with model fiber H;*

iii) *if X is a manifold and f is a map of G/H to X, then f is smooth if and only if $f \circ p : G \mapsto X$ is smooth.*

PROOF. By the preceding proof, and with the same notation, the map $\sigma : p(g \exp U) \mapsto G$ mapping $x \in g \exp U$ to $g \exp \phi_g^{-1}(x)$, (or less formally, for which $p(g \exp X)$ associates $g \exp X$) is a smooth section above $p(\exp U)$. Under these conditions, we have an explicit diffeomorphism between $p^{-1}(p(\exp U))$ and $p(\exp(U)) \times H$, given by

$$h(g) = \left(p(g), g(\sigma(p(g)))^{-1}\right).$$

Finally, if $f : G/H \mapsto X$ is such that $f \circ p$ is smooth, and if U is an open subset of G/H above which p admits a section, then

$$f_{|U} = f_{|U} \circ (p \circ \sigma_{|U}) = (f_{|U} \circ p) \circ \sigma_{|U}$$

is smooth. $\qquad \square$

What we have just seen is an example of the following situation.

Definition 4.48. *A left action of a Lie group G on a smooth manifold X is a smooth map $(g, x) \mapsto g \cdot x$ from $G \times X$ to X such that $g_1 \cdot (g_2 \cdot x) = (g_1 g_2) \cdot x$.*

Here we have an action of G on G/H given by $g \cdot x = p(gg')$ if $x = p(g')$. It is clear that the result is independent of g' and smoothness can be seen as in the Corollary 4.47 by using a local section.

The properties seen in 2.7.1 for discrete groups have analogues here. The definition of a free or effective action is the same. For a proper action, we replace finite in the Definition 2.34 with compact. A new notion appears, that of an almost effective action, which means that the subgroup

$$G_{Id} = \{g \in G : \ g \cdot x = x \text{ for all } x \in X\}$$

is a discrete subgroup of G. In practice, when we work with subgroups of linear groups, we prefer to work with almost effective actions over passing to the quotient (for example, most of the actions considered in the exercises are almost effective).

We will now see that every smooth manifold on which a Lie group acts transitively is of the form stated in Theorem 4.44.

Theorem 4.49. *Let X be a manifold equipped with a smooth and transitive action of a Lie group G having a finite number of connected components. For $a \in X$, the stabilizer*

$$G_a = \{g \in G : \ g \cdot a = a\}$$

of a is a Lie subgroup, and the map $F : g \to g \cdot a$ passes to the quotient as a diffeomorphism of G/G_a to X.

Lemma 4.50. *F is a submersion.*

PROOF. From the fact that the maps $x \mapsto g \cdot x$ are diffeomorphisms of X, the rank r of F is constant. Write $n = \dim G$, $p = \dim X$. By the rank theorem (see Exercise 10 of Chapter 1), there exists for every $g \in G$ an open subset U of G containing g, an open subset V of X containing $g \cdot a$, and diffeomorphisms ϕ, ψ of U and V to open subsets of $\mathbf{R}^n$ and $\mathbf{R}^p$ respectively such that

$$\left(\psi \circ F \circ \phi^{-1}\right)(u_1, \ldots, u_n) = (u_1, \ldots, u_r, 0, \ldots, 0).$$

This equation shows that $F(U)$ is a submanifold of X of dimension $p - r$, and therefore a negligible subset of X if $r < p$. However, the open subsets U form a covering of G, thus we may extract a countable subcover (see Section 2.8). Thus $F(G)$ is a negligible subset of X. As $F(G) = X$ by transitivity, we arrive at a contradiction. □

PROOF OF THEOREM 4.49. By the lemma, $G_a = F^{-1}(a)$ is a submanifold of dimension $n - p$.

If $h \in G_a$, we have

$$F(gh) = gh \cdot a = g \cdot (h \cdot a) = g \cdot a = F(g),$$

so F passes to the quotient, and gives a smooth map f from G/G_a to X by iii) of Corollary 4.47. As

$$T_g F = T_{p(g)} f \circ T_g p,$$

we see that f is a submersion, and therefore a local diffeomorphism for dimensional reasons, and finally a diffeomorphism as it is a bijection. □

Remark. If b is another point of X, and if $g \in G$ is such that $g \cdot a = b$, then the conjugation

$$h \longmapsto ghg^{-1}$$

gives an isomorphism between G_a and G_b, and also by passing to the quotient, a diffeomorphism of G/G_a to G/G_b.

Definition 4.51. *A smooth manifold on which a Lie group G acts transitively is called a* homogeneous space.

It is important to note that there can be many transitive actions on the same manifold. This is the case for the spheres of odd dimension for example (for a more precise statement, see Exercise 13).

Example: Grassmannians

The p-Grassmannian of a vector space E, denoted $G_p(E)$, is the set of its vector subspaces of dimension p. If $E = \mathbf{R}^n$, we write $G_{n,p}$. It is clear that the natural action of $Gl(n, \mathbf{R})$ on $\mathbf{R}^n$ gives a transitive action on $G_{n,p}$. The subgroup $T_{n,p}$ of linear maps which fix a plane P_0 generated by the first p vectors of the canonical basis is given by matrices of the form

$$\begin{pmatrix} A & B \\ 0 & C \end{pmatrix}$$

where A and C are invertible matrices of order p and $n-p$, and B is a matrix with p rows and $n-p$ columns. Then, $G_{n,p}$ is in bijection with $Gl(n, \mathbf{R})/T_{n,p}$, and inherits a smooth manifold structure.

We can also equip $\mathbf{R}^n$ with its natural inner product and let the orthogonal group act on $G_{n,p}$. The action is always transitive, as every orthonormal system of vectors may be completed to an orthonormal basis. Any $g \in O(n)$

which fixes P_0 also fixes $P_0^\perp$, and it is therefore of the form

$$\begin{pmatrix} A & 0 \\ 0 & C \end{pmatrix},$$

where $A \in O(p)$ and $C \in O(n-p)$. Thus, $G_{n,p}$ also appears as the homogeneous space $O(n)/O(p) \times O(n-p)$. With the help of Theorem 4.44, the reader is invited to check that these two representations give the same manifold structure. In passing we mention that it is useful to note that $G_{n,1} = P^{n-1}\mathbf{R}$ and that $\dim G_{n,p} = p(n-p)$.

4.7. Comments

Group actions

Actions of compact Lie groups on manifolds are well understood. See [Audin 03, Chapter 1] and [Duistermaat-Kolk 99, Chapter 2].

Haar measure

We will see in the exercises of Chapter 6 an elementary (and important) property of Lie groups: the existence of a translation invariant measure. While this measure, the so-called Haar measure, exists for every locally compact group, the case of Lie groups is much easier to treat.

Analytic structure

Unlike the case of a general manifold, on a Lie group we have at our disposal the exponential map and its local inverse. This gives a distinguished parametrization and chart. It is natural to write the group operation in this chart.

The Campbell-Hausdorff formula assures that

$$\exp X \exp Y = \exp H(X, Y),$$

where $H : \mathfrak{G} \times \mathfrak{G} \to \mathfrak{G}$ is analytic in a neighborhood of 0. Further, it is the sum of a series of *Lie monomials*, which is to say expressions using only iterated brackets of X and Y. We have

$$H(X,Y) = X + Y + \frac{1}{2}[X,Y] + \frac{1}{12}[X,[X,Y]] + \frac{1}{12}[Y,[Y,X]] + \dots$$

In particular, in a neighborhood of the identity element, multiplication is given by the bracket. One finds a crafty and (relatively) simple proof in [Stillwell 08, Chapter 7].

A few words on classification

To go further, a deep study of Lie algebras is necessary. For example, the result stated in the introduction, that every Lie group is locally isomorphic to a subgroup of the linear group, is obtained by combining the analogous result for Lie algebras (Ado's theorem, see [Postnikov 94]) and Lie's third theorem.

However, there exist Lie groups that are not isomorphic to any subgroup of a linear group: this is the case with the universal cover of $Sl(2, \mathbf{R})$. For the details, see [Doubrovine-Novikov-Fomenko 85, Chapter I, § 3]. This leads to beautiful geometry. This universal cover is also treated in a detailed way in [Godement 05, 2.7].

A milestone of the algebraic theory is the classification of simple Lie algebras, which is to say algebras with no nontrivial ideal. This classification permits us to give the classification of compact *almost simple* groups (*i.e.*, groups without a non discrete normal subgroup). Modulo finite coverings, or quotients by finite groups, we obtain the following list:

$$SO(n) \ n \neq 4; \quad SU(n); \quad Sp(n)$$

(this last group is defined in the solution to Exercise 13), and five so called exceptional groups (see [Onishchik-Vinberg 90]). One finds in [Berger 87] a purely geometric proof of the simplicity of $SO(n)$ for $n = 3$ or $n > 4$, and in [Stillwell 08, Chapter 6] an elementary proof of the simplicity of the corresponding Lie algebras. For the non simplicity of $SO(4)$, see the final part of Exercise 5.

"Infinite" Lie groups

The group of diffeomorphisms of a compact manifold, the group of diffeomorphisms preserving a volume or symplectic form are sometimes called "infinite Lie groups", where infinite refers to infinite dimension. Serious difficulties present themselves.

For example, the C^k diffeomorphisms clearly form a group. However, a simple examination of the proof of Proposition 1.43 shows that multiplication is differentiable for the C^{k-1} norm only. This problem can already be seen at the level of vector fields, as the bracket of two C^k vector fields is only C^{k-1}. In the smooth case, to have a good notion of differentiability of the composition law, we must leave the setting of normed spaces (see [Hamilton 82, p. 148] for a detailed description). Then, even if the characterization of one-parameter subgroups seen in Section 1.6 is valid in the Banach manifold setting, it is of no help.

Since a vector field X on a compact manifold M gives rise to a one-parameter group of diffeomorphisms, it is nevertheless tempting to say the group $\text{Diff}(M)$ of smooth diffeomorphisms of M is Lie group with Lie algebra $C^\infty(TM)$. After all, $C^\infty(TM)$ is a Lie algebra, and we can define the exponential as in the classical case, by posing $\exp X = \phi_1^X$. Even for the simplest compact manifold, the circle S^1, there exists diffeomorphisms arbitrarily close to the identity that are not the value at time 1 of a flow (for an example, see [Pressley-Segal 86, 3.3]). The exponential map is therefore not a local diffeomorphism at 0, which is troublesome. This has not stopped the active study of diffeomorphism groups of manifolds, but this is a another story. For two very different aspects of the study of $\text{Diff}(S^1)$, see [Hector-Hirsch 81] (for the "dynamical systems" aspect) and [Pressley-Segal 86] (for the "representation" aspect).

We note also an interesting use of Banach Lie groups to prove Lie's third theorem. See [Duistermaat-Kolk 99].

4.8. Exercises

1. *The group structure of the Artinian plane*[2]

a) Show that matrices of the form

$$\begin{pmatrix} \cosh t & \sinh t \\ \sinh t & \cosh t \end{pmatrix}$$

form a Lie group isomorphic to $\mathbf{R}$.

b) Show that this group is the identity component of the group $O(1,1)$ of matrices which preserve the quadratic form $x^2 - y^2$, and that $O(1,1)$ has four connected components.

2. *The field of quaternions*

a) Show that complex matrices of the form

$$\begin{pmatrix} u & -\overline{v} \\ v & \overline{u} \end{pmatrix}$$

form a subring of $M_2(\mathbf{C})$ where every nonzero element is invertible, and is therefore a field. This field is called the *quaternions* and is denoted $\mathbf{H}$.

b) It is traditional (and convenient) to write

$$I = \begin{pmatrix} 0 & -1 \\ 1 & 0 \end{pmatrix}; \quad J = \begin{pmatrix} i & 0 \\ 0 & -i \end{pmatrix}; \quad K = \begin{pmatrix} 0 & i \\ i & 0 \end{pmatrix}.$$

---

2. That is to say the plane equipped with the quadratic form $x^2 - y^2$. Some algebraists call it the hyperbolic plane, but we find this name rather confusing.

Denoting the $(2,2)$ identity matrix by E, show that every quaternion q may be written in a unique way in the form $aE + bI + cJ + dK$, where a, b, c, d are real, and where

$$I^2 = J^2 = K^2 = -E,$$
$$IJ = -JI = K, \quad JK = -KJ = I, \quad KI = -IK = J.$$

c) From now on we write the quaternions in this way. We often replace E, I, J and K by the corresponding lower case letters. The subset $\mathbf{R}E$ of $\mathbf{H}$ is a subring isomorphic to $\mathbf{R}$, and we identify it with $\mathbf{R}$. We write

$$\bar{q} = aE - bI - cJ - dK.$$

Check that

$$q\bar{q} = \bar{q}q = (a^2 + b^2 + c^2 + d^2),$$

(this reproves that every nonzero element of $\mathbf{H}$ is invertible), and that

$$\overline{q_1 q_2} = \bar{q}_2\,\bar{q}_1.$$

Check also that *the center* of $\mathbf{H}$ (this is to say the set of elements of $\mathbf{H}$ which commute with every element of $\mathbf{H}$) is equal to $\mathbf{R}$. We write $\operatorname{Re}(q) = a$, $\operatorname{Im}(q) = bI + cJ + dK$. A quaternion is said to be *real* if $\operatorname{Im}(q) = 0$, (which is equivalent to $\bar{q} = q$), *purely imaginary*, or more succinctly *pure* if $\operatorname{Re}(q) = 0$ (which is equivalent to $q + \bar{q} = 0$).

d) Finally we write $\|q\| = \sqrt{q\bar{q}}$. This is a norm on $\mathbf{H}$ seen as a $\mathbf{R}$-vector space as in c). Show that

$$\|q_1 q_2\| = \|q_1\|\,\|q_2\|.$$

3. *The group of multiplicative quaternions*

a) Show that the multiplicative group $\mathbf{H}^*$ of $\mathbf{H}$ is a Lie group.

b) For every quaternion q, we write

$$\exp(q) = \sum_{n=0}^{\infty} \frac{q^n}{n!}.$$

Show that the right hand side is a norm convergent series, and if q and q' commute, we have

$$\exp(q + q') = \big(\exp(q)\big)\big(\exp(q')\big).$$

c) Show that every one-parameter subgroup of the multiplicative group $\mathbf{H}^*$ is of the form

$$t \longmapsto \exp(tq).$$

Deduce that the Lie algebra of $\mathbf{H}^*$ is $\mathbf{H}$ equipped with its vector space structure over $\mathbf{R}$ and the bracket

$$[q, q'] = qq' - q'q.$$

d) Show that the multiplicative group of quaternions of norm 1 is isomorphic to $SU(2)$. State and prove the corresponding result for the Lie algebras.

4. Show that a morphism of Lie groups which is also a local diffeomorphism is a covering map.

5*. *Quaternions and rotations*

a) We identify $\mathbf{R}^3$(as a vector space) with the pure quaternions. If s is a quaternion of norm 1 and h is a pure quaternion, we define

$$\rho(s) \cdot h = sh\bar{s}.$$

Show that $\rho(s) \cdot h$ is again a pure quaternion, and that the linear map $\rho(s)$ is in $O(3)$.

b) Show that $\rho(s) \in SO(3)$, using a connectedness argument.

c) Show that

$$\rho(s)\rho(s') = \rho(ss').$$

Calculate the differential of ρ at e, and deduce that ρ is a local diffeomorphism.

d) Show that ρ is surjective, and that $\mathrm{Ker}(\rho) = \pm e$. Deduce that $SO(3)$ is isomorphic to $S^3/\pm e$, or to $SU(2)/\pm I$.

e) Show that the axis of rotation of $\rho(s)$ is given by the imaginary part of the quaternion s. To determine the *angle*, proceed as follows:

e1) Show that two conjugate rotations have the same angle.

e2) Show that for every pure quaternion t of norm 1, there exists a quaternion q of norm 1 such that $qt\bar{q} = i$ (use d)).

e3) Every quaternion s of norm 1 can be written in a unique way as $\alpha e + \beta t$, where t is a pure quaternion of norm 1, and with α and β real numbers such that $\alpha^2 + \beta^2 = 1$. Show that the angle of rotation of $\rho(s)$ depends only on α and β and compute it.

e4) Numerical application: let R_1 and R_2 be rotations of angle $2\pi/3$ with axes $(1, 1, 1)$ and $(1, 1, -1)$ respectively. Determine the rotation $R_1 \circ R_2$.

f) We may consider the map ρ_1 from $S^3 \times S^3$ to $Gl(\mathbf{R}^4)$ defined by

$$\rho_1(s, t) \cdot q = sq\bar{t}.$$

Imitating the above, show that ρ_1 is a surjective morphism of Lie groups from $S^3 \times S^3$ to $SO(4)$, and deduce that $SO(4)$ is isomorphic to $S^3 \times S^3/\pm I$.

6. *Ideals and normal subgroups*

Let G be a Lie group and let H be a normal Lie subgroup of G. Show, taking inspiration from the proof of Theorem 4.17, that the Lie algebra $\mathfrak{H}$ of H is an ideal in the Lie algebra $\mathfrak{G}$ of G. Show conversely, that a connected Lie subgroup whose Lie algebra is an ideal is a normal subgroup.

7. *Examples of exponentials*

a) Show that exp is surjective for the following groups:
- $Gl(n, \mathbf{C})$;
- $SO(n)$;
- $U(n)$;
- the group of affine transformations $x \mapsto ax + b$ of the real line such that $a > 0$.

b) Show that the subset $N \subset Gl(3, \mathbf{R})$ of matrices of the form

$$\begin{pmatrix} 1 & x & z \\ 0 & 1 & y \\ 0 & 0 & 1 \end{pmatrix}$$

is a Lie subgroup, for which the exponential map is a diffeomorphism.

c) Show that if $A \in Sl(2, \mathbf{R})$, we have

$$\mathrm{tr}(A^2) \geqslant -2$$

(use the Hamilton-Cayley theorem). Deduce the exponential map for $Sl(2, \mathbf{R})$ is not surjective, and determine its image.

8. Show that every morphism of S^1 seen as the set of complex numbers of unit modulus is of the form $t \mapsto t^k$, where $k \in \mathbf{Z}$.

9. *Comparison between $Sl(2, \mathbf{R})$ and the Lorentz group in dimension 3*

a) Show that on the vector space $Sym(2)$ of symmetric matrices of order 2, the determinant defines a quadratic form of type $(1, 2)$.

b) For $A \in Sl(2, \mathbf{R})$, and $M \in S$ we write

$$\rho(A) \cdot M = {}^t\!AMA.$$

Show that ρ defines a morphism of $Sl(2, \mathbf{R})$ to $O(1, 2)$.

c) Show that $\mathrm{Ker}\,\rho = \{\pm Id\}$, and that ρ is a covering map. Deduce the existence of an isomorphism

$$Sl(2, \mathbf{R})/\{\pm I\} \simeq SO_o(1, 2).$$

d*) Give another proof of this isomorphism by using Exercise 16 of Chapter 2.

10*. Let N be the group of Exercise 7 b). Show that the subgroup $N_{\mathbf{Z}}$ of N formed by matrices with integer coefficients is discrete, and that the manifold $N/N_{\mathbf{Z}}$ is compact.

11*. How many connected components does the pseudo-orthogonal group $O(p, q)$ have?

Hint. Let $O(p, q)$ act on the submanifold of $\mathbf{R}^{p+q}$ given by the equation

$$-x_1^2 - \cdots - x_p^2 + x_{p+1}^2 + \cdots + x_{p+q}^2 = 1.$$

12. *Universal covering of the unitary group*
a) Let

$$\widetilde{U}(n) = \{(A, t) \in U(n) \times \mathbf{R} : \det(A) = e^{2i\pi t}\}.$$

Show that the restriction of the projection on the first factor to $\widetilde{U}(n)$ is a Lie group covering, with base $U(n)$ and with kernel $I \times \mathbf{Z}$.

b) Show that $\widetilde{U}(n)$ is isomorphic to $SU(n) \times \mathbf{R}$.

(This shows, by the following exercise, that $\widetilde{U}(n)$ is the universal covering of $U(n)$.)

13. *Some homogeneous spaces*
a) Prove the following homeomorphisms:

$$S^{2n+1} \simeq U(n+1)/U(n) \simeq SU(n+1)/SU(n).$$

Deduce that the groups $SU(n)$ and $U(n)$ are connected for any n.

b) Show that $SU(n)$ is simply connected.

c) Show that the set of matrices of the form

$$\begin{pmatrix} \det(A) & 0 & \cdots & 0 \\ 0 & & & \\ \vdots & & A & \\ 0 & & & \end{pmatrix},$$

where $A \in O(n)$ forms a subgroup of $SO(n+1)$ isomorphic to $O(n)$. Abusing notation, denote this group by $O(n)$ and establish the homeomorphism

$$SO(n+1)/O(n) \simeq P^n \mathbf{R}.$$

d) With the analogous notation, establish

$$SU(n+1)/U(n) \simeq P^n \mathbf{C}.$$

e) Prove that all of these homeomorphisms are diffeomorphisms.

14*. *Orbits of a compact group action*

Let G be a compact Lie group acting differentiability on a smooth manifold M. Show that the orbits of G (this is to say the subsets of M of the form $G \cdot x$, where $x \in M$ is fixed) are submanifolds of M.

Hint. Use the rank theorem and Theorem 4.49.

We take G to be the stabilizer of a point for the natural action of $SO(n+1)$ on $P^n\mathbf{R}$. Which manifolds are diffeomorphic to the orbits of G in $P^n\mathbf{R}$? Try the same question replacing $P^n\mathbf{R}$ by $P^n\mathbf{C}$ and $SO(n+1)$ by $SU(n+1)$.

15. Show that on the vector space of polynomials, the linear operators

$$P \longmapsto P', \quad P \longmapsto XP \quad \text{and} \quad Id$$

generate a Lie algebra isomorphic to that of the Heisenberg group seen at the beginning of this chapter.

(In quantum mechanics, the position and the momentum are represented by operators – namely, in dimension 1, multiplication by x and the derivative – this is why this Lie algebra in called the Heisenberg algebra.)

16*. *Manifolds of matrices of a given rank*

Study the orbits of the action of $Gl(p, \mathbf{R}) \times Gl(q, \mathbf{R})$ on $M_{p,q}(\mathbf{R})$ given by

$$(P, Q) \cdot M = PMQ^{-1}.$$

17*. **A complex Lie group is a complex analytic manifold (see Section 2.5) equipped with a group structure such that the map $(g, h) \mapsto gh^{-1}$ is complex analytic (this is the case for $Gl(n, \mathbf{C})$ and $Sl(n, \mathbf{C})$, but absolutely not the case for $U(n)$ or $SU(n)$!). Show that every *complex* compact connected Lie group is commutative using the adjoint representation.**

18*. *Universal covering of a Lie group*

Let G be a Lie group, let $p : \widetilde{G} \to G$ be a universal covering of G and let $\widetilde{e}$ be a point of $\widetilde{G}$ such that $p(\widetilde{e}) = e$. Applying the monodromy theorem 2.45 to the map $(x, y) \mapsto p(x)p(y)$ of $\widetilde{G} \times \widetilde{G}$ to G, show that there is a unique Lie group structure on $\widetilde{G}$ with identity element $\widetilde{e}$ for which p is a morphism. Deduce that the fundamental group of a Lie group is Abelian.

Chapter 5

Differential Forms

5.1. Introduction

5.1.1. Why Differential Forms?

Does there exist a theory of integration – first for p-dimensional submanifolds of Euclidean space –, and more generally for manifolds? We can start with what we call line integrals, which is to say the circulation of a vector field V along a curve. This is classically defined as the integral

$$\int_a^b \langle V_{c(t)}, c'(t) \rangle \, dt.$$

Here, $c : [a, b] \to \mathbf{R}^n$ is a curve parametrization (in fact a piece of the parametrization as we restricted the parameter to the interval $[a, b]$) and $\langle \, , \, \rangle$ is an inner product on $\mathbf{R}^n$. Replacing the vectors $V_{c(t)}$ by *linear forms* $\alpha_{c(t)}$ has the advantage of no longer requiring the inner product. We can then integrate curves on any manifold X, the "field of linear forms" $x \mapsto \alpha_x$, for all $x \in X$, where α_x is a linear form on the tangent space $T_x X$, by writing

$$\int_c \alpha = \int_a^b \alpha_{c(t)} \big(c'(t) \big) \, dt.$$

We must of course specify the regularity of the α_x with respect to x. This question will be resolved in the same way as for vector fields.

The passage to submanifolds of dimension $p > 1$ is done in two steps. For fixed x, which is to say from an infinitesimal point of view, we reduce to an algebraic problem: to define an element of volume for vector subspaces of dimension p (the different possible tangent spaces) of a vector space of dimension n (as it happens $\mathbf{R}^n \simeq T_x \mathbf{R}^n$ or $T_x X$, if X is the n-dimensional ambient manifold). Knowing that the volume of parallelepipeds in dimension n is calculated with determinants, and that determinants are alternating n-linear forms, we are driven to introduce alternating p-linear forms, which generalize both linear forms and determinants. Then dependence with respect to x is treated as in $p = 1$.

5.1.2. Abstract

Algebraic preliminaries (alternating multilinear forms) are treated in Section 5.2. Then, in the following four sections, we study differential forms on open subsets of $\mathbf{R}^n$ at length. These are easier objects to manipulate than vector fields: every smooth map $\phi : U \to V$ (regardless of the dimensions of the spaces U and V) allows us to "pullback" differential forms on V. The pullback of a function f (a form of degree 0) is simply $f \circ \phi$, and the extension to forms is done easily using the linear tangent map of ϕ.

However the main interest in differential forms is the existence of a *natural* (which is to say that it commutes with pullback under smooth maps, the precise statement is Proposition 5.22) linear operator d which associates to each form of degree p a form of degree $p+1$. This operator is the generalization of the differential of a function (the case where $p = 0$). For example d allows us to give a unified presentation of the gradient, divergence and curl, and the algebraic properties of these operators. We deduce a remarkably concise form of the Maxwell equations in Section 5.8, this new form readily generalizes to other physical situations (see for example [Atiyah 79]).

The fundamental property of this operator is the vanishing of d^2. This is a generalization of the fact that $\partial^2_{ij} = \partial^2_{ji}$ for smooth functions. A fundamental question is the existence of a converse of this property. For example, if a form of degree 1 satisfies $d\alpha = 0$, does there exist a function such that $\alpha = df$? By the Poincaré lemma (see Section 5.6) the answer is yes (and for every degree) on open subsets diffeomorphic to $\mathbf{R}^n$. In general the answer is no. A fundamental counterexample is that of the form $\frac{x\,dy - y\,dx}{x^2 + y^2}$ on $\mathbf{R}^2 \smallsetminus \{0\}$. This corresponds to the fact that there is no way to determine the argument of a complex number in a continuous fashion. We finally note that, thanks to the fact that d commutes with pullbacks, and in particular the pullback by a diffeomorphism, the transcription of these properties to manifolds may be done painlessly.

The reader whose appetite is whetted by this long and nonetheless incomplete introduction is strongly encouraged to read (at least) the introduction of [Whitney 57]. Another beautiful reference, for physical motivation, is [Misner-Thorne-Wheeler 73].

5.2. Multilinear Algebra

5.2.1. Tensor Algebra

Let E be a vector space over a field K. The *dual space* $\mathcal{L}(E, K) = E^*$ is the vector space of K-linear maps from E to K, also called *linear forms*. Suppose E has dimension n, and suppose $(e_i)_{1 \leqslant i \leqslant n}$ is a basis. If

$$v = \sum_{i=1}^{n} v^i e_i$$

is the decomposition of a vector with respect to this basis, we denote by e^{i*} the linear form $v \mapsto v^i$, which associates to every vector its i-th coordinate. Then if $\alpha \in E^*$, we have

$$\alpha(v) = \sum_{i=1}^{n} v^i \alpha(e_i) = \sum_{i=1}^{n} \alpha(e_i) e^{i*}(v)$$

for all v. In other words the linear form α may be written as the linear combination

$$\alpha = \sum_{i=1}^{n} \alpha(e_i) e^{i*}.$$

In particular, $(e^{i*})_{1 \leqslant i \leqslant n}$ is a basis of E^*, called the *dual basis* to $(e_i)_{1 \leqslant i \leqslant n}$.

We use the *Einstein summation convention*. When we index a family of vectors or a vector field, we write a *lower* index. A good mnemonic is to think of vector fields ∂_i. When we index forms, we use an *upper* index, whether actual forms like e^{i*}, or their values on a vector such as the numbers v^i. When we decompose a vector (resp. a form) with respect to a basis, we place the indices of the coefficients in upper (resp. lower) position as we have just done. Physicists have profited from the convention that an expression where the same index appears both in upper and lower position as representing a sum over this index. For our part we will not omit the summation signs, but we will adopt the convention above for the placement of indices. This usage allows us to see at a glance whether we are working with vectors or forms.

Definition 5.1. *A linear k-form on E is any map*

$$L: \overbrace{E \times \cdots \times E}^{k \ \text{times}} \longrightarrow K$$

such that the component functions

$$x_r \longmapsto L(x_1, \ldots, x_k)$$

are linear forms on E.

The sum of two linear k-forms and the product of a linear k-form by a scalar are clearly linear k-forms.

Furthermore, if we write vectors $(x_r)_{1 \leqslant r \leqslant k}$ with respect to a given basis $(e_r)_{1 \leqslant r \leqslant k}$, we see that L is determined by n^k scalars

$$L(e_{i_1}, e_{i_2}, \ldots, e_{i_k})$$

and therefore linear k-forms form a vector space of dimension n^k.

Definition 5.2. *The* tensor product *of a linear k-form f and a linear l-form g is a linear $(k+l)$-form given by*

$$(f \otimes g)(v_1, \ldots, v_{k+l}) = f(v_1, \ldots, v_k) g(v_{k+1}, \ldots, v_{k+l}).$$

For example, if α and β are two linear forms,

$$(\alpha \otimes \beta)(u, v) = \alpha(u)\beta(v),$$

which shows by the way that $\alpha \otimes \beta \neq \beta \otimes \alpha$ if α and β are not proportional. The defining formula shows that $\otimes$ defines a bilinear map from $\mathcal{L}^k(E, K) \times \mathcal{L}^l(E, K)$ to $\mathcal{L}^{k+l}(E, K)$ for any k and l, and we have

$$f \otimes (g \otimes h) = (f \otimes g) \otimes h,$$

with the value of both sides at $(x_1, \ldots, x_{k+l+m})$ given by

$$f(x_1, \ldots, x_k) g(x_{k+1}, \ldots, x_{k+l}) h(x_{k+l+1}, \ldots, x_{k+l+m}).$$

Given a basis of E and the corresponding dual basis of E^*, this associative property allows us to introduce the n^k tensor products

$$e^{i_1 *} \otimes \cdots \otimes e^{i_p *}.$$

If we write k vectors x_i as

$$x_i = \sum_{j=1}^{n} \xi_i^j e_j$$

then

$$(e^{i_1 *} \otimes \cdots \otimes e^{i_k *})(x_1, \ldots, x_k) = \xi_1^{i_1} \xi_2^{i_2} \cdots \xi_k^{i_k}.$$

Thus, as in the simplest case of $E^* = \mathcal{L}(E, K)$, the $e^{i_1 *} \otimes \cdots \otimes e^{i_k *}$ form a basis of $\mathcal{L}^k(E, K)$. We also call this space the k-th tensor power of E^*, and we denote it by $\bigotimes^k E^*$.

Definition 5.3. *The* tensor algebra *of E^*, denoted by $T(E^*)$, is the direct sum*

$$\bigoplus_{i=0}^{\infty} \bigotimes^k E^*$$

equipped with the product obtained by extending $\otimes$ by linearity.

This is an associative algebra. **The algebraic minded reader may verify the following so-called "universal" property which characterizes $T(E^*)$: every linear map from E^* to an associative algebra A extends in a unique way to a morphism of algebras from $T(E^*)$ to A.**

5.2.2. Exterior Algebra

Definition 5.4. *A linear k-form is* alternating *if for every permutation σ of $[1, k]$ we have*

$$f(x_1, \ldots, x_k) = \epsilon(\sigma) f(x_{\sigma(1)}, \ldots, x_{\sigma(k)}),$$

where we denote the signature of σ by $\epsilon(\sigma)$. The integer k is the degree of f.

Example. Every linear form is alternating. A 2-form is alternating if and only if

$$f(x, y) = -f(y, x) \quad \text{for all } x, y \in E.$$

It is equivalent to say (if K is not of characteristic 2, for example if $K = \mathbf{R}$ or $\mathbf{C}$, which we assume hereafter) that $f(x, x) = 0$ for all x. Indeed, in this case

$$f(x + y, x + y) = 0 = f(x, x) + f(x, y) + f(y, x) + f(y, y) = f(x, y) + f(y, x).$$

By using the fact that every permutation is the product of transpositions, we can deduce the following important property.

Proposition 5.5

i) *A linear k-form is alternating if and only if*

$$f(x_1, \ldots, x_k) = 0 \quad \text{when two of the } x_i \text{ are equal.}$$

ii) *If the vectors $x_1, \ldots, x_k$ are linearly dependent,*

$$f(x_1, \ldots, x_k) = 0 \quad \text{for every alternating } k\text{-form } f.$$

PROOF. i) is a direct consequence of the above. For ii), note that if the x_i are not all zero then at least one is a combination of the others. By a permutation of the variables (f is alternating!) we may suppose x_1 is a combination of the others. Thus

$$x_1 = \sum_{i=2}^{k} \lambda^i x_i$$

and therefore

$$f(x_1, \ldots, x_k) = \sum_{i=2}^{k} \lambda^i f(x_i, x_2, \ldots, x_i, \ldots, x_k) = 0. \qquad \square$$

The vector space of alternating linear k-forms is denoted $\bigwedge^k E^*$. We have

$$\overset{1}{\bigotimes} E^* = \overset{1}{\bigwedge} E^* = E^*$$

and we declare that

$$\overset{0}{\bigotimes} E^* = \overset{0}{\bigwedge} E^* = K.$$

An important case is where $k = \dim E$: the theory of determinants says exactly that $\dim \bigwedge^n E^* = 1$, and more precisely that

$$f(x_1, \ldots, x_n) = \begin{vmatrix} \xi_1^1 & \cdots & \xi_1^n \\ \vdots & \ddots & \vdots \\ \xi_n^1 & \cdots & \xi_n^n \end{vmatrix} f(e_1, \ldots, e_n).$$

More generally

Proposition 5.6. *If $f \in \bigwedge^k E^*$ and if $(e_i)_{1 \leqslant i \leqslant n}$ is a basis of E, we have*

$$f(x_1, \ldots, x_k) = \sum_{1 \leqslant i_1 < \cdots < i_k \leqslant n} \begin{vmatrix} \xi_1^{i_1} & \cdots & \xi_1^{i_k} \\ \vdots & \ddots & \vdots \\ \xi_n^{i_1} & \cdots & \xi_k^{i_k} \end{vmatrix} f(e_{i_1}, \ldots, e_{i_k}).$$

PROOF. First, multilinearity gives

$$f(x_1, \ldots, x_k) = \sum_{i_1, \ldots, i_k} \xi_1^{i_1} \ldots \xi_k^{i_k} f(e_{i_1}, \ldots, e_{i_k}).$$

Moreover, if f is alternating, then by Definition 5.4 the only terms that remain are those where the indices i_r are pairwise distinct: the required formula is obtained by regrouping the sequences $i_1, \ldots, i_k$ corresponding to the same partition of k elements of $[1, n]$ and in applying the formula for the expansion of a determinant.   □

Corollary 5.7. $\dim \bigwedge^k E^* = \binom{n}{k}$. *In particular, $\bigwedge^k E^* = 0$ if $k > n$.*

The example of two forms of degree 1 shows that the tensor product of two alternating forms is not alternating. Nonetheless a little work allows us to equip the set of alternating forms with a multiplicative structure.

Definition 5.8. *The antisymmetrization of a linear k-form f, denoted $\mathrm{Alt}\, f$, is given by*

$$(\mathrm{Alt}\, f)(x_1, \ldots, x_k) = \frac{1}{k!} \sum_{\sigma \in \mathbf{S}_k} \epsilon(\sigma) f(x_{\sigma(1)}, \ldots, x_{\sigma(k)}),$$

where we denote the group of permutations of $\{1, 2, \ldots, k\}$ by $\mathbf{S}_k$, and the signature of the permutation σ by $\epsilon(\sigma)$.

We can easily verify that $f = \mathrm{Alt}\, f$ if and only if f is alternating.

Example. If $f \in \bigwedge^2 E^*$,

$$(\text{Alt } f)(x,y) = \frac{1}{2}\big(f(x,y) - f(y,x)\big).$$

Definition 5.9. *The exterior product of $f \in \bigwedge^k E^*$ and $g \in \bigwedge^l E^*$, denoted $f \wedge g$, is the $(k+l)$-form*

$$f \wedge g = \frac{(k+l)!}{k!\,l!}\, \text{Alt}(f \otimes g).$$

Example. If $k = l = 1$, $(f \wedge g)(x,y) = f(x)g(y) - f(y)g(x)$.

This product has the following properties, which we will assume, and whose proof (for which one can consult [Spivak 79, first volume, Chapter 7]) primarily uses properties of permutation groups.

1. *Anticommutativity:*

$$g \wedge f = (-1)^{kl} f \wedge g \quad \text{if} \quad f \in \overset{k}{\bigwedge} E^* \quad \text{and} \quad g \in \overset{l}{\bigwedge} E^*.$$

2. *Associativity:*

$$f \wedge (g \wedge h) = (f \wedge g) \wedge h \quad \text{if} \quad f \in \overset{k}{\bigwedge} E^*,\ g \in \overset{l}{\bigwedge} E^*,\ h \in \overset{m}{\bigwedge} E^*.$$

In particular $f \wedge f = 0$ if f is of odd order. This property is false for forms of even order: if α, β, γ and δ are linearly independent linear forms, and if $\omega = \alpha \wedge \beta + \gamma \wedge \delta$, we have

$$\omega \wedge \omega = 2\alpha \wedge \beta \wedge \gamma \wedge \delta \neq 0.$$

(See Exercises 1 and 2 for more details on these issues.)

Another important case is that of the product of k forms of degree 1. By induction on k and by using the properties above we can verify that

$$(f^1 \wedge \cdots \wedge f^k)(x_1,\ldots,x_k) = \begin{vmatrix} f^1(x_1) & \cdots & f^k(x_1) \\ \vdots & \ddots & \vdots \\ f^1(x_k) & \cdots & f^k(x_k) \end{vmatrix}$$

This allows a reinterpretation of Proposition 5.6 by writing

$$f = \sum_{1 \leqslant i_1 < \cdots < i_k \leqslant n} f(e_{i_1},\ldots,e_{i_k}) e^{i_1 *} \wedge \cdots \wedge e^{i_k^*}.$$

In particular, the determinant with respect to the basis $(e_i)_{1 \leqslant i \leqslant n}$ is the alternating form $e^{1*} \wedge \cdots \wedge e^{n*}$.

Definition 5.10. The exterior algebra of E^* *is the vector space*

$$\bigwedge E^* = \bigoplus_{k=0}^{\infty} \bigwedge^k E^* = \bigoplus_{k=0}^{n} \bigwedge^k E^*$$

equipped with the product above and extending $\wedge$ *by linearity.*

The fact that $\bigwedge^k E^* = 0$ if $k > \dim E$ is already a consequence of Proposition 5.5. However, the first time we wrote $\bigwedge E^*$ the dimension did not enter explicitly. We also note that $\dim \bigwedge E^* = 2^n$, and that the forms of even degree mutually commute.

Definition 5.11. *Let E and F be two vector spaces, and $f : E \to F$ a linear map. The* transpose *of f, denoted ${}^t f$, is the map from F^* to E^* given by*

$${}^t f(L) = L \circ f.$$

It is immediate that ${}^t f$ is linear, and that the map $f \mapsto {}^t f$ is also linear. Further, if E, F, G are three vector spaces, and if $f \in \mathcal{L}(E, F)$, $g \in \mathcal{L}(F, G)$, a direct application of the definition shows that

$${}^t(f \circ g) = {}^t g \circ {}^t f.$$

The definition extends immediately to the case where L is a multilinear map, by writing in a similar way

$$\big({}^t f(L)\big)(x_1, \ldots, x_k) = L\big(f(x_1), \ldots, f(x_k)\big).$$

Note that

$${}^t f(S \otimes T) = {}^t f(S) \otimes {}^t f(T),$$

and that ${}^t f(S)$ is *alternating* as soon as S is. We again have

$${}^t f(S \wedge T) = {}^t f(S) \wedge {}^t f(T).$$

If $f \in \mathcal{L}(E, E)$, the transpose ${}^t f$ also gives an endomorphism of $\bigwedge^k E^*$ for every k. For $k = \dim E$, this is an endomorphism between spaces of dimension 1, which is therefore of the form $\omega \mapsto c\omega$, where c is a scalar. If $(e_i)_{1 \leqslant i \leqslant \dim E}$ is a basis of E, by Proposition 5.6 this scalar is the *determinant* of the $f(e_i)$ with respect to the e_i, which therefore is independent of the basis chosen. We thus recover the determinant of an endomorphism. The formula ${}^t(f \circ g) = {}^t g \circ {}^t f$ gives a proof of the equation

$$\det(f \circ g) = (\det g)(\det f).$$

Remark. At the end of this voluntarily naive algebraic exposition,[1] it is good to point out it is not necessary when defining tensor or exterior algebra to have a vector space considered as a dual space. The remark following Definition 5.3 allows a direct definition of the tensor and exterior algebra of a vector space. In a more down to earth manner, it suffices to redo all of the above suppressing the stars: if $(e_i)_{1 \leqslant i \leqslant n}$ is a basis of an n-dimensional vector space, a basis of $T(E)$ is furnished by n^k symbols

$$e_{i_1} \otimes \cdots \otimes e_{i_k},$$

and we write

$$(e_{i_1} \otimes \cdots \otimes e_{i_k}) \otimes (e_{j_1} \otimes \cdots \otimes e_{j_l}) = e_{i_1} \otimes \cdots \otimes e_{i_k} \otimes e_{j_1} \otimes \cdots \otimes e_{j_l}.$$

A *tensor* is an element of $T(E)$.

The exterior algebra $\bigwedge E$ is defined in the same way. It is of course necessary and not difficult in both cases to prove that the algebraic structure obtained does not depend on the choice of basis used to define it. **A possible solution is to return to the point of view given at the beginning of the section: in finite dimensions every vector space is "naturally" isomorphic to the dual of its dual**. Finally note that every linear map from E to F extends to a map $f^\otimes$ from $\bigotimes E$ to $\bigotimes F$ and a map $f^\wedge$ of $\bigwedge E$ to $\bigwedge F$; we have

$$(f \circ g)^\otimes = f^\otimes \circ g^\otimes \quad \text{and} \quad (f \circ g)^\wedge = f^\wedge \circ g^\wedge.$$

5.2.3. Application: The Grassmannian of 2-Planes in 4 Dimensions

Let E be a vector space of dimension 4, and let ω be a nonzero element of $\bigwedge^4 E$. We define a bilinear form B on $\bigwedge^2 E$ by

$$u \wedge v = B(u, v)\omega.$$

We will show that B is symmetric, non-degenerate and of signature $(3, 3)$, and deduce that the set of 2-planes of E is in bijection with the quadric of $P^5 \mathbf{R}$ of solutions to

$$x_1 y_1 + x_2 y_2 + x_3 y_3 = 0$$

in homogeneous coordinates.

---

1. For example it is possible and conceptually preferable to give a definition "without denominators" of the exterior product.

Let (a, b, c, d) be a basis of E such that $a \wedge b \wedge c \wedge d = \frac{1}{2}\omega$. We introduce a basis $u_1, u_2, u_3, v_1, v_2, v_3$ of $\bigwedge^2 E$ by writing

$$\begin{aligned}
u_1 &= a \wedge b + c \wedge d & v_1 &= a \wedge b - c \wedge d \\
u_2 &= a \wedge c - b \wedge d & v_2 &= a \wedge c + b \wedge d \\
u_3 &= a \wedge d + b \wedge c & v_3 &= a \wedge d - b \wedge c.
\end{aligned}$$

Using the commutativity of the exterior product of even degree forms, and relations of the form

$$(a \wedge b) \wedge (a \wedge c) = -(a \wedge a) \wedge (b \wedge c) = 0,$$

we see that

$$B(u_i, u_j) = -B(v_i, v_j) = \delta_{ij} \quad \text{and} \quad B(u_i, v_j) = 0,$$

which proves the first assertion.

If now $P \subset E$ is a 2-plane, and (e, f) is a basis of E, introduce the 2-vector $e \wedge f$. If (e', f') is another basis, we have

$$e' \wedge f' = \big(\det(e', f')/(e, f) \big)(e \wedge f).$$

We associate to every 2-plane the element $[e \wedge f]$ of projective space associated to $\bigwedge^2 E$. Furthermore, by Exercise 2, a 2-vector w is of the form $e \wedge f$ if and only if $w \wedge w = 0$. We deduce the result by writing w in the form

$$x_1 u_1 + x_2 u_2 + x_3 u_3 + y_1 v_1 + y_2 v_2 + y_3 v_3.$$

We can show that the manifold in question is diffeomorphic to $S^2 \times S^2/\{I, \sigma\}$, where $\sigma(x, y) = (-x, -y)$ (see Exercise 6 of Chapter 2).

5.3. The Case of Open Subsets of Euclidean Space

5.3.1. Forms of Degree 1

We saw in Chapter 1 why the differential of a function $f \in C^\infty(U)$ (where U is an open subset of $\mathbf{R}^n$ or more generally a vector space E of dimension n) may be written in the form

$$df = \sum_{i=1}^{n} \partial_i f(x) \, dx^i.$$

If we are interested in the dependence as a function of $x \in U$, we obtain a smooth map from U to $E^* = \mathcal{L}(E, \mathbf{R})$. More generally

Definition 5.12. *A differential form of degree 1 on an open subset U of a vector space is a smooth map from U to E^*.*

We denote the set of these forms by $\Omega^1(U)$. A form α may be written

$$\sum_{i=1}^{n} \alpha_i \, dx^i,$$

$\alpha_i \in C^\infty(U)$ for all i, with the dx^i being coordinates with respect to a fixed basis (the natural basis if $E = \mathbf{R}^n$). We denote the value of α at x by α_x. A first justification of this point of view is the possibility of a definition of a line integral that does not appeal to coordinates.

Definition 5.13. *Let $I \subset \mathbf{R}$ be a closed interval, and let $c : I \to U$ be a C^1 parametrized curve. The integral of a differential form α along c is the number*

$$\int_c \alpha = \int_I \sum_{i=1}^{n} \alpha_i\big(c^i(t)\big) c'^i(t) \, dt = \int_I \alpha_{c(t)}\big(c'(t)\big) \, dt.$$

(For every t we apply the linear form $\alpha_{c(t)}$ to the vector $c'(t)$.)

Passing from vector fields to differential forms in Euclidean space; gradient

If E is equipped with an scalar product $\langle \, , \, \rangle$, which is to say if E is an inner product space, and if X is a vector field on E, we define the *circulation of X along c* as

$$\oint_c X = \int_I \langle X_{c(t)}, c'(t) \rangle \, dt.$$

The inner product defines an isomorphism between E and E^*, denoted $v \mapsto v^\flat$ (as it lowers coordinate indices, see Section 5.2). The inverse isomorphism is denoted $\alpha \mapsto \alpha^\sharp$. These isomorphisms are defined by the relations

$$\forall w \in E, \quad v^\flat(w) = \langle v, w \rangle \quad \text{and} \quad \langle \alpha^\sharp, w \rangle = \alpha(w).$$

We deduce an isomorphism between the vector space of vector fields on an open subset E and the differential forms of degree 1, which is denoted in the same way. Thus in an inner product space, the circulation of a vector field and the integral of a differential form of degree 1 along a curve are equivalent notions. However only the last one can be defined in the case where E is simply a vector space. It will be the only one to extend directly to manifolds, **whereas the circulation of a vector field is defined only with the help of Riemannian metric.**

We take advantage of this to define the *gradient* of a function f on Euclidean space. This is the vector field, denoted ∇f, defined by $\nabla f = df^\sharp$.

A classical question, notably in physics, it to know whether a vector field is a gradient vector field (or the derivative of a potential, in the language of the physicists). This question is in fact a vector field version of a more natural and general question for differential forms: given a differential form $\alpha \in \Omega^1(U)$, under what conditions is there a function $f \in C^\infty(U)$ such that $\alpha = df$? By virtue of a theorem of Clairaut/Schwarz on the interchanging of derivatives, a necessary condition is

$$\partial_i \alpha_j(x) = \partial_j \alpha_i(x) \quad \text{for all } x \in U \text{ and } i, j \in [1, n].$$

This condition is by no means sufficient.

Theorem 5.14. *On $\mathbf{R}^2 \smallsetminus \{0\}$, there is no function f of class C^1 such that the differential form*

$$\alpha = \frac{x \, dy - y \, dx}{x^2 + y^2}$$

equals df.

This is a consequence of the following lemma.

Lemma 5.15. *For every parametrized curve $c : [a, b] \to U$ such that $c(a) = c(b)$ and every function $f \in C^1([a, b], U)$ we have*

$$\int_c df = 0.$$

PROOF. It suffices to remark that

$$\int_c df = \int_a^b \sum_{i=1}^n \partial_i f\big(c(t)\big) c'^i(t) \, dt = f\big(c(b)\big) - f\big(c(a)\big). \qquad \square$$

PROOF OF THEOREM 5.14. It suffices to integrate α along the unit circle, parametrized by $c(t) = (\cos t, \sin t)$, where $t \in [0, 2\pi]$. We find $\int_c \alpha = 2\pi$. $\square$

We will not go further for the moment. The integration of forms of degree 1 (Definition 5.13) and the problem of writing a form $\alpha \in \Omega^1(U)$ as df generalizes to more general objects which we now define.

5.3.2. Forms of Arbitrary Degree

Definition 5.16. *A differential form of degree p on an open subset U of a vector space E is a smooth map from U to $\bigwedge^p E^*$.*

The vector space of forms of degree p on U is denoted $\Omega^p(U)$.

Choose a basis $(e_i)_{1 \leqslant i \leqslant n}$ for E. If $\alpha \in \Omega^p(U)$, for any $x \in U$ we obtain an alternating form α_x of degree p, which may be written

$$\alpha_x = \sum_{1 \leqslant i_1 < \cdots < i_p \leqslant n} \alpha_{i_1,\ldots,i_p}(x) e_{i_1}^* \wedge \cdots \wedge e_{i_k}^*.$$

Recalling that the linear form e^{i*} is the differential of the function $x \mapsto x^i$, we write

$$\alpha_x = \sum_{1 \leqslant i_1 < \cdots < i_p \leqslant n} \alpha_{i_1 \ldots i_p}(x)\, dx^{i_1} \wedge \cdots \wedge dx^{i_p}.$$

The algebraic operations defined in Section 5.2 for alternating forms extend naturally: for $\alpha \in \Omega^p(U)$ and $\beta \in \Omega^q(U)$, we define $\alpha \wedge \beta$ as the form of degree $p + q$ whose value at x is

$$(\alpha \wedge \beta)_x = \alpha_x \wedge \beta_x.$$

By convention $\bigwedge^0 E^* = \mathbf{R}$, and so the forms of degree 0 are smooth functions. The definition of the product of a form of degree p by a function fits this framework. Finally, the properties of anticommutativity and associativity persist in this setting.

In a way analogous to Definition 5.10, we write

$$\Omega(U) = \bigoplus_{k=0}^{\infty} \Omega^k(U) = \bigoplus_{k=0}^{\dim(U)} \Omega^k(U).$$

Any form belonging to a summand of this sum is said to be *homogeneous*.

Finally we note an obvious property, which we will use frequently in the sequel.

Lemma 5.17. $\Omega(U)$ *is generated as an algebra by functions and their differentials.*

Remark. We can say a little more, as $\Omega(U)$ is clearly generated by smooth functions and the dx^i. However the formulation of this lemma, which does not use coordinates, generalizes directly to manifolds.

Definition 5.18. *Let U and V be open subsets of a vector space, and let f be a smooth map from U to V. The* pullback *by f of $\alpha \in \Omega(V)$, denoted $f^*\alpha$, is the form on U defined by*

$$(f^*\alpha)_x = {}^t(T_x f) \cdot \alpha_{f(x)}.$$

In other words, if α is of degree p, by the definition of the transpose

$$(f^*\alpha)_x (v_1, \ldots, v_p) = \alpha_{f(x)}(T_x f \cdot v_1, \ldots, T_x f \cdot v_p).$$

If $\dim V = m$, and if $f^1, \ldots, f^m$ are the coordinates of f and if

$$\alpha_y = \sum_{1 \leqslant i_1 < \cdots < i_p \leqslant m} \alpha_{i_1, \ldots, i_p}(y) \, dy^{i_1} \wedge \cdots \wedge dy^{i_p},$$

we have

$$(f^*\alpha)_x = \sum_{1 \leqslant i_1 < \cdots < i_p \leqslant m} \alpha_{i_1, \ldots, i_p}(f(x)) \, df^{i_1} \wedge \cdots \wedge df^{i_p}.$$

Here we have replaced y by $f(x)$ and dy^i by df^i, as is common when applying the chain rule.

Examples

a) Take $U = \mathbf{R}$, $V = \mathbf{R}_+^*$ and $f(t) = \exp t$. Then $f^*(dx/x) = dt$.

b) If $U = V = \mathbf{R}^2$, and $f(r, \theta) = (r \cos \theta, r \sin \theta)$, we have

$$f^*(dx \wedge dy) = r dr \wedge d\theta.$$

c) More generally, if U and V are two open subsets of the same dimension n, then

$$f^*(dx^1 \wedge \cdots \wedge dx^n) = (\det T_x f)(dx^1 \wedge \cdots \wedge dx^n).$$

d) By the definition of a line integral, if $c : [a, b] \to U$ is a parametrized curve,

$$\int_c \alpha = \int_a^b c^*\alpha.$$

The algebraic properties of the transpose extend to f^*.

Proposition 5.19. *Pullback has the following algebraic properties:*
i) *If $f \in C^\infty(U, V)$ and if $\alpha, \beta \in \Omega(V)$, then*

$$f^*(\alpha \wedge \beta) = (f^*\alpha) \wedge (f^*\beta).$$

ii) *If $f \in C^\infty(U, V)$ and $g \in C^\infty(V, W)$, then*

$$(g \circ f)^* = f^* \circ g^*.$$

PROOF. i) is an immediate consequence of the algebraic properties of the transpose seen in Section 5.2. To show ii), we introduce a method that will be used systematically in this chapter, and is based on Lemma 5.17.

– The property is true for functions: if $\alpha \in C^\infty(W)$, we have

$$(g \circ f)^*\alpha = \alpha \circ (g \circ f) = (\alpha \circ g) \circ f = f^*(\alpha \circ g) = f^*(g^*\alpha).$$

– It is also true for forms of degree 1 by the chain rule. Indeed if $\alpha \in \Omega^1(W)$ and if $\beta = g^*\alpha$, we have

$$\beta_x(v) = \alpha_{g(x)}(T_x g \cdot v)$$

and

$$\begin{aligned}
\left(f^*\beta\right)_y(w) &= \beta_{f(y)}(T_y f \cdot w) \\
&= \alpha_{g(f(y))}\left(T_{f(y)} g \cdot (T_y f \cdot w)\right) \\
&= \alpha_{(g\circ f)(y)}\left(T_y(g \circ f) \cdot w\right).
\end{aligned}$$

Under these conditions, by applying i), we see that ii) is then true for every degree. $\qquad\qquad\square$

5.4. Exterior Derivative

The definition of the pullback seen in the previous section applies perfectly to *covariant tensors*, this is to say to the maps from an open subset U of $\mathbf{R}^n$ to $T(\mathbf{R}^{n*})$.

However, differential forms have the following specific property: it is possible to think of them as an extension of the differential of functions.

Theorem 5.20. *There exists a unique linear map* $d : \Omega(U) \to \Omega(U)$ *having the following properties:*

i) $\deg d\alpha = p + 1$ *if* $\deg \alpha = p$;

ii) *on* $\Omega^0(U)$, d *is the differential of functions;*

iii) *if* α *is homogeneous,* $d(\alpha \wedge \beta) = d\alpha \wedge \beta + (-1)^{\deg \alpha}\alpha \wedge d\beta$;

iv) $d \circ d = 0$.

PROOF. We first show uniqueness. We have $d(dx^i) = 0$ for all $i \in [1, n]$, and by using iii) we see by induction on k that

$$d\left(dx^{i_1} \wedge \cdots \wedge dx^{i_k}\right) = 0 \quad \text{for all } i_1, \ldots, i_k \in [1, n].$$

Using ii) and iii) again, we see that necessarily

$$d(f\, dx^{i_1} \wedge \cdots \wedge dx^{i_k}) = df \wedge dx^{i_1} \wedge \cdots \wedge dx^{i_k}.$$

Consequently, if α is a differential form of degree p written as

$$\alpha = \sum_{1 \leqslant i_1 < \cdots < i_p \leqslant n} \alpha_{i_1,\ldots,i_p}\, dx^{i_1} \wedge \cdots \wedge dx^{i_p},$$

we must have

$$d\alpha = \sum_{1 \leqslant i_1 < \cdots < i_p \leqslant n} d\alpha_{i_1,\ldots,i_p} \wedge dx^{i_1} \wedge \cdots \wedge dx^{i_p}. \tag{$*$}$$

We now verify such a formula holds. Properties i) and ii) are clear. To prove iii), it suffices, by using linearity, to check the case where

$$\alpha = f \, dx^{i_1} \wedge \cdots \wedge dx^{i_p} \quad \text{and} \quad \beta = g \, dx^{j_1} \wedge \cdots \wedge dx^{j_q}.$$

Then

$$d(\alpha \wedge \beta) = d(fg) \wedge dx^{i_1} \wedge \cdots \wedge dx^{i_p} \wedge dx^{j_1} \wedge \cdots \wedge dx^{j_q}$$

$$d\alpha \wedge \beta = g \, df \wedge dx^{i_1} \wedge \cdots \wedge dx^{i_p} \wedge dx^{j_1} \wedge \cdots \wedge dx^{j_q}$$

$$\alpha \wedge d\beta = f \, dx^{i_1} \wedge \cdots \wedge dx^{i_q} \wedge dg \wedge dx^{j_1} \wedge \cdots \wedge dx^{j_q}.$$

However, by the properties of the differential on functions, $d(fg) = g \, df + f \, dg$. The right hand side of the first equality then decomposes into two terms. The first is the right hand side of the second inequality, and using the commutation properties of the exterior product given in Section 5.2, the second is the right hand side of the third inequality, multiplied by $(-1)^p$.

It remains to show that $d \circ d = 0$. We first see by using $(*)$ that forms with constant coefficients have zero differential. Therefore using $(*)$ once more, we reduce to the case where we must show $d(df) = 0$ *for every function.* However

$$d(df) = \sum_{j=1}^{n} d(\partial_j f) \wedge dx^j$$

$$= \sum_{i,j=1}^{n} \partial_i(\partial_j f) \, dx^i \wedge dx^j$$

$$= \sum_{1 \leqslant i < j \leqslant n} \left(\partial_i(\partial_j f) - \partial_j(\partial_i f) \right) dx^i \wedge dx^j = 0$$

by symmetry of second derivatives. $\square$

Definition 5.21. *The operator d is called the* differential *or the* exterior derivative.

Example. For an open subset of $\mathbf{R}^3$, the differential of a form $\alpha = A \, dx + B \, dy + C \, dz$ of degree 1 is

$$d\alpha = (\partial_x B - \partial_y A) \, dx \wedge dy + (\partial_y C - \partial_z B) \, dy \wedge dz + (\partial_z A - \partial_x C) \, dz \wedge dx,$$

and that of a form $\beta = P\,dy \wedge dz + Q\,dz \wedge dx + R\,dx \wedge dy$ of degree 2 is

$$d\beta = (\partial_x P + \partial_y Q + \partial_z R)\,dx \wedge dy \wedge dz.$$

With the conventions that we have made, the formulas evoke the curl and divergence from vector calculus. We will return to this later. See also Exercise 4.

Proposition 5.22. *The exterior derivative and pullback commute. In other words, if U and V are two open subsets of a vector space and if $\varphi \in C^\infty(U, V)$, we have*

$$\varphi^*(d\alpha) = d(\varphi^*\alpha) \quad \forall \alpha \in \Omega(V).$$

PROOF. $\varphi^* d$ and $d\varphi^*$ are linear maps from $\Omega(V)$ to $\Omega(U)$, which increase a form's degree by 1. They are equal on functions of degree 0, the equality $d(\varphi \circ f) = \varphi^* df$ is equivalent to the chain rule for functions f and φ. They are also equal for forms of degree 1: due to linearity is suffices to check $f\,dg$, where $f, g \in C^\infty(V)$. However

$$\varphi^*\big(d(f\,dg)\big) = \varphi^*(df \wedge dg) = \varphi^* df \wedge \varphi^* dg$$
$$= d(\varphi^* f) \wedge d(\varphi^* g) = d\big((\varphi^* f) d(\varphi^* g)\big) = d\big(\varphi^*(f\,dg)\big).$$

Further, if α is of degree p, we have

$$(\varphi^* d)(\alpha \wedge \beta) = (\varphi^* d\alpha) \wedge \varphi^* \beta + (-1)^p \varphi^* \alpha \wedge (\varphi^* d\beta)$$
$$(d\varphi^*)(\alpha \wedge \beta) = (d\varphi^* \alpha) \wedge \varphi^* \beta + (-1)^p \varphi^* \alpha \wedge (d\varphi^* \beta).$$

With this we can proceed by induction on degree.   □

Example: the angular form

a) This point of view lets us give another proof that the form

$$\alpha = \frac{x\,dy - y\,dx}{x^2 + y^2}$$

is not exact on $\mathbf{R}^2 \setminus \{0\}$. Let S^1 be the circle defined by the equation $x^2 + y^2 = 1$, and let $i : S^1 \to \mathbf{R}^2$ be the canonical injection. By Proposition 5.22, if α is exact, then $i^*\alpha$ is too. However there cannot exist a function $f \in C^\infty(S^1)$ such that $i^*\alpha = df$: the differential of a smooth function on a compact manifold will vanish at some point (say where the maximum is attained), while $i^*\alpha$ never vanishes on S^1: a tangent vector to S^1 at a point $(\cos\theta, \sin\theta)$ is of the form $a(-\sin\theta, \cos\theta)$, and on this vector α equals a. At this stage we would prefer to work in polar coordinates. Writing $f(r, \theta) = (r\cos\theta, r\sin\theta)$, we find

$$f^*(x\,dy - y\,dx) = r\cos\theta(\sin\theta\,dr + r\cos\theta\,d\theta) - r\sin\theta(\cos\theta\,dr + r\cos\theta\,d\theta)$$
$$= r^2\,d\theta.$$

Thus $f^*\alpha = d\theta$. Alas f is not a diffeomorphism, but only a local diffeomorphism from $\mathbf{R}_+^* \times \mathbf{R}$ to $\mathbf{R}^2 \setminus \{0\}$. The fact that α is not exact says exactly that there is not an "angle function" defined on $\mathbf{R}^2 \setminus \{0\}$. Writing $d\theta$ is justified by the fact that there exists local determination of the angle given by local inverses of f, any two of which, when restricted to an open connected subset where they are defined, differ by a constant.

b) There exists a generalization of α in higher dimension. This is the *solid angle* form $\alpha_n \in \Omega^{n-1}(\mathbf{R}^n \setminus \{0\})$, defined by

$$\alpha_n = \frac{1}{\|x\|^n} \sum_{i=1}^n (-1)^{i-1} x^i \, dx^1 \wedge \cdots \wedge \widehat{dx^i} \wedge \cdots \wedge dx^n.$$

We can verify directly that $d\alpha_n = 0$. We sketch a more instructive proof. If $R \in SO(n)$, $R^*\alpha_n = \alpha_n$; and if h_t is the homothety with center 0 and with respect to t, $h_t^*\alpha_n = \alpha_n$ for $t > 0$. By Proposition 5.22, the form $d\alpha_n$ has the same properties. However, inspired by the techniques of Exercise 6, we can show that there does not exist a form of degree n on $\mathbf{R}^n \setminus \{0\}$ which is invariant under homotheties and by rotations with center 0. Thus α_n is closed.

We can generalize Theorem 5.14: there does not exist a form β in $\Omega^{n-2}(\mathbf{R}^n \setminus \{0\})$ such that $d\beta = \alpha_n$. The proof uses an integration procedure which generalizes line integrals, and will be the subject of the next chapter.

Another interpretation of differential forms is possible, where we deduce an expression for the exterior derivative d with using coordinates. For any form $\alpha \in \Omega^p(U)$, and p vector fields $X_1, \ldots, X_p$ on U we associate the function $\alpha(X_1, \ldots, X_p)$ defined by

$$\big(\alpha(X_1, \ldots, X_p)\big)(x) = \alpha_x\big((X_1)_x, \ldots, (X_p)_x\big).$$

In other words, for each x, we take the value of the alternating α_x on the vectors $(X_1)_x, \ldots, (X_p)_x$. If $\alpha = \sum_{i=1}^n \alpha_i \, dx^i$ and $X = \sum_{i=1}^n X^i \partial_i$, then

$$\alpha(X) = \sum_{i=1}^n \alpha_i X^i.$$

More generally, if

$$\alpha = \sum_{1 \leqslant i_1 < \cdots < i_p \leqslant n} \alpha_{i_1, \ldots, i_p} \, dx^{i_1} \wedge \cdots \wedge dx^{i_p},$$

and

$$X_k = \sum_{i=1}^n \xi_k^i \partial_i$$

we obtain

$$\big(\alpha(X_1,\ldots,X_p)\big)(x) = \sum_{1\leqslant i_1<\cdots<i_p\leqslant n} \alpha_{i_1,\ldots,i_p}(x) \begin{vmatrix} \xi_1^{i_1}(x) & \cdots & \xi_1^{i_p}(x) \\ \vdots & \ddots & \vdots \\ \xi_p^{i_1}(x) & \cdots & \xi_p^{i_p}(x) \end{vmatrix}$$

by applying Proposition 5.6. In particular, the coefficient $\alpha_{i_1,\ldots,i_p}$ is equal to $\alpha(\partial_{i_1},\ldots,\partial_{i_p})$.

Proposition 5.23. *If α is a differential form of degree p, the map*

$$(X_1,\ldots,X_p) \longmapsto \alpha(X_1,\ldots,X_p)$$

is an alternating linear p-form on $C^\infty(T(U))$ considered as a module over the ring $C^\infty(U)$. Conversely, if T is such a form, there exists a unique differential form of degree p such that

$$\big(T(X_1,\ldots,X_p)\big)(x) = \alpha_x\big(X_1(x),\ldots,X_p(x)\big)$$

PROOF. The first statement is a direct consequence of the definitions. Conversely, if $X_k = \sum_{i=1}^n \xi_k^i \partial_i$, linearity with respect to $C^\infty(U)$ gives

$$T(X_1,\ldots,X_p) = \sum_{1\leqslant i_1<\cdots<i_p\leqslant n} \begin{vmatrix} \xi_1^{i_1}(x) & \cdots & \xi_1^{i_p}(x) \\ \vdots & \ddots & \vdots \\ \xi_p^{i_1}(x) & \cdots & \xi_p^{i_p}(x) \end{vmatrix} T(\partial_{i_1},\ldots,\partial_{i_p}),$$

and the required form is

$$\alpha = \sum_{1\leqslant i_1<\cdots<i_p\leqslant n} T(\partial_{i_1},\ldots,\partial_{i_p})\, dx^{i_1} \wedge \cdots \wedge dx^{i_p}. \qquad \square$$

Thus, if $\alpha \in \Omega^p(U)$, the question arises to express $d\alpha$ as an alternating linear $p+1$-form on $C^\infty(TU)$.

Theorem 5.24. *If $\alpha \in \Omega^p(U)$, we have*

$$d\alpha(X_0,\ldots,X_p) = \sum_{i=0}^p (-1)^i X_i \cdot \alpha(X_0,\ldots,\widehat{X_i},\ldots,X_p) +$$

$$+ \sum_{0\leqslant i<j\leqslant p} (-1)^{i+j}\alpha([X_i,X_j],X_0,\ldots,\widehat{X_i},\ldots,\widehat{X_j},\ldots,X_p),$$

where the symbol $\,\widehat{}\,$ indicates that the corresponding term is omitted.

Example. For $p = 1$, we have

$$d\alpha(X,Y) = X \cdot \alpha(Y) - Y \cdot \alpha(X) - \alpha([X,Y]).$$

In particular, if $\alpha = df$, this form reduces to the definition of the bracket.

PROOF. The right hand side of the stated formula is $C^\infty(U)$-linear, by virtue of the relation

$$[fX, Y] = f[X, Y] - (Y \cdot f)X.$$

It then suffices to verify the formula for the ∂_i vector fields. But

$$d\alpha(\partial_{i_0}, \ldots, \partial_{i_p}) = \sum_{k=0}^{p} (-1)^k \partial_k(\alpha_{i_0, \ldots, \widehat{i_k}, \ldots, i_p}). \qquad \square$$

5.5. Interior Product, Lie Derivative

We can define the action of a vector field on differential forms, which is an "infinitesimal" version of pullback.

Definition 5.25. *The* Lie derivative *associated to a vector field X (for the moment defined only on an open subset of $\mathbf{R}^n$) is a linear map*

$$L_X : \quad \Omega^p(U) \longrightarrow \Omega^p(U)$$

which to a differential form α associates

$$\left(\frac{d}{dt}\right)(\varphi_t^* \alpha)_{|t=0},$$

where φ_t is the local one-parameter group associated to X.

It is easy to check that we obtain a differential form: for fixed $x \in U$, $(\varphi_t^* \alpha)_x$ is an alternating linear p-form depending differentiably on t, therefore is again alternating p-linear with respect to t. To see that this form depends in a smooth way on x, it suffices to check its coefficients. However

$$(L_X \alpha)_{i_1 \ldots i_p} = \frac{d}{dt} \left[\alpha_{\varphi_t(x)}(T_x \varphi_t \cdot \partial_{i_1}, \ldots, T_x \varphi_t \cdot \partial_{i_p}) \right]_{t=0}$$

is clearly smooth.

Example. For $X = \partial_i$, we of course have

$$L_X \alpha = \sum_{1 \leqslant i_1 < \cdots < i_p \leqslant n} \partial_i \alpha_{i_1, \ldots, i_p} dx^{i_1} \wedge \cdots \wedge dx^{i_p}.$$

In practice, to calculate the Lie derivative explicitly, it is very useful to use the following characterization.

Theorem 5.26. *The operator L_X is characterized by the following properties:*

i) *if $f \in C^\infty(U)$, $L_X f = df(X) = X \cdot f$;*

ii) *$L_X \circ d = d \circ L_X$, or in other words, L_X and d commute;*

iii) *for all differential forms α and β, we have*

$$L_X(\alpha \wedge \beta) = L_X \alpha \wedge \beta + \alpha \wedge L_X \beta.$$

(The last property indicates that L_X is a derivation of the algebra $\Omega(U)$.)

PROOF. We first check these properties are satisfied by L_X. This is clear for i). We obtain ii) by differentiating the identity

$$\varphi_t^* \circ d = d \circ \varphi_t^*,$$

with respect to t, obtained by applying Proposition 5.22 to the flow φ_t of X. Similarly, iii) is obtained by differentiating the identity

$$\varphi_t^*(\alpha \wedge \beta) = (\varphi_t^* \alpha) \wedge (\varphi_t^* \beta)$$

with respect to t.

Conversely, condition i) determines L_x on forms of degree 0, *i.e.*, functions, ii) and iii) determine L_X on forms of degree 1, and finally iii) determines L_X for all degrees. □

Corollary 5.27. *If X and Y are two vector fields, we have*

$$L_X \circ L_Y - L_Y \circ L_X = L_{[X,Y]}.$$

PROOF. The two sides of the equation coincide on $C^\infty(U)$ by the definition of the bracket, and both commute with d. Both sides satisfy property iii) of the preceding proposition: we saw in Section 3.4 that the bracket of two derivations is a derivation. □

Application: divergence of a vector field

We calculate $L_X(dx^1 \wedge \cdots \wedge dx^n)$ for a vector field X on $\mathbf{R}^n$. First

$$L_X(dx^1 \wedge \cdots \wedge dx^n) = \sum_{i=1}^{n} dx^1 \wedge \cdots \wedge dx^{i-1} \wedge L_X dx^i \wedge dx^{i+1} \wedge \cdots \wedge dx^n.$$

(Start with Theorem 5.26 iii) and think of the derivative as the product of n factors.) Then

$$L_X dx^i = d(L_X x^i) = d\left(\sum_{j=1}^{n} X^j \partial_j x^i \right) = dX^i.$$

Thanks to the properties of the exterior product, only the coefficient of dx^i in dX^i will occur in the i-th term of the sum above. Thus

$$L_X(dx^1 \wedge \cdots \wedge dx^n) = \left(\sum_{i=1}^{n} \partial_i X^i\right) dx^1 \wedge \cdots \wedge dx^n.$$

The function $\sum_{i=1}^{n} \partial_i X^i$ is called the *divergence* of the vector field X.

In view of the way we've obtained the divergence, we can see that it measures the infinitesimal change of volume along the flow of X. This explains its use in the equations of physics which express conservation laws. For example, in fluid dynamics, the conservation of mass of an incompressible fluid is expressed by the equation

$$\partial_t \rho + \operatorname{div}(\rho V) = 0,$$

where $\rho(x, y, z, t)$ is the mass density at a point (x, y, z) at time t, and V is the velocity at this point. The conservation of electric charge is expressed by a similar equation (see Section 5.8.2).

Theorem 5.26 gives another expression for L_X, which will be particularly useful for the sequel. For this, we will (again!) need a definition.

Definition 5.28. *The* interior product *of a differential form* α *of degree* $p > 0$ *by a vector field* X *is a form of degree* $p - 1$*, denoted* $i_X \alpha$*, given by*

$$(i_X \alpha)_x(v_1, \ldots, v_{p-1}) = \alpha_x(X_x, v_1, \ldots, v_{p-1}).$$

If $f \in \Omega^0(U) = C^\infty(U)$*, we write* $i_X f = 0$*.*

Example. If X is the radial vector field $\sum_{i=1}^{n} x^i \partial_i$,

$$i_X(dx^1 \wedge \cdots \wedge dx^n) = \sum_{i=1}^{n} (-1)^{i-1} x^i\, dx^1 \wedge \cdots \wedge \widehat{dx^i} \wedge \cdots \wedge dx^n.$$

Application: curl of a vector field

Let E be an oriented inner product space of dimension 3, and $\omega = dx \wedge dy \wedge dz$, where (x, y, z) are the coordinates with respect to a positively-oriented orthonormal basis.

The *curl* of a vector field X on E is defined by the equation

$$i_{\operatorname{curl} X} \omega = d(X^\flat).$$

As a result of the calculation which follows Definition 5.21,

$$\operatorname{curl}\left(A\frac{\partial}{\partial x} + B\frac{\partial}{\partial y} + C\frac{\partial}{\partial x}\right) = (\partial_y C - \partial_z B)\frac{\partial}{\partial x} + (\partial_z A - \partial_x C) + \frac{\partial}{\partial y}(\partial_x B - \partial_y A)\frac{\partial}{\partial z}.$$

The name comes from the following example. The instantaneous velocity vector field of a solid body moving in Euclidean space in a frame with respect to the solid is given by the formula

$$V_m = V_a + \Omega \wedge \overrightarrow{am},$$

where (for this occurrence only) $\wedge$ denotes the vector (cross) product. The vector Ω is called the *instantaneous rotation*. One verifies immediately that $\operatorname{curl} V = \Omega$.

We note that to define the divergence we need only to give an alternating form of maximal degree up to sign; to define the gradient, we need a positive-definite inner product; to define the curl we need a positive-definite inner product and an orientation (see Section 6.2) on a 3-dimensional space (physicists indicate this by saying the gradient is a polar vector and the curl is an axial vector).

The three operators that we have defined on differential forms are connected by the following formula, which we will use often.

Theorem 5.29 (Cartan's formula). *If X is a vector field on U, and if ω is a differential form, we have*

$$L_X\omega = d(i_X\omega) + i_X(d\omega).$$

PROOF. Write

$$P_X = d \circ i_X + i_X \circ d.$$

We will verify directly that P_X has all of the properties that characterize L_X. First,

$$P_X f = i_X df = df(X) = L_X f.$$

Next,

$$P_X \circ d = d \circ i_X \circ d = d \circ P_X$$

since $d \circ d = 0$. Finally, by the properties of the exterior product, if $\deg \alpha = p$, we have

$$i_X(\alpha \wedge \beta) = (i_X\alpha) \wedge \beta + (-1)^p \alpha \wedge i_X\beta.$$

Then

$$d\big(i_X(\alpha \wedge \beta)\big) = \big(d(i_X\alpha)\big) \wedge \beta + (-1)^{p-1}(i_X\alpha) \wedge d\beta + \\ + (-1)^p d\alpha \wedge i_X\beta + (-1)^{2p}\alpha \wedge \big(d(i_X\beta)\big),$$

while

$$i_X\big(d(\alpha \wedge \beta)\big) = \big(i_X(d\alpha)\big) \wedge \beta + (-1)^{p+1}d\alpha \wedge i_X\beta + \\ + (-1)^p(i_X\alpha) \wedge d\beta + (-1)^{2p}\alpha \wedge i_X d\beta.$$

Taking the sum, we find that

$$P_X(\alpha \wedge \beta) = (P_X\alpha) \wedge \beta + \alpha \wedge P_X\beta.$$

Therefore by Theorem 5.26, $P_X = L_X$.   □

We may also be interested in the derivative of $\varphi_t^*\omega$ for any t, where we denote the flow of X by φ_t. It is immediate that

$$\frac{d}{ds}\varphi_s^*\omega_{|s=t} = \frac{d}{ds}\varphi_{s+t}^*\omega_{|s=0} = \varphi_t^*\big(d(i_X\omega) + i_X(d\omega)\big).$$

We will also frequently have to evaluate the derivative of $\varphi_t^*\omega$ at t in the case where φ_t is a family of diffeomorphisms such that $\varphi_0 = Id$, which does not necessarily arise from a flow. In this case Cartan's formula still applies.

Theorem 5.30. *If $s \mapsto \varphi_s$ is a one-parameter family of diffeomorphisms with infinitesimal generator X such that $\varphi_0 = I$, and if X_t denotes the vector field on M obtained by fixing t, we have*

$$\frac{d}{ds}\varphi_s^*\omega_{|s=t} = \varphi_t^*\big(d(i_{X_t}\omega) + i_{X_t}(d\omega)\big).$$

PROOF. We first consider the case $t = 0$. The linear map $L_1 : \alpha \mapsto \frac{d}{ds}\varphi_s^*\alpha$ from $\Omega(M)$ to itself is a derivation. It coincides with L_{X_0} on functions, since

$$L_1 f = \frac{d}{ds}f\big(\varphi_s(x)\big)_{|s=0} = df \cdot X_0.$$

Next, as $d(\varphi_s^*\alpha) = \varphi_s^*(d\alpha)$, we see by differentiating with respect to s that L_1 and d commute (here we use a natural property of forms depending on a parameter, which will be justified in the next section). The derivations L_{X_0} and L_1, which coincide on functions and differentials of functions are equal on all of $\Omega(M)$.

To pass to the general case, we introduce the flow ψ_s of the vector field $(1, X)$ on $\mathbf{R} \times M$. By Section 3.7, this flow is of the form

$$(t, x) \longmapsto \big(t + s, F_s(t, x)\big),$$

with $\varphi_s(x) = F_s(0, x)$, and we remark that by Theorem 3.42

$$\varphi_{t+h}^*\alpha = \varphi_t^*(\varphi_h^{t*}\alpha), \quad \text{with } \varphi_h^t(x) = F_h(t, x).$$

The result is obtained by differentiating both sides of this equation with respect to h at $h = 0$, since by the first part, applied to the family $F_h(t, x)$ (t fixed), we have

$$\frac{d}{dh}\varphi_h^{t*}\alpha_{|h=0} = L_{X_t}\alpha.$$ □

5.6. Poincaré Lemma

5.6.1. Star-Shaped Open Subsets

We are now able to answer a question posed in Section 5.3: given a form $\alpha = \sum_{i=1}^{n} \alpha_i dx^i$ of degree 1 such that $\partial_i \alpha_j = \partial_j \alpha_i$, must there exist a function f such that $\alpha = df$? We first introduce a little vocabulary to pose this question in a more general setting.

Definitions 5.31. *A differential form of degree p on an open subset of $\mathbf{R}^n$ is said to be* closed *if $d\alpha = 0$, and said to be* exact *if there exists a form β of degree $p - 1$ such that $d\beta = \alpha$. We then say that β is a* primitive *of α.*

By Theorem 5.20, every exact form is closed, and Theorem 5.14 gives an example of a closed form of degree 1 that is not exact. We now show that on certain open subsets, we can obtain an explicit primitive for every closed form. We begin with a simple example.

Proposition 5.32. *If $\alpha = \sum_{i=1}^{n} \alpha_i \, dx^i$ is a closed form on $\mathbf{R}^n$, the function*

$$f(x) = \sum_{i=1}^{n} x^i \int_0^1 \alpha_i(tx) \, dt$$

has α as its differential.

PROOF. A direct calculation gives

$$df = \sum_{i=1}^{n} \left(\int_0^1 \alpha_i(tx) \, dt \right) dx^i + \sum_{i=1}^{n} \left(x^i \sum_{j=1}^{n} \left(\int_0^1 t \partial_j \alpha_i(tx) \, dt \right) dx^j \right).$$

The coefficient of dx^j in the second term is

$$\sum_{i=1}^{n} x^i \int_0^1 t \partial_j \alpha_i(tx) \, dt = \int_0^1 \sum_{i=1}^{n} t x^i \partial_i \alpha_j(tx) \, dt = \int_0^1 t \frac{d}{dt} \alpha_j(tx) \, dt,$$

where again, by integrating by parts,

$$\alpha_j(x) - \int_0^1 \alpha_j(tx) \, dt. \qquad \square$$

Examining the proof above highlights the role played by the following property.

Definition 5.33. *An open subset $U \subset \mathbf{R}^n$ is said to be* star-shaped *(with respect to a) if there exists $a \in U$ such that for all $x \in U$, the line segment $[a, x]$ is contained in U.*

It is the same to say that U is stable under positive homothety with center a and scaling less than 1.

Examples. A convex open subset is star-shaped (here for any $a \in U$ by the definition of convexity). On the other hand, $\mathbf{R}^n \smallsetminus \{0\}$ is not star-shaped. Figure 5.1 shows two open subsets of $\mathbf{R}^2$ diffeomorphic to the open ball, one that is star-shaped and one that is not.

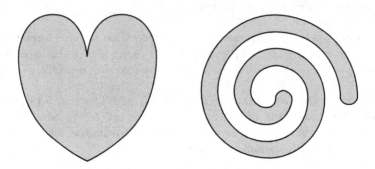

Figure 5.1: The property of being star-shaped is not invariant under diffeomorphism

It might seem curious to introduce a property that is not invariant under diffeomorphism. The answer comes from the proof of the theorem below: on a star-shaped open subset, we have an explicit representation for the primitive of any closed form. Of course, once this property is proved, we will know by Proposition 5.22 that on an open subset diffeomorphic to an star-shaped open subset (for example a ball), closed forms are exact.

Theorem 5.34 (Poincaré's lemma). *If $U \subset \mathbf{R}^n$ is a star-shaped open subset, every closed form on U is exact.*

The proof rests on the following property, which is of interest in its own right.

Lemma 5.35. *Let X be a vector field on U, whose local flow φ_t is defined on all of U for $t \in I$, where I denotes an interval. Then for every closed form α on U, the form*

$$\varphi_{t_1}^* \alpha - \varphi_{t_0}^* \alpha$$

is exact, for all $t_0, t_1 \in I$.

PROOF. We write

$$\varphi_{t_1}^* \alpha - \varphi_{t_0}^* \alpha = \int_{t_0}^{t_1} \left(\frac{d}{dt} \varphi_t^* \alpha \right) dt,$$

and remark that

$$\frac{d}{du}(\varphi_u^* \alpha)_{u=t} = \frac{d}{du}(\varphi_u^*(\varphi_t^* \alpha)])_{u=0} = L_X(\varphi_t^* \alpha).$$

Applying Cartan's formula (Theorem 5.29), and by using the fact that the pullback of a closed form is closed, we obtain

$$\frac{d}{dt}\varphi_t^* \alpha = (d \circ i_X + i_X \circ d)(\varphi_t^* \alpha) = d\big(i_X(\varphi_t^* \alpha)\big),$$

and consequently

$$\varphi_{t_1}^* \alpha - \varphi_{t_0}^* \alpha = \int_{t_0}^{t_1} \Big(d\big(i_X(\varphi_t^* \alpha)\big)\Big)\, dt = d\bigg(\int_{t_0}^{t_1}\big(i_X(\varphi_t^* \alpha)\big)\, dt\bigg). \qquad \square$$

Remark. This result persists (see Chapter 7) when replacing φ_t by a family of smooth maps depending differentiably (**and even continuously**) on t.

PROOF OF THEOREM 5.34. By applying a translation, we can suppose that U is star-shaped with respect to the origin. Considering again the calculations of the lemma for the vector field $X = x$ with flow $\varphi_t(x) = e^t x$, we obtain, for $t_0 < 0$:

$$\alpha - \varphi_{t_0}^* \alpha = d\omega,$$

where

$$\omega_x(v_1, \ldots, v_{p-1}) = \int_{t_0}^{0} \alpha_{e^t x}\big(e^t x, e^t v_1, \ldots, e^t v_{p-1}\big)\, dt.$$

It will be convenient to write $e^t = u$, and $h_\lambda(x) = \lambda x$. We then see that $\alpha - h_\lambda^* \alpha = d\beta_\lambda$, where

$$(\beta_\lambda)_x(v_1, \ldots, v_{p-1}) = \int_{\lambda}^{1} u^{p-1} \alpha_{ux}(x, v_1, \ldots, v_{p-1})\, du.$$

Allowing λ to tend to 0, we see that $\alpha = d\beta$, with

$$\beta_x(v_1, \ldots, v_{p-1}) = \int_{0}^{1} u^{p-1} \alpha_{ux}(x, v_1, \ldots, v_{p-1})\, du. \qquad \square$$

Remark. It is of course possible to verify directly that $d\beta = \alpha$ from the formula above. This is what we did for $p = 1$. The proof that we just presented explains the origin of this formula.

Example. On a star-shaped open subset with respect to 0, the primitive of $f(x)\, dx^1 \wedge \cdots \wedge dx^n$ given by this proof is

$$\beta = \sum_{i=1}^{n} (-1)^{i-1} x^i \bigg(\int_0^1 u^{n-1} f(ux)\, du\bigg) dx^1 \wedge \cdots \wedge \widehat{dx^i} \wedge \cdots \wedge dx^n.$$

5.6.2. Forms Depending on a Parameter

We have implicitly used the notion of a *differential form depending on a parameter*, which is worth further explanation.

Definition 5.36. *A one-parameter family of differential forms on an open subset $U \subset \mathbf{R}^n$ is a smooth map from $I \times U$ to $\bigwedge^p \mathbf{R}^{n*}$, where $I \subset \mathbf{R}$ is an interval.*

In other words, we have for each $t \in I$ a differential form

$$\alpha(t) = \sum_{1 \leqslant i_1 < \cdots < i_p \leqslant n} \alpha_{1 \leqslant i_1 < \cdots < i_p \leqslant n}(t, x)\, dx^{i_1} \wedge \cdots \wedge dx^{i_p},$$

with the coefficients being smooth functions on $I \times U$. We can perform (and in fact we have performed) the following operations on such a family:

a) differentiation with respect to the parameter.

We write

$$\left(\frac{d}{dt} \alpha(t) \right)_x (v_1, \ldots, v_p) = \frac{d}{dt} \big(\alpha(t)_x (v_1, \ldots, v_p) \big).$$

In coordinates, the coefficients of $\frac{d}{dt}\alpha(t)$ are obtained by differentiating those of $\alpha(t)$ with respect to t. In particular, we again obtain a one-parameter family.

b) integration with respect to the parameter.

For $a, b \in I$, we write

$$(I_a^b \alpha)_x (v_1, \ldots, v_p) = \int_a^b \alpha(t)_x (v_1, \ldots, v_p)\, dt.$$

In other words, the coefficients of this form are obtained by integrating the coefficients of $\alpha(t)$ with respect to t over $[a, b]$. By the classical results of differentiating under a summation, we see that $I_a^b \alpha \in \Omega^p(U)$.

The most important property of these operations is commuting with d, which was used in the proof Poincaré's lemma.

Lemma 5.37. *If $t \mapsto \alpha(t)$ is a one-parameter family of differential forms, the operations of differentiating and integration with respect to the parameter t commute with d. In other words*

$$d\left(\frac{d}{dt} \alpha(t) \right) = \frac{d}{dt} \big(d\alpha(t) \big)$$

$$d(I_a^b \alpha) = I_a^b (d\alpha).$$

PROOF. It suffices to prove the result for $\alpha = a(x,t)\, dx^{i_1} \wedge \cdots \wedge dx^{i_p}$. The first assertion comes from the Schwarz theorem applied to the variables t and x^i. It is more elegant to proceed as follows: knowing that α us a form on $I \times U$, we write

$$d\alpha = d^U \alpha(t) + dt \wedge \alpha'(t),$$

where d^U denotes the differential on U. Iterating we see

$$d(d\alpha) = d^U\big(d^U \alpha(t)\big) + dt \wedge \big(d^U \alpha(t)\big)' - dt \wedge d^U \alpha'(t).$$

As $d \circ d = 0$ and $d^U \circ d^U = 0$, we find

$$dt \wedge \big(d^U \alpha(t)\big)' = dt \wedge d^U \alpha'(t),$$

where $\big(d^U \alpha(t)\big)' = dt \wedge d^U \alpha'(t)$, which is what we wanted to prove. We also note that we have seen a particular case: if X is a vector field, the Lie derivative L_X commutes with d.

The second assertion comes from the theorem of differentiating under an integral, applied to the variables x^i. $\square$

5.7. Differential Forms on a Manifold

If M is a submanifold of dimension m in $\mathbf{R}^n$, and if $\alpha \in \Omega^p(\mathbf{R}^n)$, we can naturally define the restriction of α to M: for each $x \in M$, the restriction to $T_x M$ of the alternating linear p-form α_x is again an alternating linear p-form. We can define differential forms on a smooth manifold by taking inspiration from this remark: for each $x \in M$ we want an alternating linear p-form on $T_x M$. If we want to extend all of the constructions introduced for open subsets of $\mathbf{R}^n$, we must say in one way or another that α_x depends on x in a smooth fashion. We have already seen this problem for vector fields, and we proceed in the same way.

For a smooth manifold M, we denote by $\bigwedge^p T^*M$ the disjoint union

$$\coprod_{x \in M} \bigwedge^p T_x^* M.$$

Imitating what we did in Section 3.5, we equip $\bigwedge^p T^*M$ with a structure of a vector bundle over M. For each chart (U, φ), we introduce the set

$$\bigwedge^p T_x^* U = \coprod_{x \in U} \bigwedge^p T_x^* M$$

and the bijection

$$\Phi: \quad \alpha_x \longmapsto \big(\varphi(x),\, {}^t(T_x \varphi)^{-1}(\alpha_x)\big).$$

Given an atlas $(U_i, \varphi_i)_{i \in I}$ of M, we prove exactly as in Section 3.5 that there exists a unique topology on $\bigwedge^p T^*M$ such that the $\bigwedge^p T^*U_i$ are open subsets and the Φ_i are homeomorphisms. In the same way as before we see that the $(\bigwedge^p T^*U_i, \Phi_i)_{i \in I}$ form a smooth atlas.

The only modification is the calculation of $\varphi_i \circ \varphi_j^{-1}$. We are forced to take transposes and inverses of maps in the diagram

$$\mathbf{R}^n \xrightarrow{\;T_y \varphi_j^{-1}\;} T_{\varphi_j^{-1}(y)} M \xrightarrow{\;T_{\varphi_j^{-1}(y)} \varphi_i\;} \mathbf{R}^n.$$

For $(y, \omega) \in \varphi_j\big(\bigwedge^p T^*U_i \cap \bigwedge^p T^*U_j\big)$, we also find that

$$(\varphi_i \circ \varphi_j^{-1})(y, \omega) = \big(\varphi_i \circ \varphi_j^{-1}(y),\, {}^t\!\big(T_y(\varphi_i \circ \varphi_j^{-1})\big) \cdot \omega\big).$$

We again have a vector bundle, of rank $\binom{n}{p}$, equipped with local trivializations

$$\bigwedge^p T^*U_i \xrightarrow{\;\psi_i\;} U_i \times \bigwedge^p \mathbf{R}^n$$

with $\psi_i(\alpha_x) = (x, {}^t(T_x \varphi)^{-1} \cdot \alpha_x)$.

Definitions 5.38

a) *We call $\bigwedge^p T^*M$ the bundle of alternating forms on M; for $p = 1$, the bundle obtained, denoted T^*M, is called the* cotangent bundle.

b) *A differential form of degree p on a smooth manifold M is a section of π, this is to say a smooth map $\alpha : M \to \bigwedge^p T^*M$ such that for all x, $\alpha(x) \in \bigwedge^p T_x^* M$.*

As we did for vector fields, we denote the linear p-form on $T_x M$ thus obtained by α_x.

As in the case of open subsets of $\mathbf{R}^n$, the set of differential forms of degree p on M is denoted $\Omega^p(M)$, and we write

$$\Omega(M) = \bigoplus_{p=0}^{\dim M} \Omega^p(M).$$

The *exterior product* of two forms α and β of degrees p and q respectively is the form of degree $p + q$ defined by

$$(\alpha \wedge \beta)_x = \alpha_x \wedge \beta_x.$$

This is a smooth section: by using charts of the form (U, φ) and $(\bigwedge^p T^*U, \Phi)$, everything follows from the fact that transposition respects the exterior product.

The key example of differential forms are the differentials of functions. If f is smooth, we must check that the map $df : x \mapsto T_x f$ of M to T^*M is smooth. However if (U, φ) is a chart and (T^*U, Φ) is the associated chart of T^*M, we have

$$\Phi \circ df \circ \varphi^{-1}(y) = \big(y, d(f \circ \varphi^{-1})_y\big).$$

In particular, if f is one of the coordinate functions φ_i, we arrive simply at dy^i.

This remark allows us to express differential forms in the domains of charts.

Proposition 5.39. *If α is a differential form of degree p on M and if $(U, \varphi^1, \ldots, \varphi^n)$ is a chart, there exists uniquely determined smooth functions $a_{i_1 i_2 \ldots i_p}$ (with $1 \leqslant i_1 < i_2 < \cdots < i_p$) on U, such that*

$$\forall x \in U, \quad \alpha_x = \sum_{1 \leqslant i_1 < \cdots < i_p \leqslant n} a_{i_1 \ldots i_p}(x)\, d\varphi_x^{i_1} \wedge \cdots \wedge d\varphi_x^{i_p}.$$

PROOF. With the same notation $\Phi \circ \alpha \circ \varphi^{-1}(y)$ is of the form (y, ω), where $\omega \in \Omega^p\big(\varphi(U)\big)$. Thus,

$$\omega_y = \sum_{1 \leqslant i_1 < \cdots < i_p \leqslant n} \omega_{i_1 \ldots i_p}(x)\, dy^{i_1} \wedge \cdots \wedge dy^{i_p}.$$

Then $a_{i_1 i_2 \ldots i_p} = \omega_{i_1 i_2 \ldots i_p} \circ \varphi^{-1}$.   □

Remark. The functions φ^i are of no particular interest, so it may be preferable to say that on the domain of a chart U, the ring of differential forms is generated by functions and differentials of functions. The same property remains true for the whole manifold M (except that we can no longer take the differential of n functions as we could for the domain of a chart). We invite the reader to prove this, with the aid of a partition of unity (see the next chapter).

The pullback of a form under a smooth map is also defined without difficulty.

Theorem 5.40. *Let $f : M \to N$ be a smooth map, and $\alpha \in \Omega^p(N)$ a differential form. Then the family of alternating linear p-forms indexed by $x \in M$ given by*

$$\beta_x(v_1, \ldots, v_p) = \alpha_{f(x)}(T_x f \cdot v_1, \ldots, T_x f \cdot v_p)$$

is a differential form on M.

PROOF. We remark that the operation defined in the statement respects the exterior product fiber by fiber. By the preceding proposition, it suffices to check the case of the differential of a function. But if $\alpha = du$, then $\beta = d(u \circ f)$ by the chain rule.   □

Definition 5.41. *The differential form whose existence is assured by this theorem is the* pullback *of α by f, and is denoted $f^*\alpha$.*

Example. If M is a submanifold of a manifold N, for example a submanifold of Euclidean space, and if $i : M \to N$ is the canonical injection, for $\alpha \in \Omega(N)$ the form $i^*\alpha$ is called *the restriction of α to N* for obvious reasons.

In particular, the proof of the fact that $(f \circ g)^* = g^* \circ f^*$ remains valid in the case of manifolds, as it depends only on the chain rule.

Similarly, to define the exterior derivative, it suffices to see that Theorem 5.20 is valid in the manifold case. Given this importance, we reformulate the statement.

Theorem 5.42. *If M is a smooth manifold, there exists a unique smooth map $d : \Omega(M) \to \Omega(M)$ having the following properties:*

  i) $\deg d\alpha = p+1$ *if* $\deg \alpha = p$;
 ii) *on* $\Omega^0(M)$, *d is the differential of functions;*
iii) *if α is homogeneous, $d(\alpha \wedge \beta) = d\alpha \wedge \beta + (-1)^{\deg \alpha}\alpha \wedge d\beta$;*
 iv) $d \circ d = 0$.

PROOF. The arguments of the proof of Theorem 5.20 are valid when replacing Euclidean coordinates x^i with local coordinates, thanks to the lemma below. $\square$

Lemma 5.43. *If d satisfies the hypothesis of Theorem 5.42, if U is an open subset of M and if α and β are differential forms such that $\alpha_{|U} = \beta_{|U}$, then $d\alpha_{|U} = d\beta_{|U}$.*

PROOF. Let a be a point of U, V an open subset containing a such that $\overline{V} \subset U$, and let f be a function with support contained in U, equal to 1 on V. Then $f(\alpha - \beta) = 0$, and $d\big(f(\alpha - \beta)\big) = 0$ by linearity. However

$$d\big(f(\alpha - \beta)\big) = fd(\alpha - \beta) + df \wedge (\alpha - \beta).$$

At a, we have $df = 0$ by ii) and the choice of f, and $f(a) = 1$. We deduce that $(d\alpha)_a = (d\beta)_a$, and since the choice of a is arbitrary, we deduce that $d\alpha_{|U} = d\beta_{|U}$ as stated. $\square$

We may now conclude that all of the properties of the exterior derivative on forms on open subsets of $\mathbf{R}^n$ that we have seen extend to manifolds, as the proofs use only the axiomatic characterization of d obtained in Theorems 5.20 and 5.42. Thus, if $f : M \to N$ is a smooth map, we still have

$$f^* \circ d = d \circ f^*.$$

The Lie derivative and the interior product by a vector field are defined exactly as in the Euclidean case, and

$$L_X = d \circ i_X + i_X \circ d.$$

Finally we give the "manifold" version of Poincaré's lemma.

Theorem 5.44. *If M is a smooth manifold, every point $x \in M$ is contained in an open subset U having the following property: if $\alpha \in \Omega^p(U)$ is closed, there exists a form $\beta \in \Omega^{p-1}(U)$ such that $d\beta = \alpha$.*

PROOF. It suffices to choose U diffeomorphic to a ball.   □

This lemma is decidedly local. We will see in the subsequent chapters how the same problem viewed globally, *i.e.*, on all of M brings the topology of the manifold into play. We have already seen examples of this situation: we saw as a result of Theorem 5.14 that $\mathbf{R}^2 \setminus \{0\}$ is not diffeomorphic to any star-shaped open subset.

Remark. One method to study differential forms on a manifold begins with Proposition 5.23, by defining differential forms of degree p on M as the alternating linear p-forms on $C^\infty(TM)$ considered as a module over the ring $C^\infty(M)$. It is not difficult to verify, by using bump functions, that this definition is equivalent to the one we have presented. To define d, we can either take the axiomatic point of view, or *define $d\alpha$*, under the inspiration of Theorem 5.24, by the formula

$$d\alpha(X_0, \ldots, X_p) = \sum_{i=0}^{p} (-1)^i X_i \cdot \alpha\big(X_0, \ldots, \widehat{X_i}, \ldots, X_p\big) +$$

$$+ \sum_{0 \leqslant i < j \leqslant p} \alpha\big([X_i, X_j], X_0, \ldots, \widehat{X_i}, \ldots, \widehat{X_j}, \ldots, X_p\big).$$

This is a conceptually satisfying point of view (and can be especially helpful when developing a theory of differential forms in infinite dimension) but is more technically complicated. To convince yourself of this, you are invited to verify that $d \circ d = 0$ using this definition directly!

A third point a view, more down to earth, is suggested by the local expressions of vector fields seen in Section 3.5. Given an atlas (U_i, φ_i) on a manifold M, a choice of a differential form α on M is equivalent to a choice of α_i on each open subset $\varphi_i(U_i) \subset \mathbf{R}^n$, with the compatibility condition

$$\alpha_{i|\varphi_i(U_i \cap U_j)} = (\varphi_i \circ \varphi_j^{-1})^*(\alpha_{j|\varphi_j(U_i \cap U_j)}).$$

There is once inconvenience: nearly every proof will require the use of charts.

5.8. Maxwell's Equations

One of the great successes of differential forms is the simple and condensed formulation of Maxwell's equations in special relativity. We first briefly recall the "classical" (non relativistic) formulation. We account for electromagnetic phenomena with two vector fields, the electric field E and the magnetic field B. The momentum of a particle with charge q and velocity v satisfies the equation

$$\frac{dp}{dt} = q(E + v \wedge B) \quad \text{(here } \wedge \text{ denotes the vector product)}.$$

On the other hand, E and B satisfy the following system of equations

$$\operatorname{curl} E = \quad -\frac{\partial B}{\partial t} \qquad \operatorname{div} B = 0$$

$$\operatorname{curl} B = J + \frac{\partial E}{\partial t} \qquad \operatorname{div} E = \rho$$

where we have denoted the current by J and the charge density by ρ. The first and the second lines of this system of equations are called the first and second group of Maxwell's equations.

We can view the components of E and B as the six components of an alternating form of degree 2 on Minkowski space. To do this, we will give a brief description of this space. For more information, and most of all for discussion of the physical aspects, see the course by Feynman ([Feynman-Leighton-Sands 63], especially *Mainly Electromagnetism and Matter*) for example as well as the introduction to [Misner-Thorne-Wheeler 73]. This section was greatly inspired by these references.

5.8.1. Minkowski Space

Minkowski space $\mathcal{M}$ is a 4-dimensional vector space equipped with a quadratic form of signature $+---$. By analogy with the Euclidean case, we can define *pseudo-orthonormal* or *Lorentzian* bases on M, $(e_i)_{0 \leqslant i \leqslant 3}$, by the condition

$$\varphi(e_i, e_j) = 0 \text{ if } i \neq j; \quad \varphi(e_i, e_i) = -1 \text{ if } i \neq 0 ; \quad \varphi(e_0, e_0) = 1,$$

where φ is the bilinear form associated with q. If (t, x, y, z) denotes the coordinates of a vector w in with respect to such a basis, then

$$q(w) = t^2 - x^2 - y^2 - z^2.$$

Similarly,

$$\varphi(w_1, w_2) = t_1 t_2 - x_1 x_2 - y_1 y_2 - z_1 z_2.$$

Performing calculations on velocity vectors, and hence tangent vectors is done with a slight change in the point of view: by using the canonical parallelism of a vector space, we consider Minkowski space as a real vector space of dimension 4 equipped with a quadratic differential form with constant coefficients. Given a Lorentzian basis, this quadratic form can be written

$$dt^2 - dx^2 - dy^2 - dz^2.$$

The associated bilinear form defines an isomorphism between vector fields and differential forms of degree 1, which we denote by $\flat$ as in Section 5.3.1: this isomorphism is similarly defined by replacing the Euclidean inner product with φ. If (t, x, y, z) are coordinates with respect to a Lorentzian frame,

$$\flat \frac{\partial}{\partial x} = dx, \quad \flat \frac{\partial}{\partial y} = dy, \quad \flat \frac{\partial}{\partial z} = dz, \quad \text{but} \quad \flat \frac{\partial}{\partial t} = -dt.$$

5.8.2. The Electromagnetic Field as a Differential Form

Starting with the electric $E = (E_x, E_y, E_z)$ and magnetic $B = (B_x, B_y, B_z)$ fields, expressed in a given Lorentzian frame, we define a form of degree 2 by writing

$$\begin{aligned} F = &E_x \, dx \wedge dt + E_y \, dy \wedge dt + E_z \, dz \wedge dt + \\ &+ B_x \, dy \wedge dz + B_y \, dz \wedge dx + B_z \, dx \wedge dy \end{aligned} \tag{$*$}$$

(the letter F is traditional: F as in "field" or better yet, F as in Faraday).

A direct calculation shows the first group of Maxwell's equations is equivalent to the condition

$$dF = 0$$

(the vanishing of the part with dt as a factor gives the first equation, and the vanishing of the coefficient of $dx \wedge dy \wedge dz$ the second).

The second group of equations is expressed in an analogous way, given a few additional algebraic developments. On forms of degree 2 we introduce the operator $*$ by the relations

$$*(dx \wedge dt) = dy \wedge dz \qquad *(dy \wedge dz) = -dx \wedge dt$$
$$*(dy \wedge dt) = dz \wedge dx \qquad *(dz \wedge dx) = -dy \wedge dt$$
$$*(dz \wedge dt) = dx \wedge dy \qquad *(dx \wedge dy) = -dz \wedge dt$$

Of course $*$ may be defined intrinsically: there exists a "natural" extension of φ to p-forms such that,

$$\forall \alpha, \beta \in \Omega^2(M), \quad \alpha \wedge \beta = \varphi(\alpha, *\beta)\omega,$$

where ω is the volume form with equal to 1 on Lorentzian bases. We then have

$$*F = E_x \, dy \wedge dz + E_y \, dz \wedge dx + E_z \, dx \wedge dy -$$
$$- (B_x \, dx \wedge dt + B_y \, dy \wedge dt + B_z \, dz \wedge dt).$$

The second group of Maxwell's equations can then be written

$$d(*F) = 0$$

in a vacuum, and

$$d(*F) = 4\pi * \mathbf{J}$$

in the presence of electric charges. We denote by $\mathbf{J}$ the current, defined by

$$\mathbf{J} = J_x \, dx + J_y \, dy + J_z \, dz + \rho \, dt.$$

We observe that since $d^2 = 0$, the 3-form $*\mathbf{J}$ is *closed*. This property may be written

$$\operatorname{div} J + \frac{\partial \rho}{\partial t} = 0,$$

which is a statement of the conservation of electric charge.

5.8.3. Electromagnetic Field and the Lorentz Group

For a formulation independent of coordinates, it is necessary to reverse things. The Faraday tensor is a form of degree 2 on $\mathcal{M}$, subject (say in a vacuum) to the conditions

$$dF = 0 \quad \text{and} \quad d(*F) = 0,$$

which, after similar formulations do not rely on any particular coordinates.

On the other hand, electric and magnetic fields are not defined only with respect to a given Lorentzian frame $\mathcal{R}$; we can recover them by writing F under the form $(*)$ with respect to coordinates in this frame. We deduce how E and B transform in this frame. Let (t', x', y', z') be the coordinates in a new Lorentzian frame $\mathcal{R}'$, and suppose for simplification that $(y', z') = (y, z)$. By Exercise 1 of Chapter 4,

$$\begin{pmatrix} t' \\ x' \end{pmatrix} = \begin{pmatrix} \cosh u & \sinh u \\ \sinh u & \cosh u \end{pmatrix} \begin{pmatrix} t \\ x \end{pmatrix}$$

It is convenient to introduce $v = -\tanh u$: at order 1 with respect to v, we have $x' = x - vt$, and the corresponding Galilean situation is that of a uniform translation with velocity v. Comparing the expressions for F in the two frames we find

$$E'_x = E_x \qquad\qquad\qquad B'_x = B_x$$

$$E'_y = \frac{1}{\sqrt{1-v^2}}(E_y - vB_z) \qquad B'_y = \frac{1}{\sqrt{1-v^2}}(B_y + vE_z)$$

$$E'_z = \frac{1}{\sqrt{1-v^2}}(E_z + vB_y) \qquad B'_z = \frac{1}{\sqrt{1-v^2}}(B_z - vE_y).$$

Finally we see how to obtain directly the equations of motion for a charged particle from the Faraday tensor. With respect to a frame $\mathcal{R}$, the motion is given by a function

$$c : \ t \longmapsto \big(x(t), y(t), z(t), t\big).$$

We write the spatial coordinates x, y, z as a function of time constrained to this frame. The "classical" velocity in this frame must be less than the velocity of light, or in other words

$$\sqrt{x'^2(t) + y'^2(t) + z'^2(t)} < 1$$

which amounts to saying that $q\big(c'(t)\big) > 0$.

Thus the interior product of F with $c'(t)$ is given by

$$q\Big(-\big(E_x\,dx + E_y\,dy + E_z\,dz\big) - \big(v_yB_z - v_zB_y\big)\,dx -$$

$$- \big(v_zB_x - v_xB_z\big)\,dy - \big(v_xB_y - v_yB_x\big)\,dz + \big(E_xv_x + E_yv_y + E_zv_z\big)\,dt \Big)$$

Returning to the corresponding vector, we obtain

$$-\big(i_{c'(t)}F\big)^\sharp = q(E + v \wedge B) + \langle E, v\rangle \frac{\partial}{\partial t},$$

where the scalar and vector products were computed in the (Euclidean!) frame formed by $\frac{\partial}{\partial x}, \frac{\partial}{\partial y}, \frac{\partial}{\partial z}$. We recover not only the derivative of the momentum, but also the derivative of energy.

This formula is still not satisfying, as it depends on a specific frame (a physicist would say an observer). With out appealing to a frame, the motion of a particle is given by a curve in Minkowski space, every parametrization $u \mapsto \gamma(u)$ of this curve satisfies the condition $q\big(\gamma'(u)\big) > 0$. In a Lorentzian frame $\frac{\partial}{\partial x}, \frac{\partial}{\partial y}, \frac{\partial}{\partial z}, \frac{\partial}{\partial t}$ this condition may be written

$$t'(u)^2 - x'(u)^2 - y'(u)^2 - z'(u)^2 > 0.$$

This allows us to choose a parameter such that $t'(u) = 1$. This is the time with respect to the frame, up to an additive constant.

However there exists a parametrization which depends only on the curve, and not the frame, obtained by the condition

$$q\big(\gamma'(\tau)\big) = 1.$$

This parameter, which brings to mind the arc-length in the Euclidean case, is called the *proper time* of the particle. With respect to the given time with respect to the frame $\mathcal{R}$ preceding, we have

$$d\tau = \sqrt{1 - v^2}\, dt = \sqrt{1 - x'(t)^2 - y'(t)^2 - z'(t)^2}$$

The vector $\gamma'(\tau)$ is called the four-vector velocity, and $m\gamma'(\tau)$, analogous to the classical moment, the energy-momentum. The equation of motion is then written

$$-\big(i_{\gamma'(\tau)}F\big)^{\sharp} = \frac{d}{d\tau}\big(m\gamma'(\tau)\big)$$

This is of course only the start of a long story, for which we refer the reader to the books that we have already mentioned.

5.9. Comments

Differential forms and tensors

Most of the constructions of this chapter apply to *covariant tensors*. Like Definition 5.16, a covariant tensor of order p on an open subset U of a vector space E is a smooth map from U to E^*. Similarly, the construction at the beginning of this paragraph allows us to define for every manifold M the vector bundle $\bigotimes^p T^*M$ and covariant tensors of order p are sections of this bundle.

Proposition 5.23 and its proof remain valid.

In the same way we define the pullback ϕ^* by a smooth map ϕ, and the Lie derivative with respect to a vector field. Once more we have that

$$(\phi \circ \psi)^* = \psi^* \circ \phi^*.$$

Similarly, for two tensors S and T on M,

$$\phi^*(S \otimes T) = \phi^*S \otimes \phi^*T$$
$$L_X(S \otimes T) = L_X S \otimes T + L_X T \otimes S.$$

On the other hand, we can show that the differential d has no analogue (even if we only require having a linear operator on tensors that commutes with ϕ^*). This explains the overarching role of differential forms and the operator d when we study manifolds without any structure other than the differential structure.

One finds in the celebrated physics course by R. Feynmann [Feynman-Leighton-Sands 63, volume II, Chapter 31] a clear description of certain symmetric tensors of order 2 and 4 that arise in the physics of solids.

Riemannian metrics

This being said, symmetric 2-tensors, this is to say sections of the bundle $S^2 T^* M$ of symmetric bilinear forms on M, play an important role. A *Riemannian metric* on a manifold M is a symmetric tensor g whose associated quadratic form g_x is positive definite. The first example is of course Euclidean space. We then have

$$g = \sum_{i=1}^{n} dx^i \otimes dx^i,$$

with x^i being the coordinates an in orthonormal basis. Every submanifold of Euclidean space is itself equipped with a Riemannian metric: to see this it suffices to check that the restriction of a quadratic form to a vector subspace remains positive definite. More formally, if $\phi : M \to M'$ is an embedding, and if g is a Riemannian metric on M', $\phi^* g$ is a Riemannian metric on M. Applying Whitney's embedding theorem, we then see that every manifold can be equipped with a Riemannian metric.

The word metric is justified by the fact that g automatically defines a distance: we begin by defining the length of a curve $c : [a, b] \to M$ by the integral

$$\int_a^b \sqrt{g\big(c'(t), c'(t)\big)}\, dt,$$

and then the distance between two points as the infimum of the lengths of curves which join them. Warning: even in the case of submanifolds of Euclidean space, this distance *is not* in general the distance induced by the ambient space.

In the same way that smooth manifolds are natural generalizations of curves and surfaces in $\mathbf{R}^3$, we can say that Riemannian manifolds, *i.e.*, manifolds equipped with a Riemannian metric, generalize curves and surfaces *in Euclidean space*. The fundamental difference is the following. While every manifold is locally diffeomorphic to $\mathbf{R}^n$, a Riemannian manifold is not in general locally isometric to Euclidean space. Riemannian geometry studies the relationship between the metric properties of a Riemannian manifold and the topological properties of the underlying smooth manifold.

To learn more, see for example [Milnor 63, Chapter III], [Gallot-Hulin-Lafontaine 05] and [Berger 03]. One will also find in Chapter 8 the rudiments of Riemannian geometry in dimension 2.

Further, supposing that M is oriented (which is not necessary but simplifies the exposition), we can associate to g a canonical volume form ω_g, characterized by the fact that it equals 1 on every oriented orthonormal frame (this will be made precise next chapter for submanifolds of Euclidean space, equipped with the induced Riemannian metric). We can then define as in Section 5.5 a divergence operator $\mathrm{div}_g : C^\infty(TM) \to C^\infty(M)$ by means of the formula

$$L_X \omega_g = (\mathrm{div}_g X)\omega_g.$$

The second order differential operator $\Delta_g = \mathrm{div}_g(\nabla_g)$ is a natural generalization of the Laplacian $\Delta = \sum_{i=1}^n \partial_i^2$. The study of the Laplacian is another slightly more analytical aspect of Riemannian geometry (cf. the annotated bibliography at the end of the book).

Lorentzian manifolds

A Lorentzian manifold is a manifold equipped with a symmetric tensor g such that the quadratic form g_x is of signature $(\dim M - 1, 1)$ at each point. The easiest example is Minkowski space introduced in Section 5.8. The importance of these metrics comes from the fact they model spacetimes in general relativity.

There are importance differences with the Riemannian setting: a given manifold need not have a Lorentzian structure; the restriction of the tensor g to a submanifold could be either Riemannian, Lorentzian or singular.

Conversely, Riemannian, Lorentzian or more general pseudo-Riemannian manifolds (which is to say the associated quadratic form g_x is non-degenerate for all x) have three important properties in common.

1. The tensor g defines a measure (more precisely a density, see Section 6.7) on the manifold.

2. The metric defines an fiber isomorphism between TM and T^*M, which allows one to identify differential forms and vector fields. We have already used this isomorphism in the Euclidean case to define the gradient of a function. The gradient is defined in the more general pseudo-Riemannian setting.

3. The tensor g allows one to define a *connection* on TM, which is to say a directional derivative on vector fields. Geometrically this gives us the ability to compare tangent vectors at different points on the manifold.

For more details on pseudo-Riemannian geometry, see [Doubrovine-Novikov-Fomenko 90].

Symplectic Manifolds

A *symplectic manifold* is a manifold equipped with a differential form ω of degree 2, with maximum rank at each point (which already implies that the dimension is even) and which is closed. The standard example is $\mathbf{R}^{2n}$ equipped with the form

$$\omega = \sum_{i=1}^{n} dx^i \wedge dx^{i+n}.$$

To every function H on a symplectic manifold (M, ω) we can associate a *Hamiltonian*. This is the vector field X_H defined by

$$i_{X_H}\omega = dH.$$

The existence and uniqueness of X_H comes from the hypothesis on the rank of ω. Further more, as $d\omega = 0$, the Cartan formula gives

$$L_{X_H}\omega = d(i_{X_H}\omega) + i_{X_H}d\omega = ddH = 0.$$

As a consequence, X_H has remarkable dynamical properties (Poincaré recurrence theorem, cf. [Arnold 78, Chapter 3]).

On the other hand, we leave to the reader the (easy) task of showing that on the round sphere S^2 embedded in $\mathbf{R}^3$, the (Riemannian!) gradient of a coordinate function is not the Hamiltonian for *any* symplectic form on S^2 (use Exercise 7 of Chapter 3).

The interest in symplectic manifolds was first raised in mechanics: many problems from classical mechanics, for example the Kepler problem, the three body problem and more generally frictionless systems can be reduced to the study of the trajectories of a Hamiltonian vector field (see [Arnold 78] once more).

Warning. Darboux's theorem (see Exercises 14 and 17) ensures that every symplectic form may be *locally* written in the form $\sum_{i=1}^{n} dx^i \wedge dx^{i+n}$ in a certain coordinate system, but this is not a great help for the global study of symplectic manifolds and Hamiltonian vector fields.

5.10. Exercises

1. *Decomposable and indecomposable alternating forms*

An alternating p-linear form on a vector space E of dimension n is called *decomposable* if it may be written as the exterior product of p linear forms, and *indecomposable* otherwise.

a) Show that every form of degree n or $n-1$ is decomposable.

   Hint. If $\omega \in \bigwedge^{n-1} E^*$, introduce the map

$$\theta \longmapsto \theta \wedge \omega$$

   from E^* to $\bigwedge^n E^*$.

b) Let $\theta \in E^*$. Show that a p-form α may be written in the form

$$\alpha = \theta \wedge \theta'$$

   if and only if, $\theta \wedge \alpha = 0$ (we the say that α is *divisible by* θ).

c) Show that if $\alpha, \beta, \gamma, \delta$ are independent linear forms, the 2-form

$$\omega = \alpha \wedge \beta + \gamma \wedge \delta$$

   is indecomposable.

2. *Forms of degree 2; symplectic group*

Let E be a real vector space of dimension n, equipped with an alternating bilinear form ω.

a) Show that there exists a basis $\mathcal{B} = (\theta^1, \theta^2, \ldots, \theta^n)$ of E^* and an integer $p \leqslant \frac{n}{2}$ such that

$$\omega = \theta^1 \wedge \theta^{p+1} + \cdots + \theta^p \wedge \theta^{2p}$$

   and the integer p depends only on ω (we can show that if ω nonvanishing, there exists two vectors a and b and a subspace E' of E such that $E = E' \oplus \mathbb{R}a \oplus \mathbb{R}b$, with $\omega(a,b) = 1$ and $\omega(a,x) = \omega(b,x) = 0$ for all $x \in E'$).

b) Show that the codimension of the vector space

$$F = \{x \in E : \ \forall y \in E, \ \omega(x,y) = 0\}$$

   is equal to $2p$. For this reason, the integer $2p$ is called the *rank* of ω.

c) Show that p is the smallest integer such that $\omega^p \neq 0$. In particular a 2-form ω is decomposable if and only if $\omega \wedge \omega = 0$.

d) Let u be an automorphism of E. Show that u leaves ω invariant, this is to say that

$$\omega\big(u(x), u(y)\big) = \omega(x,y) \quad \forall x, y \in E$$

   if and only if the matrix A of u in the dual basis of $\mathcal{B}$ satisfies $A^t S A = S$, where S is the matrix

$$S = \begin{pmatrix} 0 & I_p & 0 \\ -I_p & 0 & 0 \\ 0 & 0 & 0 \end{pmatrix}.$$

Deduce a characterization of the elements of the *symplectic group* $Sp(n, \mathbf{R})$, defined as the group of automorphisms of $\mathbf{R}^{2n}$ that leave the form ω given by

$$\omega(x, y) = \sum_{1 \leqslant i \leqslant n} \left(x^i y^{n+i} - x^{n+i} y^i \right)$$

invariant.

e) Show that $Sp(n, \mathbf{R})$ is a $2n^2 + n$-dimensional Lie subgroup of $GL(2n, \mathbf{R})$, and determine its Lie algebra.

3*. *An application of exterior algebra to Lie groups*
Set $E = \bigwedge^2 (\mathbf{R}^{4*})$. Given a nonvanishing 4-form ω on $\mathbf{R}^4$, we consider the bilinear form B on E defined in Section 5.2.3

a) Show that every element of $GL(4, \mathbf{R})$ leaving the form ω invariant defines an automorphism of E respecting B.

b) Deduce the existence of a Lie group morphism

$$\rho : \ SL(4, \mathbf{R}) \longrightarrow O(3, 3)$$

with image the identity component of $O(3, 3)$ and with kernel $\mathbf{Z}/2\mathbf{Z}$.

c) Similarly deduce, starting with $\mathbf{C}^4$ instead of $\mathbf{R}^4$, a morphism of **complex Lie groups** from $Sl(4, \mathbf{C})$ to $SO(6, \mathbf{C})$ (we denote $O(n, C)$ the subgroup of $Gl(n, \mathbf{C})$ which leaves the quadratic form $\sum_{i=1}^{n} z_i^2$ invariant. This is a *noncompact* complex Lie group; $SO(n, \mathbf{C})$ denotes the subgroup of endomorphisms of determinant 1).

d) Equip $\mathbf{C}^4$ with a hermitian product and show that

$$SU(4)/\{\pm I\} \simeq SO(6).$$

4. *Gradient, divergence, curl*
a) Prove the following identities using the properties of exterior algebra:

$$\mathrm{curl}(\mathrm{grad}\, f) = 0$$
$$\mathrm{div}(\mathrm{curl}\, V) = 0$$
$$\mathrm{div}(fV) = f \,\mathrm{div}\, V + (\mathrm{grad}\, f) \cdot V$$
$$\mathrm{curl}(fV) = f \,\mathrm{curl}\, V + (\mathrm{grad}\, f) \wedge V$$
$$\mathrm{div}(u \wedge v) = v \cdot (\mathrm{curl}\, u) - u \cdot (\mathrm{curl}\, v).$$

Here $\wedge$ denotes the vector (cross) product.

b) Show that the local flow of a vector field V on $\mathbf{R}^n$ leaves $\omega = dx^1 \wedge \cdots \wedge dx^n$ invariant if and only if div $V = 0$.

5. *Homogeneous forms*

A differential form ω on $\mathbf{R}^n$ is said to be *homogeneous of degree* α if

$$h_t^*(\omega) = t^\alpha \omega,$$

where we have denoted the homothety with respect to t $(t > 0)$ by h_t.

a) Show that if ω is of degree k, then this amounts to saying that the coefficients are homogeneous of degree $n - \alpha$.

b) Show that the differential of a homogeneous form is homogeneous of the same degree.

6*. *Forms invariant under a group*

a) Let ω be the differential form on $\mathbf{R}^{n+1}$ equal to

$$\sum_{i=0}^n (-1)^i x^i \, dx^0 \wedge \cdots \wedge \widehat{dx^i} \wedge \cdots \wedge dx^n,$$

(the notation $\widehat{}$ indicates that the corresponding terms is omitted). Show that if $\phi \in Sl(n + 1, \mathbf{R})$, then ω is invariant under ϕ, and is (up to a constant factor) the only form of degree n having this property.

b) Let ω be a differential form of degree k on S^n such that $\phi^*\omega = \omega$ for any $\phi \in SO(n + 1)$. Show that if $0 < k < n$, then $\omega = 0$.

7. Calculate the primitive of a given closed form of degree 2 on $\mathbf{R}^n$ *explicitly*, using the formula which is provided at the end of the proof of Poincaré's lemma.

8. *Forms invariant under a Lie group*

a) If G is a Lie group of dimension n, show that the forms of degree p invariant under left translation (we call these left invariant forms) form a vector space of dimension $\binom{n}{p}$.

b) Show, without using the structure theorem of commutative Lie groups, that if G is commutative, every left invariant form is closed.

c) Take $G = Gl(n, \mathbf{R})$. Show that the coefficients of the matrix $X^{-1}dX$, where $X = (x_i^j)_{1 \leqslant i, \, j \leqslant n}$, are left invariant forms. Re-derive the expression of the bracket of $\mathfrak{G}$ by differentiating $X^{-1}dX$ and using Theorem 5.24.

d) Consider the group G of matrices of the form

$$\begin{pmatrix} a & b \\ 0 & 1 \end{pmatrix}, \quad a, b \in \mathbf{R}, \, a \neq 0.$$

Show that this is a Lie group. Determine the differential forms of degree 1 and 2 on G that are left invariant. Repeat this for right invariant forms. What can you conclude?

9. *Interior product and Lie derivative*

Prove the following identity concerning the operators i_X and L_X acting on differential forms on a manifold M:

$$i_{[X,Y]} = L_X \circ i_Y - i_Y \circ L_X.$$

10. *Complex forms*

A complex differential form of degree k on an open subset $U \subset \mathbf{R}^n$ is a smooth map from U to $\wedge \mathbf{R}^n \otimes \mathbf{C}$, where we have denoted the $\mathbf{C}$-vector space of alternating p-linear maps from $\mathbf{R}^n$ to $\mathbf{C}$ by $\wedge \mathbf{R}^n \otimes \mathbf{C}$. We write

$$dz^k = dx^k + i\, dy^k, \quad d\bar{z}^k = dx^k - i\, dy^k.$$

If df is the differential of a smooth complex-valued function, we define $\frac{\partial f}{\partial z^k}$ and $\frac{\partial f}{\partial \bar{z}^k}$ by the formula

$$df = \sum_{k=1}^{n} \left(\frac{\partial f}{\partial z^k} dz^k + \frac{\partial f}{\partial \bar{z}^k} d\bar{z}^k \right).$$

a) Express $\frac{\partial f}{\partial z^k}$ and $\frac{\partial f}{\partial \bar{z}^k}$ as a function of $\frac{\partial f}{\partial x^k}$ and $\frac{\partial f}{\partial y^k}$.

b) Show that a map (say C^1) from $\mathbf{C} = \mathbf{R}^2$ to $\mathbf{C}$ is holomorphic if and only if $\frac{\partial f}{\partial \bar{z}} = 0$. Deduce that the Jacobian of f seen as a map from $\mathbf{R}^2$ to $\mathbf{R}^2$ is equal to $|f'(z)|^2$ (note that $dz \wedge d\bar{z} = -2i\, dx \wedge dy$).

c) Let P be the half-plane consisting of complex numbers with strictly positive imaginary part. Show that maps T of the form

$$T(z) = \frac{az + b}{cz + d}, \quad \text{where } a, b, c, d \in \mathbf{R}, \ ad - bc > 0$$

are diffeomorphisms of P which leave the differential form $\omega = \frac{dx \wedge dy}{y^2}$ invariant.

11. Let ω be a differential form of degree p on $\mathbf{R}^n$, with $p < n$. Suppose that the restriction of this form to every hyperplane of the form $x^k = \text{constant}$ $(1 \leqslant k \leqslant n)$ vanishes. Show that $\omega = 0$.

12. *Forms invariant under rotation*

a) Show that the differential forms

$$\alpha = x\, dx + y\, dy \quad \text{and} \quad \beta = x\, dy - y\, dx$$

on $\mathbf{R}^2$ are invariant under the group of rotations $SO(2)$. (We can of course verify this by a direct calculation. But there are more elegant proofs.)

b) Show that every differential form of degree 1 on $\mathbf{R}^2 \smallsetminus \{0\}$ invariant under $SO(2)$ is of the form

$$f(r)\alpha + g(r)\beta,$$

where $r = \sqrt{x^2 + y^2}$, and with f and g smooth functions on $\mathbf{R}^2 \smallsetminus \{0\}$.

Which forms of degree 1 on $\mathbf{R}^2$ are invariant under $SO(2)$?

c) We now work in dimension 3. Write

$$r = \sqrt{x^2 + y^2 + z^2}$$

and denote by X the vector field

$$x\frac{\partial}{\partial x} + y\frac{\partial}{\partial y} + z\frac{\partial}{\partial z}.$$

We will now show that every form of degree 1 on $\mathbf{R}^3 \smallsetminus \{0\}$ invariant under the action of $SO(3)$ is of the form $f(r)\, dr$.

c1) Show that if α is $SO(3)$-invariant, $i_X \alpha$ is also.

c2) Show that the map

$$(t, u) \longmapsto tu \quad \text{from } \mathbf{R}^{+*} \times S^2 \text{ to } \mathbf{R}^3 \smallsetminus \{0\}$$

is a diffeomorphism.

c3) Let $\alpha \in \Omega(\mathbf{R}^3 \smallsetminus \{0\})$ be such that $i_X \alpha = 0$, and where the restriction to each sphere with center 0 vanishes. Show that $\alpha = 0$.

c4) Show that every $SO(3)$-invariant 1-form on S^2 vanishes.

c5) Deduce the conclusion.

d) Does the argument above generalize to higher dimensions?

13. Consider the sphere S^2 embedded in $\mathbf{R}^3$. Let X be the vector field

$$x\frac{\partial}{\partial x} + y\frac{\partial}{\partial y} + z\frac{\partial}{\partial z}.$$

a) Show that the restriction of the 2-form $i_X(dx \wedge dy \wedge dz)$ to the sphere never vanishes (in other words, it is a *volume form* on S^2; see the next chapter for more details). We denote this form α.

b) Show that on the complement of the equator $z = 0$,

$$\alpha = \frac{dx \wedge dy}{z}.$$

c) Let I be the stereographic projection from the north pole $N = (0,0,1)$ from $S^2 \setminus \{N\}$ to the plane $z = 0$.

Write out I and I^{-1} (for clarity, denote the coordinates of $\mathbf{R}^2 \times \{0\}$ by (u,v), and calculate $I^{-1*}\alpha$.

14*. *Darboux's theorem*

Let M be a manifold of dimension $2n$ equipped with a closed 2-form ω with rank $2n$ at each point (in other words a symplectic form). We will show that every point $a \in M$ admits a coordinate neighborhood U with coordinates $\left(x^1, \ldots, x^{2n}\right)$ such that

$$\omega_{|U} = dx^1 \wedge dx^{n+1} + \cdots + dx^n \wedge dx^{2n} \quad \text{(Darboux's symplectic theorem)}.$$

In what follows we suppose that $a = 0$ in the coordinates considered. This is not a restriction but merely simplifies the argument.

If X_1 and X_2 are two commuting vector fields, we recall (see Exercise 17 of Chapter 3) that if at a point $a \in M$, X_1 and X_2 are linearly independent, a admits an open neighborhood U with local coordinates $(x^1, \ldots, x^n)$ such that $X_{i|U} = \frac{\partial}{\partial x^i}$ for $i = 1, 2$.

a) First suppose $n = 1$. Show that if $\left(y^1, y^2\right)$ is an arbitrary system of coordinates at a, we can find a differentiable function defined in a neighborhood of a such that $\left(y^1, h\right)$ is a system of coordinates that works in a neighborhood of a.

b) We now study the general case. If f is a differential function such that $f(a) = 0$ and $df_a \neq 0$, show that there exists a unique vector field X_2 on M such that $i_{X_2}\omega = df$ and a differentiable function g such that $X_2 g = 1$ on an neighborhood of a. Show that if X_1 is the vector field defined by $i_{X_1}\omega = dg$, we have the relations

$$\omega\left(X_1, X_2\right) = 1 \quad \text{and} \quad [X_1, X_2] = 0.$$

We then consider the local coordinates $\left(y^1, \ldots, y^{2n}\right)$ on a neighborhood V of a such that $X_1 = \frac{\partial}{\partial y^1}$ at every point of V. Show that

$$\left(-f, g, y^3, \ldots, y^{2n}\right) = \left(x^1, \ldots, x^{2n}\right)$$

is a coordinate system on an open subset U' containing a such that

$$\omega_{|U'} = dx^1 \wedge dx^2 + \sum_{2 < i < j \leqslant 2n} \omega_{ij}\left(x^3, \ldots, x^{2n}\right) dx^i \wedge dx^j.$$

Deduce the result.

15*. *Darboux's theorem for contact forms*

Let M be a manifold of dimension $2n + 1$ equipped with a 1-form θ such that $\theta \wedge (d\theta)^n \neq 0$ at every point (this is called a contact form). We will show that at every point a admits a neighborhood U with local coordinates $(x^0, \ldots, x^{2n})$ such that $\theta_{|U} = dx^0 + x^1 dx^{n+1} + \cdots + x^n dx^{2n}$.

a) Show there exists a unique vector field R on M such that $i_R\theta = 1$ and $i_R d\theta = 0$ (we can consider the map from TM to T^*M defined by $X \mapsto i_X\theta \cdot \theta + i_X d\theta$).

b) If $(y^0, \ldots, y^{2n})$ is a local coordinate system at a such that $R = \frac{\partial}{\partial y^0}$, show that $d\theta$ defines a symplectic form on the submanifold defined by the equation $y^0 = 0$. Deduce the result by using the preceding exercise and modifying the coordinate y^0 as necessary.

16*. *Differential forms on $P^n\mathbf{R}$*

Let $p : \mathbf{R}^{n+1} \smallsetminus \{0\} \to P^n\mathbf{R}$ be the canonical projection.

a) Show that
$$p^* : \Omega(P^n\mathbf{R}) \longrightarrow \Omega(\mathbf{R}^{n+1} \smallsetminus \{0\})$$
is injective.

b) Show that the image of p^* is the set of forms $\omega \in \Omega(\mathbf{R}^{n+1} \smallsetminus \{0\})$ such that
$$h^*_\lambda \omega = \omega$$
$$i_X \omega = 0 \quad (X \text{ is the radial vector field}).$$

17*. *Darboux's theorem by Moser's trick*

Darboux's theorem is purely local, and so we may reduce to the case of a symplectic form on an open subset U of $\mathbf{R}^n$.

a) Let $a \in U$. Show that by a linear change of coordinates, we can reduce to the case where
$$\omega_a = \left(dx^1 \wedge dx^{n+1} + \cdots + dx^n \wedge dx^{2n}\right)_a$$

b) Write
$$\tilde{\omega} = dx^1 \wedge dx^{n+1} + \cdots + dx^n \wedge dx^{2n} \quad \text{and} \quad \omega_t = t\omega + (1-t)\tilde{\omega} \quad (0 \leqslant t \leqslant 1).$$
Show there exists an $r > 0$ such that the restriction of the ω_t to the open ball $B(a, r)$ is non-degenerate, and consequently symplectic.

c) Imitating the technique of Theorem 3.44, show the existence of a family $(\phi_t)_{0 \leqslant t \leqslant 1}$ of diffeomorphisms of $B(a, r)$ such that
$$\forall t \in [0, 1], \quad \phi_t^* \omega_t = \tilde{\omega},$$
and deduce the result.

18*. *Volume form*

a) Consider $\mathbf{R}^n$ with the "standard" volume form $\omega_0 = dx^1 \wedge \cdots \wedge dx^n$. Let r and R be two positive real numbers, with $R > r$. Show that for every $a, b \in B(0, r)$ there exists a diffeomorphism f of $\mathbf{R}^n$ such that $f(a) = b$, $f(x) = x$ if $\|x\| \geqslant R$ and $f^*\omega_0 = \omega_0$.

b) Let M be a manifold equipped with a volume form ω. Show that for all $a \in M$ there exists an open subset U containing a and a diffeomorphism of g from U to an open subset of $\mathbf{R}^n$ such that $\omega = g^*\omega_0$.

c) Conclude from a) and b) that the group of diffeomorphisms of M which preserves ω acts transitively on M.

Chapter 6

Integration and Applications

6.1. Introduction

If f is a function of a real variable with continuous derivative,

$$\int_a^b f'(t)\,dt = f(b) - f(a).$$

This is the so-called "fundamental theorem of calculus". The Russians pay homage to the two founders of differential and integral calculus by calling this the "Leibnitz-Newton formula". Stokes's theorem, which is the pivotal result of this chapter, is a generalization of this formula to higher dimensions. It gives a good example of a situation where elaboration of the statement is harder than the proof itself.

The starting point of the fundamental theorem is the convention

$$\int_a^b g(t)\,dt = -\int_b^a g(t)\,dt \quad \text{if } a \geqslant b.$$

This comes from integration on an oriented interval instead of merely an interval, which is to say an interval where one chooses a starting and ending point. If $I = [a, b]$, there are two possible orientations, depending on whether we take a or b as the origin.

It is this step – moving to an integral seen as a measure of an "oriented" object – which will be generalized. Alas, we're in a tricky situation. On $\mathbf{R}$ there is one natural orientation given by the usual order. This is no longer the case for curves in $\mathbf{R}^2$, even if they are straight lines. This is also no longer the case in vector spaces (we are only referring here to the "naive" notion of orientation, the precise details will be given in Section 6.2). There are (apparently obligatory) references to everyday life when orienting the circle (the "counterclockwise" sense) and three dimensional space (the "right hand" rule). This issue is discussed in detail in [Feynman-Leighton-Sands 63, volume I, Chapter 52]. All that we can really do is compare orientations. This is illustrated by the existence of manifolds for which a choice of coherent orientation is impossible. This impossibility can be seen "experimentally". Indeed anticipating what we will discuss formally in Section 6.2, on every oriented surface of oriented Euclidean three space, we can define a normal vector field in the following way: if p is a point on the surface and R_p is a positively oriented orthonormal frame of the tangent plane at p, n_p is the unique unit vector at p such that the frame (R_p, n_p) is positively oriented in $\mathbf{R}^3$. This is impossible for the Möbius strip, as seen in Figure 6.1.

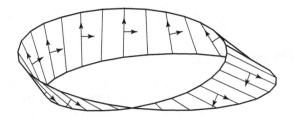

Figure 6.1: Möbius strip

We start with a discussion of these orientation problems, before starting our study of integration theory.

An important role is played by *regular domains* of a manifold. These are more or less compact subsets whose boundary is a submanifold of codimension 1 (possibly not connected, as in the case of a compact interval of $\mathbf{R}$). The fundamental theorem generalizes by replacing $[a, b]$ by a regular domain D of dimension n, $\{a, b\}$ by the boundary ∂D, and f by a differential form α of degree $n - 1$. We have

$$\int_D d\alpha = \int_{\partial D} \alpha,$$

with the bulk of the work being to define these objects. This generalization extends Green's theorem, the Divergence theorem and the classical Stokes's theorem that concerns surface with boundary in three dimensional space, where it seems we live.

Integration on manifolds and Stokes's theorem do more than just provide a more conceptual framework for the classical theorems we have mentioned. As an example, we end this chapter with two topological applications: "the hairy ball theorem" (the sphere S^2 cannot be combed without a cowlick, in other words every vector field on S^2 vanishes at some point), and Brouwer's theorem (every continuous map from a closed ball to itself has a fixed point), which can be proved without using algebraic topology.

6.2. Orientation: From Vector Spaces to Manifolds

6.2.1. Oriented Atlas

We say that two bases $(e_i)_{1 \leqslant i \leqslant n}$ and $(e_i')_{1 \leqslant i \leqslant n}$ of a real vector space E of dimension n have the same orientation if the determinant of $(e_i)_{1 \leqslant i \leqslant n}$ with respect to $(e_i')_{1 \leqslant i \leqslant n}$ is *positive*. We thus obtain an equivalence relation on the set of bases of E and a *orientation* of E is by definition a choice of one of these two equivalence classes.

Exterior algebra allows us to express this intrinsically: the vector space $\bigwedge^{\dim E} E^*$ is of dimension 1. Let $\omega \in \bigwedge^{\dim E} E^* \smallsetminus \{0\}$ be a basis of this space. Then the bases $(e_i)_{1 \leqslant i \leqslant n}$ and $(e_i')_{1 \leqslant i \leqslant n}$ define the same orientation if and only if $\omega(e_1, \ldots, e_n)$ and $\omega(e_1', \ldots, e_n')$ have the same sign. Indeed, by Proposition 5.6

$$\frac{\omega(e_1', \ldots, e_n')}{\omega(e_1, \ldots, e_n)} = \det\left((e_1', \ldots, e_n'), (e_1, \ldots, e_n)\right),$$

where the notation $\det\left((e_1', \ldots, e_n'), (e_1, \ldots, e_n)\right)$ indicates that we compute the "usual" determinant of the components of the basis $(e_1', \ldots, e_n')$ expressed in terms of the basis $(e_1, \ldots, e_n)$.

Choosing an orientation thus reduces to choosing a frame ω of $\bigwedge^{\dim E} E^*$, and the equivalence class of bases $(e_1, \ldots, e_n)$ such that $\omega(e_1, \ldots, e_n) > 0$. If we replace ω by $t\omega$, where t is a positive real number, the orientation is unchanged.

Thus, choosing an orientation is equivalent to choosing one of the two connected components of $\bigwedge^{\dim E} E^* \smallsetminus \{0\}$.

To extend the notion of orientation to manifolds, it is necessary to speak of orientations on the tangent spaces $T_x M$ varying differentiably with x. The discussion which follows is parallel to that just given for vector spaces.

Definitions 6.1

a) *An* orientation atlas *of a manifold M is any atlas $(U_i, \varphi_i)_{i \in I}$ such that the change of coordinates $\varphi_i \circ \varphi_j^{-1}$ has positive Jacobian.*

b) *An* orientable manifold *is a manifold for which there exists an oriented atlas.*

For example the atlas on the sphere S^2 consisting of stereographic projections from the north and south poles is not an oriented atlas: the transition map between charts is the inversion $x \mapsto \frac{x}{\|x\|^2}$, which reverses orientation (see Section 1.3). If we replace one of the charts by its composition with a reflection with respect to a line, we then obtain an oriented atlas. This shows that S^2 is an orientable manifold. We will soon see a more elegant proof.

These definitions are directly inspired by the discussion of oriented bases of a vector space: we want to orient $T_x M$ by means of the basis formed by the vectors $(T_{\varphi(x)} \varphi^{-1} \cdot e_k)_{1 \leqslant k \leqslant n}$, where $(e_k)_{1 \leqslant k \leqslant n}$ is the canonical basis of $\mathbf{R}^n$. Oriented atlases are defined to make this notion consistent.

Given two oriented atlases $(U_i, \varphi_i)_{i \in I}$ and $(V_j, \psi_j)_{j \in J}$, we can associate a map s from M to $\{\pm 1\}$ in the following way. Any $x \in M$ belongs to at least one of the U_i and one of the V_j, and the sign of the Jacobian of $\varphi_i \circ \psi_j^{-1}$ does not depend on the choice of i and j. Indeed, if $x \in U_{i'} \cap V_{j'}$, we have

$$\varphi_{i'} \circ \psi_{j'}^{-1} = (\varphi_{i'} \circ \varphi_i^{-1}) \circ (\varphi_i \circ \psi_j^{-1}) \circ (\psi_j \circ \psi_{j'}^{-1}).$$

The map s (as in sign!) is locally constant, therefore constant on M if connected. We say that the atlases $(U_i, \varphi_i)_{i \in I}$ and $(V_j, \psi_j)_{j \in J}$ give the same orientation if $s = 1$. We also obtain an equivalence class on oriented atlases. This discussion also shows that if m is connected there are exactly two equivalence classes.

Definitions 6.2

a) *An* orientation *of an oriented manifold is a choice of equivalence class in the oriented atlas.*

b) *An* oriented manifold *is a manifold with an orientation.*

c) *Let M and N be two manifolds oriented by oriented atlases (U_i, φ_i) and (V_j, ψ_j) respectively. A local diffeomorphism $f : M \to N$ preserves (resp. reverses) the orientation if the Jacobian of $\psi_j \circ f \circ \varphi_i^{-1}$ is positive (resp. negative) where it is defined.*

A consequence of the preceding discussion is that this condition does not depend on the choice of an atlas in the same equivalence class. We also note the analogy with the sign conditions of the determinant in linear algebra.

Oriented atlases play the same role for manifolds as bases for vector spaces seen in the opening discussion. But there is an important difference: as we will soon see, a given manifold need not possess an oriented atlas.

6.2.2. Volume Forms

Following the analogy with linear algebra, we can define orientation by means of differential forms of degree dim M.

Definition 6.3. *A* volume form *on a manifold M is a everywhere nonvanishing differential form of maximum degree* dim M.

This amounts to saying that $\omega_x(v_1, \ldots, v_n) \neq 0$ for all $x \in M$ and every basis $(v_1, \ldots, v_n)$ of T_xM.

Lemma 6.4

i) *If $\varphi : M \to N$ is a local diffeomorphism and if $\omega \in \Omega^n(N)$ is a volume form on N, then $\varphi^*\omega$ is a volume form on M.*

ii) *If $\omega \in \Omega^n(M)$ is a volume form, then every form α of degree n can be written in a unique way as $f\omega$, where $f \in C^\infty(M)$.*

PROOF.

i) If $x \in M$ and $v_1, \ldots, v_n \in T_xM$, then by the definition of the pullback,

$$(\varphi^*\omega)_x(v_1, \ldots, v_n) = \omega_{\varphi(x)}(T_xf \cdot v_1, \ldots, T_xf \cdot v_n).$$

If $v_1, \ldots, v_n$ are linearly independent, then so are their images under $T_x\varphi$ by hypothesis, and so the right hand side is nonzero as ω is a volume form, thus the left hand side is nonzero as well.

ii) The result is true for every open subset of a chart U_i as it is true for the forms $\varphi_i^{-1*}\omega$ and $\varphi_i^{-1*}\alpha$. Thus there exists a function $f_i \in C^\infty(U_i)$ such that

$$\alpha_{|U_i} = f_i\left(\omega_{|U_i}\right).$$

If $U_i \cap U_j \neq \emptyset$, the restrictions of f_i and f_j to $U_i \cap U_j$ coincide: if $x \in U_i \cap U_j$ and if $v_1, \ldots, v_n$ are n independent vectors in T_xM, then

$$\alpha_x(v_1, \ldots, v_n) = f_i(x)\omega_x(v_1, \ldots, v_n) = f_j(x)\alpha_x(v_1, \ldots, v_n).$$

By hypothesis $\alpha_x(v_1, \ldots, v_n) \neq 0$, thus f_i and f_j coincide on $U_i \cap U_j$, which shows that there exists a function $f \in C^\infty(M)$ such that $f_{|U_i} = f_i$, which is also unique since the f_i are uniquely determined. $\square$

Theorem 6.5. *A compact manifold is orientable if and only if it admits a volume form.*

PROOF. Let $\omega \in \Omega^n(M)$ be a volume form and let $(U_i, \varphi_i)_{i \in I}$ be a finite atlas of M. By the lemma,

$$\varphi_i^{-1*}\omega = f_i\, dx^1 \wedge \cdots \wedge dx^n,$$

where $f_i \in C^\infty(\varphi_i(U_i))$ is everywhere nonzero. By interchanging coordinates, we can suppose that $f_i > 0$. Then $(U_i, \varphi_i)_{i \in I}$ is an oriented atlas. Indeed from the relation

$$(\varphi_j \circ \varphi_i^{-1})^*(f_j \, dx^1 \wedge \cdots \wedge dx^n) = f_i \, dx^1 \wedge \cdots \wedge dx^n$$

we deduce that the Jacobian at $u \in \varphi_i(U_i \cap U_j)$ of $\varphi_j \circ \varphi_i^{-1}$ is the strictly positive number

$$\frac{f_i(u)}{f_j(\varphi_j \circ \varphi_i^{-1}(u))}.$$

Conversely, if M is orientable, there exists a finite oriented atlas $(U_i, \varphi_i)_{i \in I}$. For each $i \in I$, we define $\omega_i \in \Omega^n(U_i)$ by

$$\omega_i = \varphi_i^*(dx^1 \wedge \cdots \wedge dx^n).$$

To construct a volume form on the entire manifold from the ω_i, we use bump functions. By the Lemma 3.6, there exists an open covering $(V_i)_{i \in I}$ such that $\overline{V}_i \subset U_i$ for all i, and by Corollary 3.5 for each i there exists a bump function f_i with compact support in U_i and equal to 1 on V_i. We deduce the existence of an n-form α_i on M given by

$$(\alpha_i)_x = f_i(x)(\omega_i)_x \quad \text{if } x \in U_i, \quad \text{and} \quad (\alpha_i)_x = 0 \quad \text{if } x \notin \mathrm{Supp}\, f_i.$$

Then

$$\alpha = \sum_{i \in I} \alpha_i$$

is a volume form. Indeed, every $x \in M$ belongs to at least one V_{i_0} and on $\varphi_{i_0}(V_{i_0})$ the form

$$\varphi_{i_0}^{-1*}\alpha = \sum_{i:U_i \cap U_{i_0} \neq \emptyset} \varphi_{i_0}^{-1*}\alpha_i$$

$$= \left(1 + \sum_{i \neq i_0,\ U_i \cap U_{i_0} \neq \emptyset} (f_i \circ \varphi_{i_0}^{-1})\,\mathrm{Jac}(\varphi_i \circ \varphi_{i_0}^{-1})\right) dx^1 \wedge \cdots \wedge dx^n$$

is everywhere nonzero as the transition maps have positive Jacobian. $\qquad\square$

Remarks

a) With the orientation given by ω, the oriented bases $(v_1, \ldots, v_n)$ of $T_x M$ are those for which
$$\omega_x(v_1, \ldots, v_n) > 0.$$

b) This result is still true for noncompact manifolds which are countable at infinity, which is to say, as we said in Chapter 2, all reasonable manifolds. The proof that the existence of a volume form implies orientability does not use any topological assumptions on the manifold. For the converse, one must use a more refined argument than partitions of unity. (See [Spivak 79].)

c) A volume form is an everywhere nonvanishing section of the vector bundle $\bigwedge^n T^*M$. This bundle is of rank 1, so we may reformulate Theorem 6.5 by saying that a manifold (countable at infinity) is orientable if and only if the bundle $\bigwedge^n T^*M$ is trivializable.

This result is at the heart of most of the proofs of (non) orientability.

Example. The sphere S^n is orientable: the restriction to S^n of the form $\omega \in \Omega^n(\mathbf{R}^{n+1})$ defined by

$$\omega = \sum_{i=0}^{n} (-1)^i x^i \, dx^0 \wedge \cdots \wedge \widehat{dx^i} \wedge \cdots \wedge dx^n$$

is everywhere nonvanishing, since if $(v_1, \ldots, v_n)$ is a basis of $T_x S^n$,

$$\omega_x(v_1, \ldots, v_n) = \det(x, v_1, \ldots, v_n) \neq 0.$$

We can easily check that for the restriction to the open subset $U_k = \{x \in S^n : x^k \neq 0\}$ we have

$$\omega = \frac{1}{x^k} dx^0 \wedge \cdots \wedge \widehat{dx^k} \wedge \cdots \wedge dx^n.$$

On the other hand:

Theorem 6.6. *If n is even, the projective space $P^n\mathbf{R}$ is not orientable.*

PROOF. We consider $P^n\mathbf{R}$ as the quotient of S^n by the group $\{Id, \sigma\}$, where $\sigma(x) = -x$. We denote the corresponding covering map by $p : S^n \to P^n\mathbf{R}$. If $\alpha \in \Omega^n(P^n\mathbf{R})$ is a volume form, then $\omega = p^*\alpha$ is a volume form on S^n. Then

$$\omega = f\omega_0,$$

where ω_0 is the volume form of S^n of the example above, and $f \in C^\infty(S^n)$ is nowhere vanishing.

Since $p \circ \sigma = p$, we have $\sigma^*\omega = \omega$. On the other hand, $\sigma^*\omega_0 = (-1)^{n+1}\omega_0 = -\omega_0$ if n is even. However, we also have

$$\sigma^*\omega = \sigma^*(f\omega_0) = -(f \circ \sigma)\omega_0.$$

Consequently, f attains both signs on S^n, and so must vanish somewhere, which is impossible. So $P^n\mathbf{R}$ is not orientable if n is even. $\qquad\square$

On the other hand, we can see that odd-dimensional real projective spaces are orientable by examining homogeneous coordinate charts. However it is simpler and more instructive to show that the volume form ω_0 of S^n, which this time is σ-invariant, passes to the quotient as a volume form on $P^n\mathbf{R}$. This comes from the following

Proposition 6.7. *Let Γ be a discrete group acting properly and smoothly on a manifold X, and let $p : X \to X/\Gamma$ be the covering map to the quotient. If ω is a differential form on X such that*

$$\gamma^*\omega = \omega \quad \forall \gamma \in \Gamma,$$

then there exists a unique differential form α on X/Γ such that $p^*\alpha = \omega$. Further, if ω is a volume form, then α is too.

PROOF. We recall that the linear tangent map at p is invertible. For $x \in X/\Gamma$ and $v_1, \ldots, v_k$ we set

$$\alpha_x(v_1, \ldots, v_k) = \omega_y\big((T_y p)^{-1} \cdot (v_1)_x, \ldots, (T_y p)^{-1} \cdot (v_k)_x\big).$$

With the hypothesis on ω, the right hand side does not depend on the choice of y in $p^{-1}(x)$: changing y amounts to replacing it by some $\gamma(y)$, and by the properties of covering maps,

$$(T_{\gamma(y)}p)^{-1} = T_y\gamma \circ (T_y p)^{-1}.$$

To show this defines a differential form, it suffices to check this for restrictions to a open subsets $U \subset X/\Gamma$ satisfying the defining property of coverings. In other words we may assume that U is of the form $p(V)$, where V is an open set in X such that the $\gamma(V)$ are mutually disjoint. Then $p_{|V}$ is a diffeomorphism, and by the definition of α, we have

$$\alpha_{|U} = (p_{|V})^{-1*}(\omega_{|V}).$$

This formula also shows that α is a volume form if ω is. $\qquad\square$

Example. The projective spaces $P^{2n+1}\mathbf{R}$ are orientable as we have just seen; tori are orientable.

6.2.3. Orientation Covering

We can take up the discussion of Section 6.2.1 from another point of view, by "putting together" the orientations. An orientation at a point x of M is given by a connected component of $\bigwedge^n T_x^*M \smallsetminus \{0\}_x$, and the set of orientations of T_xM can be seen as the quotient of $\bigwedge^n T^*M \smallsetminus M$ by the multiplicative action of $\mathbf{R}_+^*$ on each fiber (namely, we take away from $\bigwedge^n T^*M$ the union of the $\{0\}_x$ as x varies over M, which is to say the image of the *zero section* of the vector bundle $\bigwedge^n T^*M$, which is identified with M). It is possible to give this quotient a manifold structure. This type of construction is important, but as we will not use it in this book, we prefer to skip it by introducing a Riemannian metric on M.

Such a metric allows us to define a norm on $\bigwedge^n T_x^*M$: we declare that for each orthonormal basis $(e_{1,x}, \ldots, e_{n,x})$ of T_xM, the covector $e_{1,x}^* \wedge \cdots \wedge e_{n,x}^*$ has norm 1. We denote by $\widetilde{M}$ the fibration of unit spheres of $\bigwedge^n T^*M$, in other words

$$\widetilde{M} = \{\omega_x : \ x \in M \text{ and } \|\omega_x\|_x = 1\}.$$

Of course, the unit sphere of $\bigwedge^n T_x^*M$, which is a vector space of dimension 1, consists of two points. The map $p : \widetilde{M} \to M$ defined by $p(\omega_x) = x$ is therefore a covering of order 2. The interchange of two elements of each fiber, denoted σ, defines an action of the group of two elements on $\widetilde{M}$, such that $M/\{I, \sigma\} \simeq M$.

We define a differential form Ψ of degree n on $\widetilde{M}$ by

$$\Psi_{\omega_x}(v_1, \ldots, v_n) = \omega_x(T_{\omega_x}p \cdot v_1, \ldots, T_{\omega_x}p \cdot v_n)$$

for n tangent vectors to $\widetilde{M}$ at ω_x.

Theorem 6.8. *The form Ψ is a volume form on $\widetilde{M}$ such that $\sigma^*\Psi = -\Psi$; the manifold M is orientable if and only if the covering map $p : \widetilde{M} \to M$ is trivial, or if and only if $\widetilde{M}$ is not connected.*

PROOF. Let $a \in M$, and let U be a trivializing open subset for p. Then $p^{-1}(U) = W_1 \cup W_2$, where the W_i are disjoint and $p_{|W_i}$ is a diffeomorphism p_i from W_i to U. The inverse diffeomorphism is a local section with norm 1 of $\bigwedge^n T^*M$, and therefore a volume form on U. Suppose for example that $\omega_a \in W_1$, and let $\overline{\omega} \in \Omega^n(U)$ be the corresponding volume form. Then $\Psi_{|W_1} = p_1^{-1*}\overline{\omega}$, which proves that Ψ is an n-form which never vanishes. The relation $\sigma^*\Psi = -\Psi$ is an algebraic consequence of the definition.

Next, if the covering is trivial, it admits a global section q and $q^*\Psi$ is a volume form on M. Conversely, if α is a volume form on M, the form $\frac{\alpha}{\|\alpha\|}$ defines a section of p.

Finally, by Theorem 2.44, the total space of a covering of degree 2 with connected base is connected if and only if the covering is nontrivial. $\square$

One can show that changing the Riemannian metric results in replacing M with a diffeomorphic manifold. We also remark that by Theorem 2.44, a simply connected manifold is orientable.

Definition 6.9. *We call $\widetilde{M}$ the* orientation covering *of M.*

Examples. The orientation covering of the Klein bottle is the torus T^2, that of the Möbius strip is the cylinder, and that of the projective plane is the sphere S^2.

6.3. Integration on Manifolds; A First Application

6.3.1. Integral of a Differential Form of Maximum Degree

On $\mathbf{R}^n$ there exists a natural measure, the Lebesgue measure, which up to a multiplicative factor, is the unique translation invariant measure. On the other hand, there is not a "natural" measure on a manifold in general. However we saw in Section 2.8 a natural notion of a negligible set. This should be compared with the existence of *a natural family of measures* on any manifold which are given by differential forms of maximum degree.

As always when integrating, it is convenient to add hypothesis to avoid problems of noncompactness.

Definition 6.10. *The* support *of a differential form α is the closure of the set of points x where $\alpha_x \neq 0$.*

We denote the set of compactly supported differential forms by $\Omega_0(M)$.

We begin with open subsets of $\mathbf{R}^n$. Once and for all, we suppose *these sets are oriented by the form $dx^1 \wedge \cdots \wedge dx^n$*, which amounts to saying that we orient $\mathbf{R}^n$ by means of its natural basis.

Definition 6.11. *Given a form $\alpha \in \Omega_0^n(U)$, where U is an open subset of $\mathbf{R}^n$, we call the expression*

$$\int_U f \, dx^1 \ldots dx^n \quad if \quad \alpha = f \, dx^1 \wedge \cdots \wedge dx^n$$

the integral of α, *denoted $\int_U \alpha$.*

It seems we have pulled a sleight of hand! In reality, the following property shows interest in this point of view.

Proposition 6.12. *Let $\varphi : U \to V$ be a diffeomorphism and let $\alpha \in \Omega_0^n(V)$. Then*

$$\int_U \varphi^* \alpha = \pm \int_V \alpha$$

depending on whether φ preserves or reverses the orientation.

PROOF. We have $\alpha = f\, dx^1 \wedge \cdots \wedge dx^n$ and

$$\varphi^*\alpha = (f \circ \varphi)\varphi^*(dx^1 \wedge \cdots \wedge dx^n) = (f \circ \varphi)\mathrm{Jac}(\varphi)\, dx^1 \wedge \cdots \wedge dx^n.$$

However by the change of variable formula for multiple integrals,

$$\int_V f\, dx^1 \dots dx^n = \int_U (f \circ \varphi)|\mathrm{Jac}(\varphi)|\, dx^1 \dots dx^n. \qquad \square$$

This invariance under diffeomorphism opens the door to generalization to manifolds. The idea is to decompose a form as a sum of forms supported in the domains of charts, and to integrate each. This decomposition uses the following notion.

Definition 6.13. *Given a covering $(U_i)_{i \in I}$ of a manifold M, a partition of unity subordinate to the cover is a family of nonnegative smooth functions $(p_i)_{i \in I}$ such that*

$$\mathrm{Supp}\, p_i \subset U_i \quad and \quad \sum_{i \in I} p_i = 1.$$

Proposition 6.14. *For every finite open covering of compact manifold, there exists a subordinate partition of unity.*

PROOF. We again use Corollary 3.5 and Lemma 3.6: there exists an open covering $(V_i)_{i \in I}$ such that $\overline{V_i} \subset U_i$ for all i, and for each i a bump function q_i equal to 1 on V_i with support contained in U_i. Now the function

$$q = \sum_{i \in I} q_i$$

is strictly positive everywhere, and it suffices to take $p_i = \frac{q_i}{q}$. $\qquad \square$

We are now in good shape to integrate differential forms on a compact oriented manifold.

Theorem 6.15. *If M is an oriented compact manifold of dimension n, there exists a unique linear map INT from $\Omega^n(M)$ to $\mathbf{R}$ such that for every form α with support contained in an open subset of a chart (U, φ), we have*

$$INT(\alpha) = \int_{\varphi(U)} \varphi^{-1*}\alpha$$

if φ preserves the orientation.

PROOF. Let $(U_i, \varphi_i)_{i \in I}$ be an atlas (we can take it to be finite since M is compact) compatible with the orientation of M, and let $(p_i)_{i \in I}$ be a subordinate partition of unity. Then every form $\alpha \in \Omega^n(M)$ may be written

$$\alpha = \sum_{i \in I} p_i \alpha = \sum_{i \in I} \alpha_i \qquad (*)$$

where $\operatorname{Supp} \alpha_i \subset U_i$. Then, necessarily

$$INT(\alpha) = \sum_{i \in I} \int_{\varphi_i(U_i)} \varphi_i^{-1*} \alpha_i \,,$$

which shows the uniqueness of INT.

To establish existence, we must show that the result we obtain does not depend on the choices made. Let (V_j, ψ_j) be another finite atlas compatible with the orientation and let $(q_j)_{j \in J}$ be a corresponding subordinate partition of unity. We may then write

$$\alpha = \sum_{(i,j) \in I \times J} p_i q_j \alpha.$$

The support of the form $\alpha_{ij} = p_i q_j \alpha$ is contained in $U_i \cap V_j$ (in particular $\alpha_{ij} = 0$ if $U_i \cap V_j = \emptyset$), and it suffices to check for these forms that the two results coincide. This result is Proposition 6.12.   □

Definition 6.16. *The map $INT(\alpha)$ is called the integral of ω over the oriented manifold M, and is denoted $\int_M \omega$.*

An immediate consequence of the proof of Theorem 6.15 is that for an n-form $\alpha(t)$ depending continuously (resp. smoothly) on a parameter t, the function $t \mapsto \int_M \alpha(t)$ is continuous (resp. smooth) as we can apply classical results of integral calculus. We also see that Proposition 6.12 extends to any oriented manifold.

6.3.2. The Hairy Ball Theorem

We give a spectacular application of this result.

Theorem 6.17. *If n is even, then every vector field on S^n admits a zero.*

PROOF. Consider S^n as the set of vectors of unit norm in $\mathbf{R}^{n+1}$ with the standard inner product. Then a vector field on S^n can be identified with a map X from S^n to $\mathbf{R}^{n+1}$ such that $\langle x, X(x) \rangle = 0$. If X is everywhere nonvanishing, we can suppose that $\|X(x)\| = 1$ for all x by replacing the vector field by $X/\|X\|$. Define

$$f_\epsilon(x) = x + \epsilon X(x).$$

We thus obtain a smooth map from S^n to the sphere $S^n\left(\sqrt{1+\epsilon^2}\right)$ (here we have denoted the sphere of radius r and center 0 in $\mathbf{R}^{n+1}$ by $S^n(r)$).

Lemma 6.18. *There exists an $\epsilon_0 > 0$ such that f_ϵ is a diffeomorphism for $\epsilon < \epsilon_0$.*

PROOF. Define $\omega \in \Omega^n(\mathbf{R}^{n+1})$ by

$$\omega = \sum_{i=0}^{n} (-1)^i x^i \, dx^0 \wedge \cdots \wedge \widehat{dx^i} \wedge \cdots \wedge dx^n.$$

We will abuse notation and denote any restriction of ω to $\omega_{|S^n(r)}$ by ω. Orient $S^n(r)$ by ω. The formula giving f_ϵ shows that $f_\epsilon^* \omega$ is a differential form which depends polynomially on the parameter ϵ. More precisely,

$$f_\epsilon^* \omega = \omega + \epsilon \alpha(\epsilon),$$

where $\alpha(\epsilon)$ is a form which depends on ϵ polynomially, and is thus continuous *a fortiori*. By the properties of volume forms, we can write, with the same regularity conditions

$$f_\epsilon^* \omega = (1 + \epsilon g_\epsilon)\omega,$$

which shows that $f_\epsilon^* \omega$ is a volume form if ϵ is sufficiently small, and that f_ϵ is an immersion.

We now show that f_ϵ is injective for ϵ sufficiently small. If not, there exists a sequence ϵ_k approaching 0, and distinct points $x_k, y_k \in S^n$ such that

$$x_k + \epsilon_k X(x_k) = y_k + \epsilon_k X(y_k),$$

and therefore

$$\frac{x_k - y_k}{\|x_k - y_k\|} = -\epsilon_k \frac{X(x_k) - X(y_k)}{\|x_k - y_k\|}$$

However the left hand side has norm 1, while the right hand side tends to zero by the mean value theorem (applied to the function $x \mapsto X(x/\|x\|)$ on an appropriate annulus).

Under these conditions, f_ϵ is an injective local diffeomorphism from $S^n(1)$ to $S^n(\sqrt{1 + \epsilon^2})$, and thus a diffeomorphism since its image, necessarily open and closed, is the whole sphere $S^n(\sqrt{1 + \epsilon^2})$. □

END OF THE PROOF OF THEOREM 6.17. With the notations and orientation conventions of the lemma,

$$\int_{S^n(r)} \omega = r^{n+1} \int_{S^n} \omega = c_n r^{n+1}.$$

(We will calculate c_n later, its precise value is not important here.) On the other hand,

$$\int_{S^n(r)} \omega = \int_{S^n} f_\epsilon^* \omega.$$

We already know by the proof of the lemma that this last expression is a polynomial in ϵ. For ϵ sufficiently small we then have,

$$c_n(1 + \epsilon^2)^{\frac{n+1}{2}} = P(\epsilon),$$

which is clearly impossible if n is even. □

Remarks

a) This argument is due to J. Milnor. Next chapter we will see a more classical proof using degree theory.

b) We could have avoided the lemma above by invoking the existence of a natural topology on the set of C^1 maps of a compact manifold to itself for which $\mathrm{Diff}(M)$ is *open* (see [Hirsch 76, Chapter 2] for the details). Conversely, we note in passing that the set of homeomorphisms of a compact metric space K is not open in $C^0(K,K)$ as we have already seen for $K = [0,1]$ or $K = S^1$.

6.4. Stokes's Theorem

6.4.1. Integration on Compact Subsets

Exactly as in the classical case of Lebesgue measure, we will now define the integral of differential forms of degree n on a compact subset of an oriented manifold of dimension n. We begin with the case where the compact subset K is included in the open domain of a chart U. If φ is the corresponding chart, and if $\varphi^{-1*}\alpha = f\,dx^1 \wedge \cdots \wedge dx^n$, we of course define

$$\int_K \alpha = \int_{\varphi(K)} f\,dx^1 \ldots dx^n.$$

By Proposition 6.12, the result is independent of the chart chosen.

In the general case, if $(U_i, \varphi_i)_{i\in I}$ is an atlas of M, we can cover K by a finite number of open subset U_i, take a partition of unity subordinate to this cover (this requires a slight modification of Proposition 6.14 which we leave to the reader) and proceed as in the proof of Theorem 6.15.

Definition 6.19. *The number thus obtained is the* integral of the form α *over the compact subset K.*

Of course, if K and L are two disjoint compact subsets, $\int_{K\cup L} \alpha = \int_K \alpha + \int_L \alpha$. Better still:

Proposition 6.20. *If K and L are two compact subsets with negligible intersection,*

$$\int_{K\cup L} \alpha = \int_K \alpha + \int_L \alpha.$$

PROOF. If $K \cup L$ is included in the domain of a chart, this result is the same as for the Lebesgue integral in $\mathbf{R}^n$. If not, the preceding discussion allows us to reduce to this case through partitions of unity. □

As an immediate consequence we obtain the following result, whose statement is a little heavy, but is very useful.

Proposition 6.21. *Let M be a compact manifold, let $(D_i)_{i \in I}$ be a finite family of compact subsets of $\mathbf{R}^n$, and for each i, let φ_i be a smooth map with values in M, defined on an open subset containing D_i, and whose restriction to $\text{int}(D_i)$ is a parametrization preserving the orientation. Suppose that*

i) $M \smallsetminus \left(\bigcup_{i \in I} \varphi_i \big(\text{int}(D_i) \big) \right)$ *is negligible.*

ii) $\varphi_i \big(\text{int}(D_i) \big) \cap \varphi_j \big(\text{int}(D_j) \big) = \emptyset$ *if $i \neq j$. Then*

$$\int_M \alpha = \sum_{i \in I} \int_{D_i} \varphi_i^* \alpha.$$

Example. Consider the spherical coordinate map

$$F : \ (\theta, \varphi) \longmapsto (\cos \theta \cos \varphi, \sin \theta \cos \varphi, \sin \varphi)$$

from $\mathbf{R}^2$ to S^2 (spherical coordinates). Then F is a diffeomorphism from $(0, 2\pi) \times (-\frac{\pi}{2}, \frac{\pi}{2})$ to S^2 minus the "origin meridian" $\theta = 0$.

For the orientation of S^2 given by F, if $\alpha \in \Omega^2(S^2)$, we have

$$\int_{S^2} \alpha = \int_{[0, 2\pi] \times [-\frac{\pi}{2}, \frac{\pi}{2}]} F^* \alpha.$$

(See Exercises 6 and 7 for examples of calculations.)

6.4.2. Regular Domains and Their Boundary

Definition 6.22. *A compact $D \subset M$ is a* regular domain *if it equals the closure of its interior and if the boundary is either empty or an embedded submanifold of dimension $n - 1$.*

This submanifold will be denoted ∂D, and called the *boundary* of D.

Examples

a) A compact manifold is a regular domain with empty boundary.

b) A closed ball is a regular domain in Euclidean space; the hemisphere $x^0 \geqslant 0$ is a regular domain of the sphere.

c) In a normed space E, the closed annulus

$$C(r_1, r_2) = \big\{ x \in E : \ r_1 \leqslant \|x\| \leqslant r_2 \ (\text{with } r_1 < r_2) \big\}$$

is a regular domain whose boundary has two connected components, the spheres with center 0 and radii r_1 and r_2 respectively. More generally, for any integer k, there exist regular domains whose boundary has k connected components.

d) A square or closed half-disk in the plane are not regular domains: they are equal to the closure of their interior, however their boundaries are not submanifolds.

We now make the structure of a regular domain more precise.

Lemma 6.23. *Let $D \subset M$ be a regular domain. Then:*

i) *D has a finite number of connected components, each of which is a regular domain.*

ii) *For every $p \in \partial D$, there exists an open subset U containing p and a chart $\varphi : U \to \mathbf{R}^n$ such that*

$$\varphi(U \cap D) = \{x \in \varphi(U) : x^1 \leqslant 0\}.$$

PROOF. We leave i) as an exercise. It remains to show ii). By the definition of submanifold, there exists an open U containing p and a chart φ such that

$$\varphi(U \cap \partial D) = \{x \in \varphi(U) : x^1 = 0\}.$$

Furthermore, we can always choose U and φ so that $\varphi(U)$ is an open ball with center 0. Then the open subset $\varphi(\operatorname{int}(D) \cap U)$ of $\varphi(U) \smallsetminus \{x^1 = 0\}$ has empty boundary. Therefore it is either equal entirely to $\varphi(U) \smallsetminus \{x^1 = 0\}$ or to one of the connected components of this open set. But the first alternative is excluded as $\varphi(\operatorname{int}(D) \cap U)$ has a nonempty boundary in $\varphi(U)$. By modifying φ by changing x^1 to $-x^1$, we may suppose that

$$\varphi(\operatorname{int}(D) \cap U) = \{x \in \varphi(U) : x^1 < 0\}.$$

By taking the closure, we deduce that

$$\varphi(D \cap U) = \{x \in \varphi(U) : x^1 \leqslant 0\}. \qquad \square$$

Theorem 6.24. *The boundary of a regular domain of an orientable manifold is itself orientable.*

PROOF. Fix an orientation of the ambient manifold M. If $p \in \partial D$, there exists an open subset U containing p and a chart $\varphi : U \to \mathbf{R}^n$ satisfying the conditions of the previous lemma. Further, by changing x^2 to $-x^2$, we can suppose that the chart φ preserves orientation.

The choices being made for φ, we introduce the map φ_1 from $U \cap \partial D$ to $\mathbf{R}^{n-1}$ which associates to $p \in U \cap \partial D$ the coordinates $x^2, \ldots, x^n$. We verify that we thus obtain an oriented atlas of ∂D. Indeed let (V, ψ) be another chart having the same properties, and denote by $y^1, \ldots, y^n$ the coordinates of ψ. At a point of $\varphi(U \cap V \cap \partial D)$, the matrix of the differential of the transition map $\psi \circ \varphi^{-1}$ is of the form

$$
\begin{pmatrix}
\frac{\partial y^1}{x^1} & 0 & \cdots & 0 \\
\frac{\partial y^2}{\partial x^1} & & & \\
\vdots & & \left(\frac{\partial y^i}{\partial x^j} \right)_{2 \leqslant i,\, j \leqslant n} & \\
\frac{\partial y^n}{\partial x^1} & & &
\end{pmatrix}
$$

(in fact this is the matrix of a linear map which transforms the hyperplane $x_1 = 0$ to the hyperplane $y_1 = 0$), with $\frac{\partial y^1}{\partial x^1} > 0$. Under these conditions, the determinant of the matrix

$$
\left(\frac{\partial y^i}{\partial x^j} \right)_{2 \leqslant i,\, j \leqslant n},
$$

which is the differential of the transition map $\psi_1 \circ \varphi_1^{-1}$, is positive. $\qquad \square$

Orientation of the boundary

In this proof we have given an orientation to ∂D. We call the *oriented boundary* of D the manifold ∂D equipped with this orientation. This orientation can also be defined using frames. For $p \in \partial D$, let $n_p \in T_p D$ be a non-tangential vector to ∂D. We say that n_p is an *outward-pointing vector* if for every curve $c : [0,1] \to D$ such that $c(0) = p$ and $c'(0) = n_p$, $c(t) \notin D$ for $t > 0$ sufficiently small. Alternatively by what we have said above $d\varphi_p^1(n_p) > 0$ for one and therefore all charts φ seen in the preceding theorem. Thus a frame $v_1, \ldots, v_{n-1}$ of $T_p \partial D$ is positively-oriented if and only if the frame $n_p, v_1, \ldots, v_{n-1}$ of $T_p D$ is positively-oriented.

In the case of a regular domain of Euclidean space, **and more generally of an oriented Riemannian manifold**, it is easy to define this orientation of the boundary by means of a volume form. Indeed we have a natural choice of n_p by taking the *normal* to $T_p \partial D$. We then define a vector field along ∂D, which is to say a smooth map N from ∂D to $T\mathbf{R}^n$ such that $N_p \in T_p \partial D$ for all p. Let ω be the volume form equal to 1 for all positively-oriented orthonormal frames (of course, if $x^1, \ldots, x^n$ with respect to a positively-oriented orthonormal basis, $\omega = dx^1 \wedge \cdots \wedge dx^n$). We define a volume form σ on ∂D by

$$
\alpha_a(v_1, \ldots, v_{n-1}) = \omega_a(N_a, v_1, \ldots, v_{n-1}).
$$

The orientation of ∂D is then defined by σ. We note that with a light abuse of notation we can write $\alpha = i_N(\omega_{|\partial D})$.

Examples

a) Consider the disk with center 0 and radius a in $\mathbf{R}^2$ equipped with the natural Euclidean norm. The boundary is the circle of center 0 and radius a. In a neighborhood of a point p on the circle, we take polar coordinates (r, θ). In these coordinates, the volume form $dx \wedge dy$ may be written $r dr \wedge d\theta$, and with the notations of Lemma 6.23, we can take $\varphi_1 = r - a, \varphi_2 = \theta$. The orientation of the circle is given by $d\theta$. In other words, $\theta \mapsto (a \cos \theta, a \sin \theta)$ preserves the orientation. We say that the circle is oriented counterclockwise.

b) We can reconsider the discussion for a circular annulus $C(a, b)$, where $a < b$. As a result of the above, the circle of radius b is oriented in the trigonometric sense. Conversely, in a neighborhood of the circle of radius a we must take $\varphi_1 = a - r$, thus $\varphi_2 = -\theta$. The orientation is thus the opposite to the trigonometric orientation.

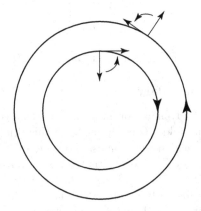

Figure 6.2: Oriented boundary of an annulus

In the same way, if D is a closed annulus of Euclidean space, ∂D has two connected components which are spheres with *opposite* orientations.

c) Figure 6.3 can be deduced in the same way as in b).

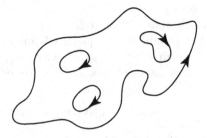

Figure 6.3: Another oriented boundary

d) Let M be a compact submanifold of codimension 1 in $\mathbf{R}^n$. Then by Alexander's theorem (cf. [Bredon 94]), $\mathbf{R}^n \setminus M$ has two connected components, one of which is bounded. The closure of the bounded component is a regular domain with boundary M. In particular, M is orientable. For $n = 2$, this result is called Jordan's theorem. It is already nontrivial (see [Berger-Gostiaux 88, 9.2]) or [Do Carmo 76, 5.7].

e) The case where $\dim D = 1$ merits special attention as the preceding proof does not apply directly. Now ∂D consists of a finite number of points (in fact 2 if D is connected). An "orientation" of a point is a choice of sign $\pm$ (the forms of degree 0 are functions which are constant here). A boundary point p is assigned the sign $+$ if in a neighborhood of p, D is defined by $x \leqslant 0$, where x is a local coordinate compatible with the orientation of D, and is assigned $-$ otherwise. Notice $\{b\} - \{a\}$ is the oriented boundary of $[a, b]$ in $\mathbf{R}$.

6.4.3. Stokes's Theorem in All of Its Forms

Theorem 6.25 (Stokes). *Let D be a regular domain of an oriented manifold M of dimension n, ∂D is oriented boundary, and let $\alpha \in \Omega^{n-1}(M)$. Then*

$$\int_{\partial D} \alpha_{|\partial D} = \int_D d\alpha.$$

Example. If D is the interval $[a, b] \subset \mathbf{R}$, then α is a function f, and with the preceding convention we have

$$\int_{\partial [a,b]} f = f(b) - f(a),$$

and we recover the fundamental theorem of calculus:

$$f(b) - f(a) = \int_a^b f'(x) \, dx.$$

PROOF. First, take an open set V containing D whose closure is compact. We can cover V by a finite number of domains of charts U_i such that if U_i intersects ∂D, then $U_i \cap \partial D = \{x, x^1 = 0\}$, the orientation of ∂D being given on $U_i \cap \partial D$ by $\varphi_i^*(dx^2 \wedge \cdots \wedge dx^n)$. If f_i is a partition of unity subordinate to this cover, we have $\alpha = \sum_{i \in I} f_i \alpha$, and it suffices to prove the result for the $\alpha_i = f_i \alpha$, which is to say for forms supported within the open subset U_i.

Thus let α be such that $\operatorname{Supp} \alpha \subset U_i$. There are three cases to consider:

1) If $\operatorname{Supp} \alpha \subset M \setminus D$, then α vanishes on ∂D, and $d\alpha$ vanishes on D. The result is then clear.

2) If $U_i \subset D \setminus \partial D$, then α vanishes on ∂D, and

$$\int_D d\alpha = \int_{\varphi_i(U_i)} \varphi_i^{-1*}(d\alpha) = \int_{\varphi_i(U_i)} d(\varphi_i^{-1*}\alpha).$$

As $\varphi_i^{-1*}\alpha$ has compact support, this integral is again equal to

$$\int_{\mathbf{R}^n} d(\varphi_i^{-1*}\alpha).$$

However $\varphi_i^{-1*}\alpha$ may be written

$$\sum_{k=1}^n f_k \, dx^1 \wedge \cdots \wedge \widehat{dx^k} \wedge \cdots \wedge dx^n,$$

and

$$d(\varphi_i^{-1*}\alpha) = \left(\sum_{k=1}^n (-1)^{k+1} \frac{\partial f_k}{\partial x^k} \right) dx^1 \wedge \cdots \wedge dx^n.$$

Therefore

$$\int_{\mathbf{R}^n} d(\varphi_i^{-1*}\alpha) = \sum_{k=1}^n (-1)^{k+1} \int_{\mathbf{R}^n} \frac{\partial f_k}{\partial x^k} \, dx^1 \ldots dx^n.$$

By Fubini's theorem, the k-th term may be written

$$\int_{\mathbf{R}^{n-1}} \left(\int_{-\infty}^{+\infty} \frac{\partial f_k}{\partial x^k} \, dx^k \right) dx^1 \ldots \widehat{dx^k} \ldots dx^n.$$

However because f_k has compact support, we have

$$\int_{-\infty}^{+\infty} \frac{\partial f_k}{\partial x^k} \, dx^k = 0$$

for all k, which shows that $\int_D d\alpha = 0$ and proves the theorem in this case.

3) It remains to examine the case where U_i intersects ∂D. In this case,

$$\int_D d\alpha = \int_{\varphi_i(D \cap U_i)} d(\varphi_i^{-1*}\alpha).$$

With the notations of 2) and the hypotheses on the support, this integral is equal to

$$\int_{\mathbf{R}_-^n} \left(\sum_{k=1}^n (-1)^{k+1} \frac{\partial f_k}{\partial x^k} \right) dx^1 \ldots dx^n,$$

where $\mathbf{R}_-^n = \{x \in \mathbf{R}^n, x^1 \leqslant 0\}$. We calculate this integral as in 2) by integrating first the k-th term with respect to x^k. If $k \geqslant 2$, we obtain 0 for the same reasons as before. Finally

$$\int_D d\alpha = \int_{\mathbf{R}^{n-1}} \left(\int_{-\infty}^0 \frac{\partial f_1}{\partial x^1} dx^1 \right) dx^2 \ldots dx^n$$

$$= \int_{\mathbf{R}^{n-1}} f_1(0, x^2, \ldots, x^n)\, dx^2 \ldots dx^n$$

As $\varphi_i(\partial D \cap U_i) = \varphi_i(U_i) \cap \{x, x^1 = 0\}$, with the choice of orientation of ∂D this last integral is equal to $\int_{\partial D} \alpha$.   □

Corollary 6.26. *If α is a differential form of degree $n - 1$ on a compact oriented manifold M of dimension n,*

$$\int_M d\alpha = 0.$$

In particular, a volume form on a compact manifold is never exact.

Remark. Compactness is an essential hypothesis for the last assertion. For example, on $\mathbf{R}^n$, one has

$$d(x^1 dx^2 \wedge \cdots \wedge dx^n) = dx^1 \wedge \cdots \wedge dx^n.$$

One of the interests in Stokes's theorem is the unification of several integral formulas in dimension 2 and 3.

a) Green's theorem.

This is the case where D is a regular domain in $\mathbf{R}^2$, and α is a differential form of degree 1. One has

$$\int_{\partial D} P\, dx + Q\, dy = \int_D (\partial_x Q - \partial_y P)\, dx\, dy.$$

The plane is oriented by $dx \wedge dy$, and we have subtly replaced $dx \wedge dy$ by the Lebesgue measure $dxdy$.

We will not dwell further on this elementary example. However, it shows that weakening the hypothesis made on D is a natural question: it is easy to check directly that Green's formula holds for a square! For Stokes's theorem in a very general setting, see [Whitney 57, Chapter 3].

b) Divergence theorem.

Here D is a regular domain of $\mathbf{R}^3$. If $\omega = dx \wedge dy \wedge dz$ and if X is a vector field on $\mathbf{R}^3$, recall (see Section 5.5) that

$$d(i_X \omega) = L_X \omega = (\operatorname{div} X)\omega.$$

Stokes's theorem gives

$$\int_{\partial D} i_X \omega = \int_D (\operatorname{div} X) \omega \qquad (*)$$

where if $X = P \partial_x + Q \partial_y + R \partial_z$,

$$\int_{\partial D} P\, dy \wedge dz + Q\, dz \wedge dx + R\, dx \wedge dy = \int_D (\partial_x P + \partial_y Q + \partial_z R)\, dx \wedge dy \wedge dz.$$

Once written, this formula uses only the volume form ω and not the inner product structure. It can also be reinterpreted in terms of flux (in the language of physicists) or a surface integral (in the language of the mathematicians) of a vector field. To do this, we introduce on $S = \partial D$ the outward normal vector field N and the volume form $\sigma = i_N \omega$ introduced in 6.4.2 (see Figure 6.4). The flux of the vector field X across S, denoted $\int_S X$, is by definition the integral

$$\int_S \langle X, N \rangle \sigma.$$

By noting that

$$X = \langle X, N \rangle N + Y, \quad \text{where } Y \text{ is tangent to } S,$$

we see that

$$i_X \omega_{|S} = i_{\langle X, N \rangle N} \omega_{|S} = \langle X, N \rangle \sigma.$$

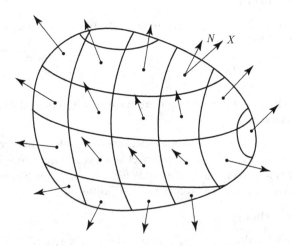

Figure 6.4: Flux across a surface

The formula $(*)$ above then gives

$$\int_S X = \int_D (\operatorname{div} X)\, dx \wedge dy \wedge dz \quad \text{where } S = \partial D.$$

c) "Classical" Stokes's theorem.

This concerns regular domains of surfaces in $\mathbf{R}^3$. We again denote such a domain by S, and suppose its boundary $\partial S = C$ is connected. Then for $\alpha \in \Omega^1(\mathbf{R}^3)$, we have

$$\int_C \alpha = \int_S d\alpha.$$

This formula can be rewritten in terms of a vector field if we utilize the Euclidean metric. Let X be a vector field and $\alpha = X^\flat$ the differential form associated by the metric (cf. 5.3.1). Recall the curl of X is defined by

$$i_{\operatorname{curl} X}\omega = d(X^\flat)$$

Taking b) into account, Stokes's formula can then be written (see Figure 6.5)

$$\oint_C X = \int_S \operatorname{curl} X \quad \text{if } C = \partial S.$$

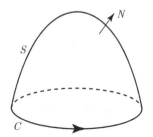

Figure 6.5: Flux and circulation

6.5. Canonical Volume Form of a Submanifold of Euclidean Space

We will see that on Euclidean space, the inner product furnishes a natural measure on submanifolds. First recall what happens in $\mathbf{R}^n$. The Lebesgue measure is characterized up to a factor by its invariance under translation.

To determine it completely, it suffices to know the measure of a reference set, for example a parallelepiped

$$P_0 = \left\{ \sum_{i=1}^{n} t^i e_i, \ 0 \leqslant t^i \leqslant 1 \right\},$$

where $(e_i)_{1 \leqslant i \leqslant n}$ is a basis of $\mathbf{R}^n$.

Then, if we normalize Lebesgue measure by declaring that $\mathrm{vol}(P_0) = 1$, the volume of a parallelepiped constructed by n vectors $a_1, \ldots, a_n$, is the absolute value of the determinant of the system $a_1, \ldots, a_n$ with respect to the basis $(e_i)_{1 \leqslant i \leqslant n}$.

This normalization of the Lebesgue measure on $\mathbf{R}^n$ does not give a normalization of vector subspaces of dimension $k < n$. To convince ourselves of this, take the case $n = 2$, $k = 1$. Imagine that area determines a normalization of the Lebesgue measure on lines. Then every linear transformation of $\mathbf{R}^2$ which preserves area and leaves the subspace $\mathbf{R} \times \{0\}$ invariant conserves the Lebesgue measure of this subspace. This is impossible because these linear transformations are of the form

$$\begin{pmatrix} a & b \\ 0 & 1/a \end{pmatrix}.$$

Everything changes if $\mathbf{R}^n$ is equipped with an inner product, for example the canonical inner product

$$\langle x, y \rangle = \sum_{i=1}^{n} x^i y^i.$$

Now the reference set is a parallelepiped constructed from an orthonormal basis, and if $E \subset \mathbf{R}^n$ is a vector subspace of dimension k, we can now take the reference set for E to be a parallelepiped constructed from an orthonormal basis of E. The following algebraic property allow us to perform explicit calculations.

Lemma 6.27 (Gram determinants). *If $e_1, \ldots, e_n$ is an orthonormal basis, for any vectors $a_1, \ldots, a_n$, we have*

$$\left(\det \left((a_1, \ldots, a_n), (e_1, \ldots, e_n) \right) \right)^2 = \det \left(\langle a_i, a_j \rangle_{1 \leqslant i, \, j \leqslant n} \right).$$

PROOF. We may suppose that $\langle \, , \, \rangle$ is the canonical inner product on $\mathbf{R}^n$, and $e_1, \ldots, e_n$ the canonical basis. Then if

$$a_i = \sum_{k=1}^{n} \alpha_i^k e_k,$$

we have

$$\langle a_i, a_j \rangle = \sum_{k=1}^{n} \alpha_i^k \alpha_j^k.$$

Thus the matrix

$$\langle a_i, a_j \rangle_{1 \leqslant i,\, j \leqslant n}$$

is the product of the matrix of coordinates of a_i by its transpose. $\square$

It remains to transport these constructions to submanifolds of $\mathbf{R}^n$.

Theorem 6.28. *Let M be an oriented submanifold of $\mathbf{R}^n$ of dimension p. Then there exists a unique volume form σ on M such that if $v_1, \ldots, v_k$ is a positively-oriented orthonormal basis of $T_x M$, then*

$$\sigma_x(v_1, \ldots, v_k) = 1.$$

If U is an open subset of $\mathbf{R}^p$ and $F : U \to \mathbf{R}^n$ is a local parametrization of M compatible with the orientation, we have

$$F^* \sigma = \sqrt{\det((\langle \partial_i F, \partial_j F \rangle)_{1 \leqslant i,\, j \leqslant p})}\ dx^1 \wedge \cdots \wedge dx^p.$$

PROOF. The condition in the statement determines $\alpha_x \in \bigwedge^n T_x^* M$ in a unique way from the uniqueness of σ. Now, if $\partial_1, \ldots, \partial_p$ is the canonical basis of $T_u \mathbf{R}^p$ and F is a local parametrization of M to an open subset U containing u, by Lemma 6.27 we have

$$\sigma_{F(u)}(T_u F \cdot \partial_1, \ldots, T_u F \cdot \partial_p) = \sqrt{\det((\langle \partial_i F, \partial_j F \rangle)_{1 \leqslant i,\, j \leqslant p})},$$

with the sign $+$ if F preserves orientation. This proves the second part of the statement, and also the existence of σ, as we have obtained smooth local expressions. $\square$

Definition 6.29. *The volume form above is the* canonical volume form *of M. Note that it depends both on an inner product of $\mathbf{R}^n$ as well as a choice of orientation (it changes sign if we change the orientation of M or of $\mathbf{R}^n$).*

Examples

a) We saw in Section 3.8 that a submanifold of dimension 1 in $\mathbf{R}^n$ is given by an embedding c of $\mathbf{R}$ or of S^1. Such an embedding determines an orientation of C and the corresponding volume form is equal at the point $c(t)$ to $\|c'(t)\|\, dt$.

b) Let S^n be the unit sphere in $\mathbf{R}^{n+1}$. We can choose an orientation of S^n by deciding that a basis $v_1, \ldots, v_n$ of $T_x S^n$ is positively-oriented if and only if the basis $x, v_1, \ldots, v_n$ of $\mathbf{R}^{n+1}$ is positively-oriented. The canonical volume form is then given by

$$\sigma_x(v_1, \ldots, v_n) = \det(x, v_1, \ldots, v_n),$$

where the determinant is taken with respect to a positively-oriented orthonormal basis. We note that σ is the restriction of

$$i_x(dx^0 \wedge \cdots \wedge dx^n) = \sum_{i=0}^{n} (-1)^i x^i \, dx^0 \wedge \cdots \wedge \widehat{dx^i} \wedge \cdots \wedge dx^n.$$

to S^n.

Remark: the volume form of a Riemannian manifold

**More generally, if M is a oriented manifold equipped with a Riemannian metric g, we associate to g a volume form ω_g by deciding that

$$(\omega_g)_x(v_1, \ldots, v_n) = 1$$

for positively-oriented orthonormal frames $(v_1, \ldots, v_n)$ of $T_x M$. By the same argument as in Theorem 6.28, if $F : U \to M$ is a local parametrization of M and if

$$F^* g = \sum_{i,j=1^n} g_{ij} \, dx^i \otimes dx^j,$$

then

$$F^* \omega_g = \sqrt{\det(g_{ij})_{1 \leqslant i, \, j \leqslant n}} \, dx^1 \wedge \cdots \wedge dx^n. **$$

Definition 6.30. *The* volume *of a compact submanifold M of $\mathbf{R}^n$ is the integral over M of "its" canonical volume form σ.*

We put quotes because σ depends on the orientations of M and of $\mathbf{R}^n$. But it is clear that $\int_M \sigma$ does not depend on these. For $\dim M = 1$ we speak of the *length*, and for $\dim M = 2$ the *area*.

Remark. We can define the volume (finite or infinite) of a noncompact manifold M as the upper bound of the integrals of σ taken over regular domains of M.

By using Proposition 6.21, we see by an easy calculation that

$$\text{area}(S^2) = \int_0^{2\pi} \int_{-\frac{\pi}{2}}^{\frac{\pi}{2}} \cos \phi \, d\theta \, d\phi = 4\pi.$$

We move on to higher dimensions.

Proposition 6.31. *The volume of the unit sphere in $\mathbf{R}^{n+1}$ is equal to*

$$\frac{2\pi^{\frac{n+1}{2}}}{\Gamma(\frac{n+1}{2})}.$$

PROOF. We will calculate the integral

$$\int_{\mathbf{R}^{n+1}} e^{-\|x\|^2} dx^0 \wedge \cdots \wedge dx^n$$

in two ways. On one hand, by Fubini's theorem, this integral equals I^{n+1}, where

$$I = \int_{-\infty}^{\infty} e^{-t^2} dt.$$

On the other hand, if $F : \mathbf{R}^+ \times S^n \to \mathbf{R}^{n+1}$ is given by $F(r, u) = ru$, we have

$$F^*(dx^0 \wedge \ldots \wedge dx^n) = r^n\, dr \wedge \sigma,$$

where σ is the canonical volume form of S^n (to see this, it suffices to evaluate the two forms above at (r, u) on $\frac{\partial}{\partial r}$ and an orthonormal basis $v_1, \ldots, v_n$ of $T_u S^n$). This formula shows that F is a diffeomorphism on $\mathbf{R}^{n+1} \smallsetminus \{0\}$. Then

$$\int_{\mathbf{R}^{n+1}} e^{-\|x\|^2} dx^0 \wedge \cdots \wedge dx^n = \int_{\mathbf{R}^+ \times S^n} e^{-r^2} r^n\, dr \wedge \sigma = \mathrm{vol}(S^n) \int_0^{\infty} e^{-r^2} r^n\, dr.$$

However

$$\int_0^{\infty} e^{-r^2} r^n\, dr = \frac{1}{2} \int_0^{\infty} e^{-t} t^{\frac{n-1}{2}}\, dt = \frac{1}{2}\Gamma\left(\frac{n+1}{2}\right).$$

As $\mathrm{vol}(S^1) = \int_{S^1} x\, dy - y\, dx = 2\pi$, for $n = 1$ we obtain $I = \sqrt{\pi}$ and the result follows. $\qquad\square$

In particular, $\mathrm{vol}(S^{2p-1}) = \frac{2\pi^p}{(p-1)!}$. We also note that by Stokes's theorem

$$\int_{S^n} \sigma = \int_{B^{n+1}(1)} (n+1)\, dx^0 \wedge \cdots \wedge dx^n,$$

and thus

$$\mathrm{vol}(B^{n+1}(1)) = \frac{1}{n+1}\, \mathrm{vol}(S^n).$$

The preceding calculation used a result analogous to Fubini's theorem, which we state as follows:

Lemma 6.32. *Let X and Y be two compact oriented manifolds, and let $X \times Y$ be their product equipped with the product orientation (see Exercise 1).*

Let $p : X \times Y \to X$ and $q : X \times Y \to Y$ be the canonical projections. Then if $\alpha \in \Omega^{\dim X}(X)$ and $\beta \in \Omega^{\dim Y}(Y)$, we have

$$\int_{X \times Y} p^* \alpha \wedge q^* \beta = \left(\int_X \alpha \right) \left(\int_Y \beta \right).$$

PROOF. Let $(U_i, \varphi_i)_{i \in I}$ (resp. $(V_j, \psi_j)_{j \in J}$) be a finite oriented atlas of X (resp. of Y). Then $(U_i \times V_j, \varphi_i \times \psi_j)_{(i,j) \in I \times J}$ is an oriented atlas of $X \times Y$, which by definition gives the *product* orientation by the atlases on the factors. Now if $(f_i)_{i \in I}$ (resp. $(g_j)_{j \in J}$) is a partition of unity subordinate to the cover $(U_i)_{i \in I}$ (resp. $(V_j)_{j \in J}$), then $(f_i g_j)_{(i,j) \in I \times J}$ is a partition of unity subordinate to the cover $(U_i \times V_j)_{(i,j) \in I \times J}$ of $X \times Y$. It suffices to prove the property for forms $f_i g_j p^* \alpha \wedge q^* \beta$. We have thus reduced to the classical Fubini theorem. $\square$

We finish this paragraph with spectacular result, the *isoperimetric inequality*.

Theorem 6.33.[1] *Let D be a regular domain of n-dimensional Euclidean space. Then we have the inequality*

$$\frac{\left(\operatorname{vol}(\partial D) \right)^n}{\left(\operatorname{vol}(D) \right)^{n-1}} \geqslant \frac{\left(\operatorname{vol}(S^{n-1}) \right)^n}{\left(\operatorname{vol}(B(0,1)) \right)^{n-1}}.$$

Further, the equality occurs if and only if D is a ball.

For example, for $n = 2$, we have

$$\left(\operatorname{length}(\partial D) \right)^2 \geqslant 4\pi \operatorname{area}(D),$$

with the equality attained if and only if D is a disk.

In proving this, one might think that a relation between the volume of a domain and that of its boundary involves the use of Stokes's theorem. There is indeed such a proof, but this is the most recent of all. It is due to M. Gromov, from the 1980s. See [Berger-Gostiaux 88, 6.6.9].

6.6. Brouwer's Theorem

In this paragraph, we will see a topological application of Stokes's theorem.

Theorem 6.34. *Let U be an open subset of $\mathbf{R}^n$ containing the closed unit ball B^n. Then there does not exist a smooth map of U to $\partial B^n = S^{n-1}$ whose restriction to S^{n-1} is the identity.*

---

1. Statement only, not used in the sequel.

PROOF. Let $F = (f^1, \ldots, f^n)$ be such a map. Then the restriction to S^{n-1} of the differential forms

$$x^1 \, dx^2 \wedge \cdots \wedge dx^n \quad \text{and} \quad f^1 \, df^2 \wedge \cdots \wedge df^n$$

coincide. In particular,

$$\int_{S^{n-1}} x^1 \, dx^2 \wedge \cdots \wedge dx^n = \int_{S^{n-1}} f^1 \, df^2 \wedge \cdots \wedge df^n.$$

Apply Stokes's theorem to each side. On the one hand

$$\int_{S^{n-1}} x^1 \, dx^2 \wedge \cdots \wedge dx^n = \int_{B^n} dx^1 \wedge dx^2 \wedge \cdots \wedge dx^n = \mathrm{vol}(B^n) \neq 0.$$

On the other hand

$$\int_{S^{n-1}} f^1 \, df^2 \wedge \cdots \wedge df^n = \int_{B^n} df^1 \wedge df^2 \wedge \cdots \wedge df^n.$$

However, by hypothesis $\sum_{i=1}^n (f^i)^2 = 1$, and so $\sum_{i=1}^n f^i \, df^i = 0$, from which it follows

$$df^1 \wedge \cdots \wedge df^n = 0,$$

a contradiction. □

Remark. Methods of algebraic topology permit a proof of a C^0 version of this result: there does not exist a continuous map from B^n to S^{n-1} whose restriction to S^{n-1} is the identity. However, as we will see, differential methods also allow us to prove purely topological results, such as the following result due to Brouwer.

Theorem 6.35. *Every continuous map from a closed ball in Euclidean space to itself admits a fixed point.*

PROOF. We will reduce to the preceding result. First, an approximation result will allow us to return to the smooth case.

Lemma 6.36. *For every continuous map of the closed unit ball B^n to itself, and for each $\epsilon > 0$, there exists numbers r_1 and r_2 with $r_1 < 1 < r_2$ and a smooth map g from the open ball $B(r_2)$ to the open ball $B(r_1)$ such that*

$$\forall x \in B^n, \quad \|f(x) - g(x)\| < \epsilon.$$

PROOF. By the Stone-Weierstrass theorem applied to the components of f, there exists a a map $P = (p^1, \ldots, p^n)$, where the p^i are polynomials such that

$$\sup_{x \in B^n} \|f(x) - P(x)\| < \frac{\epsilon}{4}.$$

We check $g = P/(1 + \frac{\epsilon}{2})$ satisfies the required conditions. First, for $x \in B^n$ we have

$$\|g(x)\| \leqslant \frac{1 + \frac{\epsilon}{4}}{1 + \frac{\epsilon}{2}} = 1 - \frac{\epsilon}{4 + 2\epsilon}.$$

By virtue of uniform continuity of g (say on $B(3/2)$), we can then find a $r_2 > 1$ such that $g(x) < 1 - \frac{\epsilon}{8}$ if $\|x\| < r_2$. Further, if $x \in B^n$ we have

$$\|f(x) - g(x)\| \leqslant \|f(x) - P(x)\| + \left\|\left(1 - \frac{1}{1 + \frac{\epsilon}{2}}\right)P(x)\right\| < \epsilon. \qquad \square$$

PROOF OF THEOREM 6.35. We argue by contradiction. Let h be a continuous map without fixed point from the closed ball B^n to itself. Then by compactness of the ball, the number

$$\delta = \inf_{x \in B^n} \|h(x) - x\|$$

is strictly positive. Apply the lemma to h with $\epsilon = \frac{\delta}{2}$, and let g be the smooth map we obtain. Now if $x \in B^n$, we have

$$\|x - g(x)\| > \|x - f(x)\| - \|f(x) - g(x)\| > \frac{\delta}{2}$$

thus $g(x) \neq x$. If $x \in B(r_2) \setminus B^n$, as $g(x) \in B(r_1)$, we have $g(x) \neq x$. This lets us construct a map k from $B(r_2)$ to S^{n-1} whose restriction to S^{n-1} is the identity: the straight line which joins x and $g(x)$ (which is well defined because $g(x) \neq x$) cuts S^{n-1} at two distinct points, and we take $k(x)$ to be the point of intersection such that $g(x)$ is exterior to the segment $[x, k(x)]$ (see Figure 6.6).

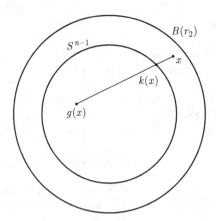

Figure 6.6: Brouwer's theorem by contradiction

Explicitly, $k(x)$ is of the form

$$g(x) + t\big(x - g(x)\big) \quad \text{where } t \geqslant 0,$$

which shows that t is the unique positive root of the second degree equation

$$\|x - g(x)\|^2 t^2 + 2\langle g(x), x - g(x)\rangle t + \|g(x)\|^2 - 1 = 0,$$

and that k is smooth. By applying Theorem 6.34, we obtain a contradiction.

$\square$

Remark. **Of course there exist purely topological proofs of these results which use cohomology of the ball and sphere. The proof we have given has the advantage of being much more elementary. In reality, it is not far from the other proof: we will see in the next chapter how cohomology in maximal dimension can be interpreted in terms of integration.**

6.7. Comments

Integration on chains: towards algebraic topology

We can of course integrate forms of degree p on regular domains of p-dimensional submanifolds. Stokes's theorem

$$\int_{\partial D} \alpha_{|\partial D} = \int_D d\alpha$$

can be seen as a duality formula between operators $\alpha \mapsto d\alpha$ and $D \mapsto \partial D$. However if we want to consider ∂ as a linear map, we must work in a more general setting than domains with boundary: for each p we introduce the vector space $\mathcal{C}_p(M)$ of *p-chains*, namely of formal linear combinations $\sum \lambda_i S_i$ (say with real coefficients) of submanifolds with boundary of dimension p.[2] We then define $\partial : \mathcal{C}_p(M) \to \mathcal{C}_{p-1}(M)$ by posing

$$\partial\Big(\sum \lambda_i S_i\Big) = \sum \lambda_i \partial S_i.$$

We have $\partial \circ \partial = 0$ simply because the boundary of a manifold with boundary is a smooth manifold (without boundary).

We are thus driven to introduce in each dimension the *homology groups*

$$H_p(X) = \mathrm{Ker}(\partial_p)/\mathrm{Im}(\partial_{p-1})$$

---

2. We are cheating a bit in this summary.

which play a dual role to the de Rham groups $H^p(X)$ defined at the beginning of the next chapter. More precisely, by Stokes's theorem, the map

$$(c, \alpha) \longmapsto \int_c \alpha,$$

where c is a chain of dimension p and α is a form of degree p, passes to the quotient and defines a bilinear map from $H_p(X) \times H^p(X)$ to $\mathbf{R}$, and if the manifold X is compact, one may prove that this bilinear form is non-degenerate.

For more details, see [Greenberg-Harper 81], or [Dieudonné 88] once more.

Non orientable manifolds

A Möbius strip, made with scissors and glue does have an area! To have a theory of integration valid on non orientable manifolds, we introduce the notion of *density*. A density is a measure δ on X such that for every chart (U, φ), the pushforward $\varphi_* \delta$ is of the form $f_\varphi \mu$, where μ denotes the Lebesgue measure of $\mathbf{R}^n$ and f_φ is a strictly positive smooth function at each point, with the compatibility condition

$$f_\psi = f_\varphi \circ (\varphi \circ \psi^{-1}) |\mathrm{Jac}(\varphi \circ \psi^{-1})|$$

if the domains of φ and ψ intersect. Imitating the proof of Theorem 6.5, we see there always exists densities on compact manifolds. To define integration with respect to a density, we simply repeat the details of section 6.3. For more details on densities, see [Berger-Gostiaux 88, 3.3].

6.8. Exercises

1. *Orientability and oriented atlases*

a) Show that for every manifold M, the tangent bundle TM is orientable.

b) Show that the product of two orientable manifolds is orientable if and only if each factor is orientable.

2. *Orientability and volume forms*

a) Let X be a submanifold of $\mathbf{R}^n$ of the form $f^{-1}(0)$, where $f : \mathbf{R}^n \mapsto \mathbf{R}$ is a smooth map whose differential at every point $x \in X$ is surjective. Show that X is orientable.

b) Same question for a submanifold of X of $\mathbf{R}^n$ of the form $f^{-1}(0)$, where f is a smooth map from $\mathbf{R}^n$ to $\mathbf{R}^p$ having the same property.

c) Is a submanifold of an orientable manifold orientable?

d) Does an orientable submanifold of an orientable manifold have a "natural" orientation (like the boundary of a regular domain of a oriented manifold)?

3. *The Klein bottle*

a) Show that the quotient K of $T^2 = S^1 \times S^1$ by the group with two elements $\{I, \sigma\}$, where

$$\sigma(\theta, \varphi) = (-\theta, \varphi + \pi)$$

is a non orientable manifold (this manifold is called the *Klein bottle*). Using spherical coordinates, construct a smooth map from X to S^2.

b) Show that the formulas

$$x(u, v) = \left(r + \cos \frac{u}{2} \sin v - \sin \frac{u}{2} \sin 2v\right) \cos u$$

$$y(u, v) = \left(r + \cos \frac{u}{2} \sin v - \sin \frac{u}{2} \sin 2v\right) \sin u$$

$$z(u, v) = \qquad \sin \frac{u}{2} \sin v + \cos \frac{u}{2} \sin 2v$$

(with $r \geqslant 2$) define an immersion of the Klein bottle into $\mathbf{R}^3$.

Sketch this map. Do you recognize one of the designs on the cover of this book?

4. Equip the unit sphere $S^n \subset \mathbf{R}^{n+1}$ with the volume form

$$\sigma = \sum_{i=0}^{n} (-1)^i x^i \, dx^0 \wedge \cdots \wedge \widehat{dx^i} \wedge \cdots \wedge dx^n,$$

and calculate (using convenient values of the function Γ, and taking inspiration from the calculation of the volume of S^n) the integrals

$$I(p_0, \ldots, p_n) = \int_{S^n} (x^0)^{p_0} \ldots (x^n)^{p_n} \sigma.$$

5. Let ω be a differential form of maximal degree on a compact manifold M, and let $X \in C^\infty(TM)$ be a vector field. Show that

$$\int_M L_X \omega = 0.$$

6. Calculate the volume of the domain given by the equation

$$\frac{x^2}{a^2} + \frac{y^2}{a^2} + \frac{z^2}{c^2} \leqslant 1$$

(this is the interior of an ellipsoid of revolution).

7. *Archimedes's formula*

Let ω be the volume form $x\,dy \wedge dz + y\,dz \wedge dx + z\,dx \wedge dy$ on the sphere $S^2 \subset \mathbf{R}^3$.

a) Using spherical coordinates, determine an explicit primitive of ω on $S^2 \smallsetminus \{S \cup N\}$ which is invariant under the rotations about the north-south axis (we denote the north and south poles N and S respectively).

Application. Calculate the area of a "spherical segment"

$$\Sigma_{h,k} = \big\{(x,y,z) \in S^2 : \ h \leqslant z \leqslant k\big\}.$$

b) Determine an explicit primitive of ω on $S^2 \smallsetminus \{S\}$ invariant under the rotations about the north-south axis.

8. *Haar measure of a Lie group*

a) Show that on a Lie group G of dimension n, there exists a nonzero left invariant (resp. right invariant) differential form of degree n that is unique up to a multiplicative factor, and that this form is a volume form. By abuse of terminology, we call the measure defined by this form "the" left (or right) *Haar measure* on G.

b) What is the Haar measure on $\mathbf{R}$? on $\mathbf{R}^*$? on $\mathbf{C}$? on $\mathbf{T}^n$?

c) Show that the Haar measure (either left or right) on $Gl(n,\mathbf{R})$ is defined by the form

$$(\det X)^{-n} \bigwedge_{1\leqslant i\leqslant j\leqslant n} dx_{i,j}.$$

9*. *Compact subgroups of the linear group*

Let G be a compact subgroup of $Gl(n,\mathbf{R})$. Using the Haar measure on G, show that there exists a positive definite quadratic form q on $\mathbf{R}^n$ such that

$$\forall g \in G, \ \forall x \in \mathbf{R}^n, \quad q(gx) = q(x).$$

Deduce that there exists a $g \in Gl(n,\mathbf{R})$ such that

$$gGg^{-1} \subset O(n).$$

10. *Modulus of a Lie group*

a) Let G be a Lie group, and let ω be a left Haar measure on G. Show that $R_g^*\omega$ is also a left Haar measure. Deduce the existence of a smooth function (denoted mod and called the *(Haar) modulus* of G), such that

$$R_g^*\omega = \text{mod}(g)\omega, \quad \forall g \in G.$$

Show that mod is a morphism of G to $\mathbf{R}^*$, and to $\mathbf{R}_+^*$ if G is connected.

b) Show that $\mathrm{mod}(g) = \pm 1$ if G is compact.

c) Show that $\mathrm{mod}(g) = \det(\mathrm{Ad}\, g^{-1})$. Calculate mod for $G = A(1, \mathbf{R})$.

d) A Lie group is called *unimodular* if $\mathrm{mod}(g) = \pm 1$, in other words if the left and right Haar measures coincide. We say in b) that every compact group is unimodular, and in Exercise 9 that $Gl(n, \mathbf{R})$ is unimodular. Show that a connected Lie group is unimodular is and only if the endomorphism $Y \mapsto [X, Y]$ of $\mathfrak{G}$ is trace free for all $X \in \mathfrak{G}$. Application: show that the Heisenberg group and $Sl(n, \mathbf{R})$ are unimodular, but that the group of affine isomorphisms $x \mapsto ax + b$ of the real line is not. **More generally, nilpotent groups and connected semi-simple groups are unimodular for the same reason. (See for example [Hall 03, C.4].)**

11**. *Cauchy-Crofton formula*

Represent the set of oriented lines of the Euclidean plane by $S^1 \times \mathbf{R}$, by associating to each oriented line its "Euler equation" (see Section 2.1.1)

$$x \cos \theta + y \sin \theta - p = 0.$$

a) Show that the differential form $dp \wedge d\theta$ is invariant under the natural action of the group of affine isometries of the plane.

b) Let C be a closed curve of the plane with length L parametrized by arc length s. Let F be the map from $[0, L] \times [0, \pi]$ to $S^1 \times \mathbf{R}$ which associates (s, φ) with the line which passes through the point with arc-length parameter s and making an angle of φ with the (oriented) tangent to the curve at this point. Show that

$$F^*(dp \wedge d\theta) = \sin \varphi ds \wedge d\varphi.$$

c) Deduce for almost every line D, the set $D \cap C$ is finite, and that

$$\int_{S^1 \times \mathbf{R}} \mathrm{card}(D \cap C)\, dp \wedge d\theta = 2L.$$

There are many formulas of this type, cf. [Santaló 76].

12. *Center of mass; Guldin's theorem*

a) Let D be a regular domain of $\mathbf{R}^n$ seen as affine space. Choose an origin, inner product and an orientation, and write

$$m(D) = \frac{\left(\int_D x^i \omega\right)_{1 \leqslant i \leqslant n}}{\int_D \omega},$$

where ω is the volume form defined by the inner product and orientation. Show that $m(D)$ depends neither on the origin, metric nor orientation.

b) Verify that the map F from $S^1 \times \mathbf{R}_+^* \times \mathbf{R}$ to $\mathbf{R}^3$ given by

$$(r, \theta, z) \longmapsto (r \cos \theta, r \sin \theta, z)$$

is a diffeomorphism to $\mathbf{R}^2 \setminus \{0\} \times \mathbf{R}$. If ω is the volume form of $\mathbf{R}^3$ for its canonical inner product structure, calculate $f^*\omega$.

c) Let D be a regular domain of $\mathbf{R}_+^* \times \mathbf{R}$, and a the distance of $m(D)$ to the second coordinate axes. Show that

$$\mathrm{vol}\left(f(D)\right) = 2\pi a \cdot \mathrm{area}(D)$$

(we say that $f(D)$ is the "domain of revolution" given by D).

13. *Archimedes's theorem*

We consider the regular domain D as a solid body immersed in liquid with mass density ρ. Space is measured with respect to an orthonormal frame (i, j, k), the surface of the liquid represented by the plane (i, j) and a constant gravitation force given by $-gk$. Then, by the fundamental principle of hydrostatics, the infinitesimal weight at $p \in \partial D$ is $-zg\rho\sigma n$, where σ is the volume form of ∂D, z is the distance of the surface of the liquid and n is an outward unit normal to ∂D. Then the resulting pressure forces and their moments in $m \in \mathbf{R}^3$ are given by

$$\int_{\partial D} -\rho g z n \sigma \quad \text{and} \quad \int_{\partial D} (\overrightarrow{pm} \wedge zn) \rho g \sigma,$$

where once and for all $\wedge$ denotes the cross product. Show that

$$\int_{\partial D} -\rho g z n \sigma = g\rho \, \mathrm{vol}(D) k$$

and

$$\int_{\partial D} (\overrightarrow{pm} \wedge zn) \rho g \sigma = -g\rho \, \mathrm{vol}(D)\left(\overrightarrow{m(D)m} \wedge k\right).$$

14. Let D be a regular domain of $\mathbf{R}^n$. The *normal derivative* of a smooth (or even C^1) function on D, denoted $\frac{\partial f}{\partial n}$ is a function on ∂D given by $\langle \nabla f, N \rangle$, where N is the outward unit normal.

a) Show that if f and g are two C^2 functions C^2 on D, and if $\sigma_{\partial D}$ denotes the canonical volume form on ∂D, we have

$$\int_D (f\Delta g - g\Delta f)\, dx^1 \wedge \cdots \wedge dx^n = \int_{\partial D} \left(f\frac{\partial g}{\partial n} - g\frac{\partial f}{\partial n}\right)\sigma_{\partial D}$$

(write the left hand side as the integral of a divergence).

b) Show that if f is a harmonic function on D (which is to say if $\Delta f = 0$) that vanishes on ∂D, then $f = 0$ everywhere. Try the same question supposing that $\frac{\partial f}{\partial n}$ vanishes on ∂D.

15. *Tubular neighborhood of a curve*

This exercise uses several elementary results about parametrized curves. Let $c : \mathbf{R}/L\mathbf{Z} \to E$ be an embedded C^2 curve in the Euclidean plane of length L, parametrized by arc-length.

Show that the area of a tubular neighborhood $V_r(c)$ (see Exercise 24 of Chapter 3) is equal to $2rL$.

**More generally, if M is a submanifold of dimension p in $\mathbf{R}^n$ equipped with the standard Euclidean norm, we have

$$\mathrm{vol}\left(V_r(M)\right) = \sum_{k=0}^{[\frac{p}{2}]} a_k r^{n-p+2k},$$

where $a_0 = \mathrm{vol}(M)\,\mathrm{vol}\left(B^{n-p}(1)\right)$, and the a_k only depend on the Riemannian geometry of M. For an explanation of these invariants, see [Berger-Gostiaux 88, 6.9.8] and the references therein.** See also Theorem 8.24.

Cohomology and Degree Theory

7.1. Introduction

In the preceding chapters we saw several ways to show that two open subsets of $\mathbf{R}^n$, and more generally two manifolds, are not diffeomorphic.

For example, Theorem 5.14 shows that on $\mathbf{R}^2 \smallsetminus \{0\}$ there exist forms which are not exact, thus this open subset of $\mathbf{R}^2$ is not diffeomorphic to $\mathbf{R}^2$. The same argument works for $\mathbf{R}^n \smallsetminus \{0\}$. The "solid angle" form is closed but not exact because its restriction to a sphere with center 0 is a volume form which has nonzero integral.

Another type of argument is possible: the complex exponential furnishes a nontrivial covering of $\mathbf{R}^2 \smallsetminus \{0\}$ by $\mathbf{R}^2$, and so we see that $\mathbf{R}^2 \smallsetminus \{0\}$ is not simply connected. This argument is in some sense better as we see that $\mathbf{R}^2$ and $\mathbf{R}^2 \smallsetminus \{0\}$ are not homeomorphic, but in some sense worse because the argument does not extend to higher dimensions.

The de Rham cohomology spaces, which form the subject of this chapter, take into account the failure of the Poincaré lemma: in each degree these spaces are the quotient of the vector space of closed forms by the subspace of exact forms. These are the simplest invariants which allow one to show that two manifolds are not diffeomorphic.

We begin by describing what happens in the maximal dimension. The point of departure is a well known but under-appreciated analytical fact: the primitives of a periodic function are periodic if and only if the integral of this function over a period is zero. While this is a criteria for exactness on closed forms on the circle, this property extends to compact connected orientable manifolds. The integral of forms of degree $n = \dim X$ on such a manifold gives an isomorphism between the de Rham space $H^n(X)$ of maximum degree and $\mathbf{R}$ (Theorem 7.9).

This information, which appears relatively weak if we only consider manifolds, becomes much richer as soon as we are interested in maps between manifolds. Indeed the correspondence which associates to a manifold X its de Rham space $H^k(X)$ of degree k is what mathematicians call a functor. If $f : X \to Y$ is a smooth map, we obtain a linear map $f^* : H^k(Y) \to H^k(X)$ (this is passing to the quotient in the inverse image seen in Chapter 5). This functor is called "contravariant" as it reverses the direction of the arrow (for functorial machinery, which is useful but must not be abused, see the beginning of [Douady 05]) or [MacLane 71].

For compact orientable manifolds of the same dimension, in maximum degree, f^* is a linear map between spaces of dimension 1, this is to say multiplication by a number. This number which is *a priori* any real number, is in fact an *integer*, called the degree, which also has a geometric interpretation.

Very roughly, the degree of a map f from a manifold X to a manifold Y of the same dimension is the number of solutions of the equation $f(x) = y$ (X and Y here are assumed to be connected and compact). We really want to say that such a number depends continuously on y *and* f. Earlier we say in Chapter 2 (see Theorem 2.14) that on the regular values of f, the cardinality of $f^{-1}(y)$ is finite and locally constant. But this point of view is naive: the drawing below of a graph of a function from $\mathbf{R}$ to $\mathbf{R}$ (easily embeddable as $S^1 = P^1\mathbf{R}$), shows that one must take into account the orientation (here, of the fact that f can be increasing or decreasing). As the picture suggests, the number of solutions of the equation $f(x) = y$ does not depend on the regular

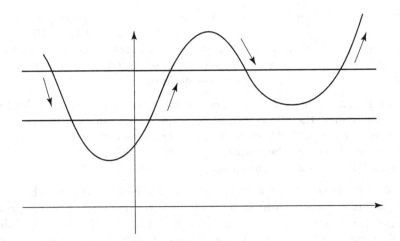

Figure 7.1: Equation $f(x) = y$

value y only if we impose the sign $+$ for the x if f preserves orientation, and the sign $-$ if x reverses orientation. This is stated carefully in Theorem 7.18.

The continuity of the degree with respect to f uses the first point of view, which gives an expression by means of differential forms. As it is simultaneously continuous and an integer, the degree is constant on every continuous family of smooth maps from a given manifold to another (this is invariance by homotopy, see Section 7.4.3). This property has important applications: fundamental theorem of algebra, the index of a vector field at zero, the linking number of curves in $\mathbf{R}^3$.

We then describe cohomology in any degree. The calculation here relies on combining two ideas:

- replace the space under study by a simpler space having the same cohomology (a star-shaped open subset with respect to a point) by remarking that cohomology depends only on the "homotopy type" (the details are given in Section 7.7);

- using a convenient decomposition of the space under study into subspaces with known cohomology.

An example of this technique is the Mayer-Vietoris sequence, explained in Section 7.8.

Finally, we mention that de Rham cohomology spaces are in fact topological invariants, which is to say invariants under homeomorphism between differentiable manifolds (see for example [Bott-Tu 86, II.9] for a precise statement and proof).

7.2. De Rham Spaces

If X is a manifold, let $F^k(X)$ be the vector space of closed differential forms of degree k. As $d \circ d = 0$, the vector space $d\Omega^{k-1}(X)$ of exact forms of degree k is a vector subspace of $F^k(X)$ by the Poincaré lemma. The quotient vector space of closed forms by exact forms measures the "failure of exactness" of closed forms of degree k on X.

Definitions 7.1

a) *The quotient*

$$F^k(X)/d\Omega^{k-1}(X)$$

  is a vector space called the k-th cohomology space of X and is denoted $H^k(X)$.

b) If $\alpha \in F^k(X)$, its image $[\alpha]$ under the quotient map is called the cohomology class of α.

c) Two closed forms are said to be cohomologous if they define the same cohomology class, which is to say their difference is exact.

Remarks

a) One may prove if X is compact then the spaces $H^k(X)$ are finite dimensional (see for example [Bott-Tu 86], and Exercise 21 for an idea).

b) There exist other notions of cohomology, see [Greenberg-Harper 81]. The subject of this chapter is de Rham cohomoloy, named in honor of Georges de Rham, a Swiss mathematician of the past century. We will omit his name hereafter.

Examples

a) As there do not exist forms of degree -1, the group $H^0(X, \mathbf{R})$ is equal to $F^0(X)$: it consists of functions with zero differential, which is to say functions that are locally constant. In other words, $H^0(X, \mathbf{R}) \simeq \mathbf{R}^c$, where c is the number of connected components of X.

b) $H^p(X)$ is always zero if $p > \dim(X)$.

c) The Poincaré lemma says exactly that if U is a star-shaped open subset then $H^p(U) = 0$ if $p > 0$.

d) We come to the simplest compact manifold, the circle S^1, which will be convenient to consider as $\mathbf{R}/\mathbf{Z}$. Now a differential form $f(t)\, dt$ on $\mathbf{R}$ passes to the quotient as a form on S^1 if and only if the function f is periodic with period 1, and is exact if and only if the function f has a periodic primitive with period 1. However if f is periodic with period 1, the integral

$$\int_t^{t+1} f(u)\, du$$

is independent of t. Indeed, if we regard this as a function of t, its derivative is $f(t+1) - f(t) = 0$! The value of this integral is precisely the integral of the quotient form on S^1. In other words

Theorem 7.2. A form $\omega \in \Omega^1(S^1)$ is exact if and only if it has zero integral, and the integration map

$$\alpha \longmapsto \int_{S^1} \alpha$$

induces an isomorphism from $H^1(S^1)$ to $\mathbf{R}$. This isomorphism depends only on the choice of orientation of the circle.

This result extends, as we will now see, to every compact oriented manifold. We can interpret it as converse to Stokes's theorem for compact manifolds (see Corollary 6.26).

7.3. Cohomology in Maximum Degree

We start with an "elementary" property concerning the space $\mathbf{R}^n$, which is analogous to the previous result.

Theorem 7.3. *A differential form $\alpha \in \Omega^n(\mathbf{R}^n)$ with compact support admits a primitive if and only if*

$$\int_{\mathbf{R}^n} \alpha = 0$$

We proceed by induction on n. The result is a consequence of the following lemma, which is a version of the theorem "with parameters", and allows the induction.

Lemma 7.4. *Let $u \mapsto \alpha(u)$ be a family of differential forms of degree n with compact support contained in $(0,1)^n$ that depends differentiably on a parameter $u \in U \subset \mathbf{R}^k$. Then, if*

$$\int_{(0,1)^n} \alpha(u) = 0,$$

there exists a family $u \mapsto \beta(u)$ of compactly support forms of degree $n-1$ such that

$$\forall u \in U, \quad \beta(u) = d\alpha(u).$$

PROOF. We proceed by induction on n. For $n = 1$, the result is immediate: $\alpha(u)$ may be written as $f(x, u)dx$, where $f \in C^\infty((0,1) \times U)$. Further, there exists a closed interval $[a(u), b(u)] \subset (0,1)$ such that $f(x, u) = 0$ if $x \notin [a(u), b(u)]$, and by hypothesis

$$\int_{(0,1)} f(x, u)\, dx = \int_{a(u)}^{b(u)} g(x, u)\, dx = 0.$$

So the function

$$g(x, u) = \int_0^x f(t, u)\, dt$$

gives a family of primitives of $\alpha(u)$ satisfying the required conditions.

Now suppose the property is true for $n-1$, and let $\alpha(u)$ be a family of n-forms on $(0,1)^n$ satisfying the conditions of the lemma. Then

$$\alpha(u) = f(x^1, \ldots, x^n, u)dx^1 \wedge \cdots \wedge dx^n$$

may be written $\beta(x^n, u) \wedge dx^n$, where $\beta(x^n, u)$ is a form of degree $n-1$ on $(0,1)^{n-1}$ parametrized by $(x^n, u) \in (0,1) \times U$. Let $\sigma \in \Omega^{n-1}((0,1)^{n-1})$ be a compactly supported form such that $\int_{(0,1)^{n-1}} \sigma = 1$. We write

$$\overline{\beta}(x^n, u) = \left(\int_{(0,1)^{n-1}} f(x^1, \ldots, x^n, u)\, dx^1 \ldots dx^{n-1} \right)\sigma.$$

By construction, the form $\beta(x^n, u) - \overline{\beta}(x^n, u)$ has compact support in $(0,1)^{n-1}$ and zero integral, and by the induction hypothesis,

$$\beta(x^n, u) - \overline{\beta}(x^n, u) = d\big(\gamma(x^n, u)\big),$$

where $\gamma(x^n, u)$ is a family of forms of degrees $n-1$ parametrized by (x^n, u).

Denote by $\gamma'(u)$ the family of forms (depending on the parameter u) on $(0,1)^n$ whose restriction to the plane $x^n = x_0^n$ is $\gamma(x_0^n, u)$. In other words, $\gamma'(u)$ is obtained taking $\gamma(x^n, u)$ and allowing x^n to be regarded as a *parameter* instead of a *variable*. Then, with an obvious abuse of notation for the left hand side, we have

$$d\gamma'(u) \wedge dx^n = d\gamma(x^n, u) \wedge dx^n$$

and consequently (by omitting the parameter u),

$$\alpha = \overline{\beta} \wedge dx^n + d\gamma' \wedge dx^n.$$

The first term may be written $\sigma \wedge F(x^n, u)dx^n$, where the function $F(\cdot, u)$ has zero integral. It is therefore of the form $\sigma \wedge dG$, where the function G is compactly supported in $(0,1)$. This is, up to sign, the differential of a compactly supported form $G\sigma$ (since $d\sigma = 0$). Similarly, the second term is the differential of $x^n d\gamma'$ up to sign, which is also of compact support as we can see from its definition. □

Remark. The careful reader will see several implicit uses of the projections $p : (x^1, \ldots, x^n) \mapsto (x^1, \ldots, x^{n-1})$, inclusions $i_{x^n} : (x^1, \ldots, x^{n-1}) \mapsto (x^1 \ldots, x^{n-1}, x^n)$, and the pullbacks of forms by these maps. We made these omissions for clarity instead of rigor.

The property stated below for compact manifolds is deduced by using successive applications of this theorem to suitable open sets.

Theorem 7.5. *Let X be a compact connected oriented manifold of dimension n. A differential form of degree n on X is exact if and only if its integral vanishes.*

PROOF. The fact that an exact form of maximum degree on a compact oriented manifold has vanishing integral follows from Stokes's theorem. To prove the converse, we will need "reference formulas" for the forms σ which appeared in Lemma 7.4.

Lemma 7.6. *For every open subset U of an oriented manifold X, there exits a form $\sigma \in \Omega^n(X)$ with compact support contained in U such that $\int_X \sigma = 1$.*

PROOF. Let $\varphi : U \to \mathbf{R}^n$ be a chart preserving the orientation, and f a smooth function with compact support contained in $\varphi(U)$ and with integral equal to 1. It suffices to take σ as the form which equals $\varphi^*(f \, dx^1 \wedge \cdots \wedge dx^n)$ on U and 0 on $X \smallsetminus \varphi^{-1}(\mathrm{Supp}(f))$. $\qquad\square$

The "only if" direction of Theorem 7.5 is a consequence of the following result, which has its own uses:

Proposition 7.7. *Under the same hypothesis as in Theorem 7.5, let α and $\sigma \in \Omega^n(X)$. Suppose that $\mathrm{Supp}(\sigma)$ is included in an open subset of a chart, and that $\int_X \sigma = 1$. Then there exists a form $\beta \in \Omega^{n-1}(X)$ and a real number t such that*

$$\alpha - t\sigma = d\beta.$$

Further, $t = \int_X \alpha$.

By covering X by a finite number of open charts and taking a partition of unity subordinate to this cover, by additivity we reduce to the case where α has support contained in an open subset of a chart. If this open subset is the same as the corresponding one for σ, we are finished by Theorem 7.3. In general this is not the case, so we make a sequence of corrections to reduce to this case.

Lemma 7.8. *Let X be a connected topological space, and $(V_i)_{i \in I}$ an open covering of X. Then for all $x, y \in X$, there exists a finite sequence $V_{i_0}, \ldots, V_{i_p}$ of open subsets of the covering such that*

$$x \in V_{i_0}, \;\; y \in V_{i_p} \;\; and \;\; V_{i_k} \cap V_{i_{k+1}} \neq \emptyset \;\; if \;\; 0 \leqslant k < p.$$

PROOF. For $x \in X$, denote by $C(x)$ the set of $y \in X$ for which the property stated is satisfied by the pair (x, y). It is clear that $C(x)$ is an nonempty open set and that $C(x) = C(x')$ as soon as $C(x) \cap C(x') \neq \emptyset$. Thus, the $C(x)$ form a partition of X into open subsets, and therefore $C(x) = X$. $\qquad\square$

END OF THE PROOF OF THEOREM 7.5. By the lemma, we are in the following situation: there exists a sequence $U_0, \ldots, U_p$ of open subsets of charts, such that

$$\operatorname{Supp}\alpha \subset U_0; \quad \operatorname{Supp}\sigma \subset U_p; \quad U_k \cap U_{k+1} \neq \emptyset \ \text{ if } \ 0 \leqslant k < p.$$

We denote the corresponding charts by φ_k, and suppose these charts preserve the orientation and that $\varphi_k(U_k) \subset (0,1)^n$. By Lemma 7.6, there exists for every k $(0 \leqslant k < p)$ a differential form σ_k of degree n with support contained in $U_k \cap U_{k+1}$ such that $\int_X \sigma_k = 1$. By Theorem 7.3, there exists a differential form β_0 with compact support contained in $\varphi_0(U_0)$ such that

$$\varphi_0^{-1*}\alpha - t\varphi_0^{-1*}\sigma_0 = d\beta_0,$$

where

$$t = \int_X \alpha = \int_{\varphi_0(U_0)} \varphi_0^{-1*}\alpha.$$

We write $\varphi_0^{-1*}\beta_0 = \gamma_0$. The form $\gamma_0 \in \Omega^{n-1}(U_0)$ has compact support contained in U_0. It thus extends to a form on all of X that vanishes on $M \smallsetminus \operatorname{Supp}(\gamma_0)$, and we again have

$$\alpha - t\sigma_0 = d\gamma_0.$$

Similarly, by applying Theorem 7.3 to $\varphi_k(U_k)$ for $0 \leqslant k < p$, we see that there exists a form β_k with compact support contained in $\varphi_k(U_k)$ such that

$$\varphi_k^{-1*}\sigma_k - \varphi_k^{-1*}\sigma_{k+1} = d\beta_k,$$

and by the preceding argument, a form $\gamma_k \in \Omega^{n-1}(X)$ such that

$$\sigma_k - \sigma_{k+1} = d\gamma_k.$$

For the same reasons, we have also

$$\sigma_p - \sigma = d\gamma_p.$$

Finally, we find that

$$\alpha - t\sigma = d\left(\gamma_0 + t\sum_{k=0}^{p} \gamma_k\right). \qquad\qquad \square$$

Putting everything together, we arrive at the following fundamental result.

Theorem 7.9. *If X is a compact connected orientable manifold of dimension n, $H^n(X)$ is of dimension 1; further, given an orientation of X, the integration map*

$$\alpha \longmapsto \int_X \alpha$$

from $\Omega^n(X)$ to $\mathbf{R}$ passes to the quotient as an isomorphism between $H^n(X)$ and $\mathbf{R}$.

PROOF. It suffices to show the second part. The map $\alpha \mapsto \int_X \alpha$ passes to the quotient thanks to Stokes's theorem; Theorem 7.5 says exactly that the map obtained by passing to the quotient is injective, and Lemma 7.6 says that it is surjective. □

Corollary 7.10. *If X is a compact non orientable manifold of dimension n, $H^n(X) = 0$.*

PROOF. It suffices to give the proof when X is connected. We use the terminology of Section 6.2.3. Let $p : \widetilde{X} \to X$ be the orientation covering, and let $\alpha \in F^n(X)$. As $\widetilde{X}$ is compact, *connected*, and orientable, there exists a real number t and an $(n-1)$-form β on $\widetilde{X}$ such that

$$p^*\alpha = t\Psi + d\beta.$$

Taking the pullback under σ, and recalling that $p \circ \sigma = p$, we obtain

$$p^*\alpha = -t\Psi + d(\sigma^*\beta).$$

Thus

$$p^*\alpha = \frac{d(\beta + \sigma^*\beta)}{2}.$$

However the form $\beta + \sigma^*\beta$ is σ-invariant, and therefore of the form $p^*\gamma$. Thus $\alpha = \frac{d\gamma}{2}$ since p^* is injective. □

One may prove that the maximum degree cohomology of a noncompact manifold is zero (see for example [Bott-Tu 86, p. 87]). We will not use this result.

7.4. Degree of a Map

7.4.1. The Case of a Circle

We will again consider two points of view, by considering S^1 as the unit circle in the Euclidean plane (or the set of complex numbers of unit modulus), and as the quotient $\mathbf{R}/2\pi\mathbf{Z}$. The identification between these two models is given by the quotient map $\theta \mapsto p(\theta) = (\cos\theta, \sin\theta)$, which we also view as $\theta \mapsto p(\theta) = e^{i\theta}$.

Definition 7.11. *Let $f : \mathbf{R} \to S^1$ be a continuous function. A* lift *of f is a continuous function $F : \mathbf{R} \to \mathbf{R}$ such that $f = p \circ F$.*

Theorem 7.12. *Liftings always exist; two liftings of the same function differ by an integer multiple of 2π.*

PROOF. First we consider the case of a function f defined on a compact interval $[a, b]$. As f is uniformly continuous, there exists a $\eta > 0$ such that if $|x - y| < \eta$ then $\|f(x) - f(y)\| < 1$ (here, we take the standard Euclidean norm of $\mathbf{R}^2$, or, what amounts to the same thing, the modulus in $\mathbf{C}$). Thus the image of every interval of length less than η is contained in a semicircle, which is to say an open subset trivializing the cover p. By decomposing $[a, b]$ into a union of contiguous intervals with length less than η, we may define f piece by piece by composition with a well chosen local inverse of p. In this procedure, the only arbitrary choice is of $F(a)$, subject to the single constraint $e^{iF(a)} = f(a)$.

If f is defined on $\mathbf{R}$, we obtain a lift by taking appropriate lifts of restrictions of f to the intervals $[n, n + 1], n \in \mathbf{Z}$.

If F_1 and F_2 are two lifts of the same function, we have

$$e^{i\left(F_1(x) - F_2(x)\right)} = 1,$$

so that $F_1(x) - F_2(x)$ is always an integer multiple of 2π. However a continuous function on $\mathbf{R}$ with integer values is constant. □

Remark. This result is a special case of the monodromy theorem, namely Theorem 2.45.

Now if g is a continuous map from S^1 to S^1, we can apply the above to $f = g \circ p$.

Definition 7.13. *The* degree *if a continuous map g from S^1 to S^1 is the number $\frac{F(2\pi) - F(0)}{2\pi}$, where F is a lift of $f = g \circ p$ (this number does not depend on the lift by Theorem 7.12).*

As $p\left(F(2\pi)\right) = p\left(F(0)\right)$, this number is an *integer*. We note also the following property, which makes the link with the definition of degree in higher dimensions.

Proposition 7.14. *If g is of class C^1,*

$$\deg(g) = \frac{1}{2\pi} \int_{S^1} g^*(d\theta).$$

PROOF. Noting that the angular form is the restriction of $\mathrm{Im}\frac{dz}{z}$ to S^1, we see in the notation of Definition 7.13, that

$$g^*(d\theta) = \int_0^{2\pi} f^*(d\theta)$$

$$= \int_0^{2\pi} F^*(p^*d\theta)$$

$$= \int_0^{2\pi} F'(\theta)\, d\theta.$$

Everything relies on the fact that despite the notation $d\theta$ is not exact on S^1, while $p^*(d\theta)$ is indeed exact on $\mathbf{R}$. □

By examining the proof of Theorem 7.12 closely, one can check that the degree is continuous in the topology of uniform convergence and therefore constant on each homotopy class of maps. We leave as a further exercise to the reader to prove that two continuous maps form S^1 to S^1 of the same degree are homotopic (or see [Berger-Gostiaux 88]).

We now consider higher dimensions.

7.4.2. Definition and Basic Properties in the General Case

To take full advantage of cohomology, it is necessary to use the fact that the correspondence

$$\text{manifold} \longrightarrow \text{cohomology vector space}$$

is "functorial", which is to say that to every smooth map between manifolds, there is a natural association of a linear map $h^k(f)$ between cohomology spaces of degree k.

Proposition 7.15. *Let X and Y be two manifolds, and let $f : X \to Y$ be a smooth map. Then, for every integer k, the map*

$$f^* : \Omega^k(Y) \longrightarrow \Omega^k(X)$$

sends closed forms to closed forms, exact forms to exact forms, and passes to the quotient as a linear map

$$h^k(f) : H^k(Y) \longrightarrow H^k(X).$$

If Z is a third manifold, and $g : Y \to Z$ a smooth map,

$$h^k(g \circ f) = h^k(f) \circ h^k(g).$$

In particular, if f is a diffeomorphism, $h^k(f)$ is a vector space isomorphism.

PROOF. These assertions are immediate consequences of the relations

$$f^* \circ d = d \circ f^* \quad \text{and} \quad (g \circ f)^* = f^* \circ g^*. \qquad \square$$

This property is particularly nice if we consider the n-th cohomology space of two n-dimensional manifolds.

Corollary 7.16. *Let X and Y be two compact connected oriented manifolds of the same dimension n, and let $f : X \to Y$ be a smooth map. Then there exists a real number, denoted $\deg(f)$, such that*

$$\forall \alpha \in \Omega^n(Y), \quad \int_X f^*\alpha = \deg(f) \int_Y \alpha.$$

PROOF. Let $\sigma \in \Omega^n(Y)$ be such that $\int_Y \sigma = 1$. Then there exists a form $\beta \in \Omega^{n-1}(Y)$ such that

$$\alpha - \left(\int_Y \alpha \right) \sigma = d\beta.$$

Then

$$f^*\alpha - \left(\int_Y \alpha \right) f^*\sigma = f^*(d\beta) = d(f^*\beta).$$

By integrating we obtain

$$\int_X f^*\alpha = \left(\int_Y \alpha \right)\left(\int_X f^*\sigma \right),$$

and the stated property with $\deg(f) = \int_X f^*\sigma$. $\qquad \square$

Remark. Clearly this property can be stated in a more conceptual manner at the level of cohomology of degree n. Denoting $\int_X$ (resp. $\int_Y$) the isomorphism between $H^n(X)$ (resp. $H^n(Y)$) and $\mathbf{R}$ given by integration of forms of degree n, and a the map $x \mapsto ax$ from $\mathbf{R}$ to $\mathbf{R}$, we have a commutative diagram

$$
\begin{array}{ccc}
H^n(Y) & \xrightarrow{\; h^n(f) \;} & H^n(X) \\
{\scriptstyle \int_Y} \downarrow & & \downarrow {\scriptstyle \int_X} \\
\mathbf{R} & \xrightarrow{\; \deg(f) \;} & \mathbf{R}
\end{array}
$$

These spaces are of dimension 1, and by using the isomorphisms with $\mathbf{R}$ given by integration, $h^n(f)$ becomes a linear map from $\mathbf{R}$ to $\mathbf{R}$, which is to say multiplication by a real number, namely $\deg(f)$.

Definition 7.17. *The number $\deg(f)$ is called the degree of f.*

Remark. If $X = Y$, and if we give the domain and range the same orientation, the degree of $f : X \to Y$ is independent of choice of orientation.

In fact we will see that $\deg(f)$ is an *integer*.

Theorem 7.18. *Let X and Y be two compact oriented manifolds of the same dimension, and $f : X \to Y$ a smooth map. For every regular value y of f, we have:*

$$\deg f = \sum_{x \in f^{-1}(y)} \mathrm{or}_x f$$

where $\mathrm{or}_x f = +1$ if $T_x f$ preserves the orientation, and $\mathrm{or}_x f = -1$ otherwise. In particular

  i) $\deg f$ *is an integer, which changes sign if we change the orientation of X or of Y;*

  ii) *if f is not surjective, $\deg f = 0$;*

  iii) *the parity of the integer $\mathrm{card}\big(f^{-1}(y)\big)$ is the same for every regular value y of f.*

PROOF. Let y be a regular value, which exists by Sard's theorem (Theorem 2.49). By Theorem 2.14, $f^{-1}(y)$ is finite, and there exists an open subset V containing y such that if $f^{-1}(y) = \{x_1, \ldots, x_k\}$, we have

$$f^{-1}(V) = \bigcup_{i=1}^{k} U_i,$$

where the U_i are mutually disjoint, $x_i \in U_i$ and $f_{|U_i}$ is a diffeomorphism from U_i to V. Now let σ be a differential form of degree $n = \dim Y$, with support contained in V, and such that

$$\int_Y \sigma = \int_V \sigma = 1.$$

By the definition of degree, $\deg f = \int_X f^* \sigma$. However, clearly

$$\int_X f^* \sigma = \sum_{i=1}^{k} \int_{U_i} f^* \sigma,$$

and

$$\int_{U_i} f^* \sigma = \pm \int_V \sigma,$$

depending on whether the diffeomorphism $f_{|U_i}$ preserves or reverses orientation. If $y \notin f(X)$, we repeat the above with a form σ with support contained in the open subset $V = Y \setminus f(X)$, in which case $f^* \sigma = 0$. $\qquad\square$

Example. Let Y be a compact orientable manifold and $p : X \to Y$ a covering map. By Lemma 6.4, X is also orientable, and the two manifolds can be equipped with orientations that are preserved by p. Under these conditions, if X is compact, the degree of p that we have just defined is the same as the degree of p of the covering map.

7.4.3. Invariance of the Degree under Homotopy; Applications

Since the degree is an integer, we can expect its invariance under a continuous deformation. The notion of homotopy, which generalizes homotopy of loops (see Definition 2.41) allows us to make this remark precise.

Definition 7.19. *Two smooth maps f and g from a manifold X to a manifold Y are said to be* homotopic *if there exists a continuous map $H : X \times [0,1] \to Y$ (called a* homotopy *between f and g) such that*

i) $\forall x \in X, H(x,0) = f(x), H(x,1) = g(x)$;

ii) *the map $x \mapsto H(x,t)$ is smooth for all t.*

One can prove (see [Dieudonné 72]) that if there exists a continuous map H satisfying i), there also exists a smooth homotopy H' on $X \times [0,1]$.

Examples

a) Two maps with values in a convex open subset or a star-shaped open subset are always homotopic.

b) A more interesting case is that of two smooth maps f and g from a manifold X to S^n such that

$$\forall x \in X, \quad \|f(x) - g(x)\| < 2,$$

where we have denoted the Euclidean norm of $\mathbf{R}^{n+1}$ by $\| \ \|$, and so this condition says that $f(x)$ and $g(x)$ are never diametrically opposite. Then f and g are homotopic. As the segment $[f(x), g(x)]$ never intersects the origin, we have the homotopy

$$H(x,t) = \frac{tf(x) + (1-t)g(x)}{\|tf(x) + (1-t)g(x)\|}.$$

In fact, this is a general property.

Theorem 7.20. *Let X and Y be two compact manifolds, and let d be a metric defining the topology of Y. Define*

$$\rho(f,g) = \sup_{x \in X} d\big(f(x), g(x)\big).$$

There exists an $r > 0$ such that any two maps f, g from X to Y such that $\rho(f, g) < r$ are homotopic.

PROOF. First observe that if d and d' are two metrics defining the topology of Y, the identity map from (Y, d) to (Y, d') is *uniformly continuous*, so it suffices to prove the result for *a* particular metric. We embed Y into some R^N (see Theorem 3.7) and we take d to be the distance induced by the Euclidean distance on R^N.

Recall from the tubular neighborhood theorem (see Exercise 24 of Chapter 3) that there exists an $r > 0$ such that the open subset of $\mathbf{R}^N$ defined by

$$V_r(Y) = \{y \in \mathbf{R}^N : d(y, Y) < r\}$$

has the following properties:

1) for every $y \in V_r(Y)$ there exists a unique point of Y (which we denote $p(y)$ such that $d\big(y, p(y)\big) = d(y, Y)$;
2) the map $y \mapsto p(y)$ so defined is smooth.

Now let f and g be two smooth maps from X to Y such that $\rho(f, g) < r$. Then every point of the segment $[f(x), g(x)]$ in $\mathbf{R}^N$ belongs to $V_r(Y)$. It then suffices to take

$$H(t, x) = p\big(tf(x) + (1 - t)g(x)\big)$$

for the required homotopy.   □

This result shows that the homotopy classes form *open* subsets of $C^\infty(X, Y)$ in the topology of uniform convergence. It is not surprising then that we can associate *discrete* invariants to a homotopy class.

Theorem 7.21. *If f and g are two homotopic smooth maps between two compact connected orientable manifolds of the same dimension, then* $\deg(f) = \deg(g)$.

PROOF. If H is a homotopy between f and g, write $f_t(x) = H(x, t)$. Let $\sigma \in \Omega^n(Y)$ be a form of maximum degree and integral 1. Then the family of forms $f_t^*\sigma$ depends continuously on the real parameter t, and by the remark which follow definition 6.16 the function

$$t \longmapsto \int_X f_t^*\sigma$$

is continuous. On the other hand, its value at t is $\deg(f_t)$, which is to say an integer. However a continuous function with integer values defined on an interval is constant. In particular, $\deg(f) = \deg(f_0) = \deg(f_1) = \deg(g)$.   □

Corollary 7.22. *If X is a compact orientable manifold, the identity map from X to X is not homotopic to a constant map (or even to a non-surjective map).*

Remark. We will see in the appendix a more general version of this result, which includes compact non orientable manifolds. However the proof appeals to a version of Sard's theorem not given in this book. We emphasize that compactness is essential, as the counterexample of $\mathbf{R}^n$ shows.

This invariance of the degree under homotopy has numerous applications. We will only see a few of the main ones. We begin with another proof of Theorem 6.17, called the "hairy ball theorem".

Theorem 7.23. *If n is even, every vector field on S^n has a zero.*

PROOF. As in 6.17, we consider S^n embedded in $\mathbf{R}^{n+1}$. A vector field is identified with a map $X : S^n \to \mathbf{R}^{n+1}$ such that $\langle x, X(x) \rangle = 0$. If X is everywhere nonvanishing, we can replace X by $\frac{X}{\|X\|}$ to obtain a vector field of norm 1. Define

$$H(x,t) = (\cos \pi t)x + (\sin \pi t)X(x).$$

As $\|H(x,t)\| = 1$, we have indeed defined a homotopy between $H(x,0) = x$ and $H(x,1) = -x$. The map $x \mapsto -x$ has degree $(-1)^{n+1}$, from which it follows

$$(-1)^{n+1} = 1,$$

and we conclude n is odd.   □

Next, we note that an approximation argument allows us to define the degree for maps which are only continuous.

Lemma 7.24. *Let X and Y be two compact manifolds. The set of smooth maps from X to Y is dense in the set of continuous maps with the topology of uniform convergence.*

PROOF. We proceed as in Theorem 7.20 and reuse the terminology and notation given there. Let $V_r(Y)$ be a tubular neighborhood of Y in a Euclidean space $\mathbf{R}^N$, and let $\epsilon < r$. If $f : X \to Y$ is continuous, then by the Stone-Weierstrass theorem, there exists a smooth function $g : X \to \mathbf{R}^N$ such that $d(f,g) < \epsilon$. Then $p \circ g$ is a smooth map from X to Y which is within ϵ of f.   □

Now, with the same notation as Theorem 7.20, if $f : X \to Y$ is smooth, every smooth approximation within $\frac{r}{2}$ (and we have just seen that there are such maps) are homotopic and therefore have the same degree. We define this number to be the degree of f.

7.4.4. Index of a Vector Field

By making our study of the zeros of a vector field more precise, we can refine the hairy ball theorem. To every isolated zero x of a vector field ξ, we may associate an integer called the index of ξ at x in the following way: we choose a chart (U, φ) such that $\varphi(x) = 0$ and that x is the only zero of ξ contained in U. The index, denoted $\mathrm{ind}_x \xi$, is the degree of the map

$$y \longmapsto \frac{\eta(y)}{\|\eta(y)\|} \quad \text{from } S^{n-1}(\epsilon) \text{ to } S^{n-1}, \qquad (**)$$

where $\eta = \varphi_* \xi$, $S^{n-1}(\epsilon)$ denotes a small sphere with center 0 and radius ϵ contained in $\varphi(U)$.

Theorem 7.25. *The degree of the map* $(**)$ *is independent of both ϵ and the chart.*

PROOF. The first part is clear. For the second, it suffices to show that the degree of a map analogous to $(**)$ defined by a diffeomorphism f from an open ball $B(0, r) \subset \mathbf{R}^n$ is independent of f. We first note that f is homotopic to $T_0 f$ through a family of diffeomorphisms by Exercise 3, and that under these conditions, the maps

$$y \longmapsto \frac{f_* \eta}{\|f_* \eta\|} \quad \text{and} \quad y \longmapsto \frac{A_* \eta}{\|A_* \eta\|}$$

(here $T_0 f = A$) are themselves homotopic. We have therefore reduced to the case of a linear map. If $\det A > 0$, A is homotopic to the identity through a family of invertible linear maps because $Gl^+(n, \mathbf{R})$ is connected.

Finally, if f is a orientation-preserving diffeomorphism from $B(0, r)$ to its image, the maps $\frac{f_* \xi(y)}{\|f_* \xi(y)\|}$ and $\frac{\xi(y)}{\|\xi(y)\|}$ have the same degree. If f reverses orientation, the same connectedness argument allows us to reduce to the case of a reflection σ with respect to a hyperplane. But then

$$\sigma_* \xi = \sigma \circ \xi \circ \sigma^{-1},$$

and the same relation persists for $\frac{\xi}{\|\xi\|}$. The degree remains the same by Lemma 7.30 below. $\qquad\square$

Example in two dimensions

This being a local question, it suffices to work in a neighborhood of 0 in $\mathbf{R}^2$. Figures 7.2 and 7.3 show vector fields that vanish at the origin.

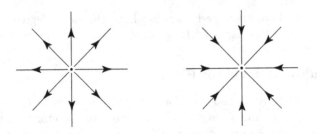

Figure 7.2: Index 1: a source and sink

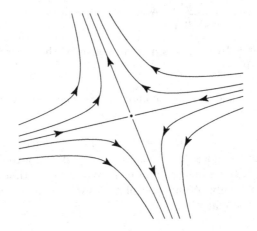

Figure 7.3: Index −1

Figure 7.2 corresponds for example to linear vector fields $X \mapsto AX$, where the eigenvalues of A are real and of the same sign (strictly positive for a source, strictly negative for a sink).

Figure 7.3 represents a zero of index -1. We obtain this for a linear vector field when the eigenvalues are real and of opposite sign.

Another example: gradient vector fields

We now calculate the index of the gradient of a function at a critical point. Here the answer is independent of the metric.

Theorem 7.26. *Let f be a smooth function on a manifold M and let a be a non-degenerate critical point of index d. Then for every Riemannian metric on M, the index of the vector field ∇f at a is equal to $(-1)^d$.*

PROOF. This is a local question, so we may reduce to the case where M is an open subset of $\mathbf{R}^n$ and $a = 0$. The proof of the result follows from a series of remarks.

1. The argument of the preceding theorem shows that if ϕ is a diffeomorphism from U to its image such that $\phi(0) = 0$, ∇f and $\nabla (f \circ \phi)$ have the same index. We can thus apply the Morse lemma (Lemma 3.44) to reduce to the case where

$$f(x) = -x_1^2 - \cdots - x_d^2 + x_{d+1}^2 + \cdots + x_n^2.$$

2. The Euclidean gradient is then

$$\nabla f = -2 \sum_{i=1}^{d} x^i \frac{\partial}{\partial x^i} + 2 \sum_{i=d+1}^{n} x^i \frac{\partial}{\partial x^i}$$

and a direct calculation shows the index of this vector field is $(-1)^d$.

3. Any two Riemannian metrics on the same manifold are homotopic, simply take the convex combination $tg_1 + (1-t)g_2, t \in [0,1]$. In particular a Riemannian metric on U is homotopic to $\sum_{i=1}^{n}(dx^i)^2$ thus the gradients corresponding to the same function are homotopic. □

The Poincaré-Hopf theorem (see [Spivak 79, Chapter 11], or the following chapter for the surface case) assures that if X is a compact manifold, and ξ is a vector filed on X having only isolated zeros,

$$\chi(X) = \sum_{x \in Z(\xi)} \text{ind}_x(\xi)$$

where $\chi(X)$ is the Euler-Poincaré characteristic of X, defined by

$$\chi(X) = \sum_{k=0}^{\dim X} (-1)^k \dim H^k(X, \mathbf{R}),$$

and $Z(\xi)$ is the set of zeros of ξ. Consequently, the hairy ball theorem (Theorem 7.23) is true for every compact manifold with nonzero Euler characteristic (we will soon see that $\chi(S^{2p}) = 2$).

Applying Theorem 7.26 we obtain the following special case:

Let X be a compact manifold, and f a smooth function on X whose critical points are non-degenerate (and therefore there are only finitely many by the Morse lemma). Then

$$\chi(X) = \sum_{x \in \text{crit}(f)} (-1)^{\text{ind}_x(f)}.$$

7.5. Fundamental Theorem of Algebra: Revisited

7.5.1. Two Proofs of the Fundamental Theorem of Algebra Using Degree Theory

To calculate the degree of a map, we have two methods at our disposal (see the proof of Theorem 7.18): we may either use the definition directly, or consider the inverse image of a regular point. In the latter case, it is important to know whether the orientation is preserved or not. An important case is that of *holomorphic functions*: we saw in Section 1.2 that if f is a holomorphic function defined on an open subset $U \subset \mathbf{C}$, its Jacobian at z, regarded as a function on an open subset of $\mathbf{R}^2$ is $|f'(z)|^2$, so that the orientation is preserved at every regular point.

From this property we can deduce two proofs of the fundamental theorem of algebra. The first is very close to that of Chapter 2, the second openly uses invariance under homotopy.

We proceed as in Section 2.4, associating to each polynomial P of one complex variable the map $f : S^2 \to S^2$ defined by

$$f(x) = i_N^{-1}\big(P(i_N(x))\big) \quad \text{if } x \neq N, \text{ and } f(N) = N.$$

Theorem 7.27. *If P is nonconstant, then f and therefore P are surjective maps.*

PROOF. We saw in Section 2.4 that P and consequently f have only a finite number of singular points if P is nonconstant. Let x be a regular value of f such that $x \neq N$ (the argument above dispenses with using Sard's theorem; it turns out N is a singular value, but this is not important here). Then f preserves the orientation at every point of $f^{-1}(x)$: it suffices to see the representation of f in the chart i_N. We have then reduced to showing that P preserves orientation, which is true by Section 1.2.2. As a consequence $\deg(f) > 0$. In particular f and P are surjective. $\square$

The second proof stems from the following property.

Theorem 7.28. *The degree of f is equal to the degree of the polynomial P.*

PROOF. For $t \in [0,1]$, we write

$$h(z,t) = a_0 z^n + t(a_1 z^{n-1} + \cdots + a_n)$$

and

$$H(x,t) = i_N^{-1}\big(h(i_N(x),t)\big) \quad \text{if } x \neq N \text{ and } H(N,t) = N.$$

Then the calculation done in Section 2.4 shows that H defines a homotopy between f and the map g defined analogously starting with the polynomial $Q(z) = z^n$. To calculate the degree of g we proceed as in Theorem 7.18. Given a regular value $x = i_N^{-1}(1)$ for example, we know that $g^{-1}(x)$ has n elements, since the equation $z^n = 1$ has n solutions. By Section 1.2.2, $\deg(g) = n$, and by invariance of homotopy, $\deg(f) = n$. $\qquad\square$

Corollary 7.29. *There exists maps from S^2 to S^2 of every degree.*

The antipodal map is of degree -1, it suffices to apply the following lemma, which is important in its own right.

Lemma 7.30. *Let X, Y, Z be three compact connected oriented manifolds of the same dimension, and let $f \in C^\infty(X, Y)$ and $g \in C^\infty(Y, Z)$. Then*

$$\deg(g \circ f) = \big(\deg(g)\big)\big(\deg(f)\big).$$

PROOF. Let σ be a form on Z of maximum degree and integral 1. Then

$$\deg(g \circ f) = \int_X (g \circ f)^* \sigma = \int_X \big(f^*(g^*\sigma)\big)$$
$$= \deg(f) \int_Y g^*\sigma = \big(\deg(f)\big)\big(\deg(g)\big). \qquad\square$$

7.5.2. Comparison of the Different Proofs of the Fundamental Theorem of Algebra

It is instructive to compare these proofs of the fundamental theorem of algebra with each other, and with other existing proofs. In the final analysis, each proof puts into play a very simple yet fundamental property of $\mathbf{C}$.

In Chapter 2, the key role is played by connectedness of $\mathbf{C}$ with a finite set of points removed, which is grossly false for $\mathbf{R}$.

In 7.27 above, it was the fact that the map $z \mapsto az$ from $\mathbf{C}$ to $\mathbf{C}$ preserves orientation (if $a \neq 0$), which is again false for false for $\mathbf{R}$.

In 7.28, we used the preceding property and a homotopy argument. This homotopy argument, when used directly, shows that P has nonzero degree, and is therefore surjective. The same homotopy argument allows us to show that a polynomial with real coefficients, extended to infinity is homotopic to the map $x \mapsto \pm x^n$ extended in the same way. Unfortunately this map is of degree ± 1 if n is odd (the case where the intermediate value theorem amply suffices), and 0 is n is even.

We now discuss proofs that use the theory of holomorphic functions. One of these starts with the Cauchy integral formula: if f is a meromorphic function on the disk $D(0,r)$, with oriented boundary $C(r)$, one may show that

$$\int_{C(r)} \frac{f'(z)}{f(z)} \, dz = 2i\pi \big(Z(f) - P(f)\big),$$

where $Z(f)$ and $P(f)$ denote the number of zeros and the number of poles of f in D. If f is a polynomial of degree n, we have $P(f) = 0$. On the other hand, if r is sufficiently large, we can write

$$\frac{f'(z)}{f(z)} = \frac{n}{z} + \frac{\epsilon(z)}{z} \quad \text{with} \quad \lim_{|z| \mapsto \infty} \epsilon(z) = 0.$$

For r sufficiently large we have

$$\left| \int_{C(r)} \frac{\epsilon(z)}{z} \, dz \right| < \frac{1}{2},$$

thus

$$\frac{1}{2i\pi} \int_{C(r)} \frac{f'(z)}{f(z)} \, dz = \frac{1}{2i\pi} \int_{C(r)} \frac{n}{z} \, dz = n$$

since these two integrals are integers. This argument is essentially *the same* as the argument used to prove the invariance of the degree under homotopy. We also note that the Cauchy integral formula puts into play the connectedness of $\mathbf{C} \setminus \{0\}$ in a way we do not necessarily think of.

Another proof uses Liouville's theorem for entire functions: if P has no zero, the function $\frac{1}{P}$ is holomorphic and bounded on P, and therefore constant. It is difficult to find a shorter argument. But the proof of Liouville's theorem rests on the Cauchy integral formula. We also note that the fact that $\frac{1}{P}$ is bounded rests on a comparison of $P(z)$ and z^n which we also used in the preceding proof... as in Theorem 7.28.

We next consider "elementary" proofs. One consists of a proof by contradiction that $\inf |P(z)| = 0$. A compactness argument proves that the minimum is attained, and after a change of variable we can suppose the minimum is at 0. Then

$$P(z) = a_n \left(1 + \sum_{k=1}^{n} \frac{a_{n-k}}{a_n} z^k \right) = a_n \big(1 + Q(z)\big),$$

where if P is nonconstant, $Q(z)$ has a principal part of cz^k as z tends to zero. Taking z sufficiently small so that cz^k is a negative real number, we can then make $|P(z)|$ less than $|a_n|$, obtaining a contradiction. Here the ability to take n-th roots of complex numbers plays the key role.

The other elementary proof by contradiction uses the maximum amount of algebra. Replacing P by $\overline{P}P$ we see that it suffices to do the proof for polynomials with real coefficients. We write the degree of the polynomial as $n = 2^k m$ where m is odd, and an ingenious argument using symmetric polynomials allows us to proceed by induction on k (see [Samuel 08, p. 53]; this argument is independent of the rest of the book). It is only for the base case (for $k = 0$!) that we need a topological argument.

All of these proofs have in common a *connectedness* argument. The final four proofs are due to Gauss. One can find further proofs in [Douady 05].

●

7.6. Linking

The object of this paragraph is in some sense opposite to that of Section 7.4: rather than give mathematical proofs of phenomena conforming to our perception of space, we instead "define" (incompletely anyway, given the difficulty of the subject) a notion suggested by this perception. It seems that the first questions that were posed by linking stemmed from questions in electromagnetism. (See Exercise 8.)

It will be useful to define a curve as an embedding of S^1 into $\mathbf{R}^3$, a pair of disjoint curves as two embeddings $f, g : S^1 \to \mathbf{R}^3$ such that $f(s) \neq g(t)$ for all s and t in S^1.

Definition 7.31. *Two pairs of disjoint curves* (f_1, f_2) *and* (g_1, g_2) *are said to be* homotopic *if there exists smooth maps*

$$H_i : \ S^1 \times [0,1] \longrightarrow \mathbf{R}^3 \quad (i = 1, 2)$$

such that

$$H_i(u, 0) = f_i(u), \quad H_i(u, 1) = g_i(u),$$

and for all fixed $t \in [0,1]$ *the two maps*

$$u \longmapsto H_i(u, t) \quad (i = 1, 2)$$

define a pair of disjoint curves.

Remark. Intuitively, we allow deformations of the curves, but we may not cut them. The fact that at each instant the curves are disjoint is essential.

Definition 7.32. *A pair of curves is said to be* trivial *is they are homotopic to the pair*

$$\left(t \longmapsto (\cos \pi t, \epsilon \sin \pi t, 0), \ u \longmapsto (3 + \cos \pi u, \epsilon' \sin \pi u, 0)\right),$$

with $\epsilon, \epsilon' = \pm 1$, *or in other words to a pair formed by two disjoint circles in the plane* $z = 0$. *A nontrivial pair of curves is said to be* linked.

Example. The pair (f, g) formed by two circles defined by

$$f(s) = (\cos s, \sin s, 0)$$
$$g(t) = (1 + \cos t, 0, \sin t) \tag{$*$}$$

(see Figure 7.4) is linked. To see this, it suffices to extract from this pair of curves a quantity invariant under homotopy.

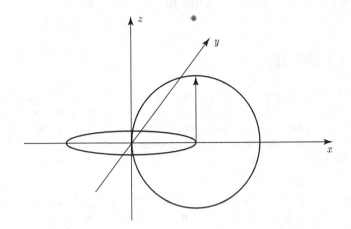

Figure 7.4: Two linked circles

Definition 7.33. The linking number $E(f, g)$ of a pair of curves f, g *is the degree of the map F from $S^1 \times S^1$ to S^2 given by*

$$F(s, t) = \frac{f(s) - g(t)}{\|f(s) - g(t)\|}.$$

Note that if we replace f (for example) by $f \circ \varphi$, where $\varphi \in \mathrm{Diff}(S^1)$, we have

$$E(f \circ \varphi, g) = \pm E(f, g)$$

depending on whether φ preserves or reverses the orientation. We note also that

$$E(f, g) = -E(g, f).$$

Remark. Let C and C' be two compact oriented submanifolds of dimension 1 in $\mathbf{R}^3$. Using the fact that these manifolds are diffeomorphic to S^1 (Theorem 3.45), we define the linking number $E(C, C')$ of C and C' as $E(f, g)$ for parametrizations compatible with the orientations. By an abuse of language, we call a compact oriented submanifolds of dimension 1 a curve. Warning: the parametrizations arise explicitly when we consider homotopies by pairs of curves.

The following property is an immediate consequence of the preceding definitions and the invariance of the degree under homotopy.

Proposition 7.34. *If the pairs of curves* (f, g) *and* (f_1, g_1) *are homotopic,* $E(f, g) = E(f_1, g_1)$.

We deduce that the pair of curves of the Example $(*)$ is linked. Indeed,

$$F^{-1}(0, 0, 1) = \left\{ \left(0, \frac{3\pi}{2} \right) \right\}.$$

It suffices then to verify that the point $(0, \frac{3\pi}{2})$ is regular. Setting $t = \frac{3\pi}{2} + t'$, we see that the projection of $F(s, t)$ onto the plane $\{z = 0\}$ admits the asymptotic expansion

$$(-t', s) + O(s^2 + t'^2)$$

in a neighborhood of $(0, \frac{3\pi}{2})$.

This shows that $(0, \frac{3\pi}{2})$ is a regular point of F and that F preserves the orientation at this point (here S^2 is oriented as usual with its standard volume form). Thus, $E(f, g) = 1$.

We can also understand linking in a more geometric way, by regarding the intersections of one of the curves with a surface bounded by the other. The first step is the following.

Theorem 7.35. *If a curve C is the oriented boundary of a surface Σ that does not intersect C', then $E(C, C') = 0$.*

PROOF. Let

$$\delta : \ \mathbf{R}^3 \times \mathbf{R}^3 \setminus \Delta \longrightarrow S^2 \quad \text{be given by} \quad \delta(x, y) = \frac{x - y}{\|x - y\|}$$

(here we denote the diagonal by Δ), and let ω be the standard volume form on S^2. By the definition of linking number, we have

$$E(C, C') = \frac{1}{4\pi} \int_{C \times C'} \delta^* \omega,$$

and therefore by Stokes's theorem

$$E(C, C') = \frac{1}{4\pi} \int_{\Sigma \times C'} d\delta^* \omega = \frac{1}{4\pi} \int_{\Sigma \times C'} \delta^* d\omega = 0. \qquad \square$$

On the contrary, if two curves are linked, our drawings suggest that every surface that bounds one curve also meets the other curve. We can make this precise in the following way.

Theorem 7.36. *Let C and C' be two closed curves in $\mathbf{R}^3$, and Σ be a surface whose oriented boundary is C. Suppose that $\Sigma \cap C'$ is finite, and that for all x in this intersection we have*

$$T_x\Sigma \cap T_xC' = \{0\}.$$

Then

$$E(C, C') = \sum_{x \in \Sigma \cap C'} \mathrm{or}_x(\Sigma \cap C'),$$

where $\mathrm{or}_x(\Sigma \cap C') = +1$ if the union of a positively-oriented basis of $T_x\Sigma$ and a positively-oriented basis of T_xC' is a positively-oriented basis of $\mathbf{R}^3$, and $\mathrm{or}_x(\Sigma \cap C') = -1$ otherwise.

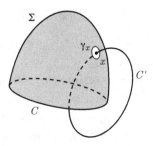

Figure 7.5: If $C = \partial\Sigma$ and C' are linked, Σ intersects C'

PROOF (SKETCH). We proceed exactly as in the proof of the preceding theorem. Given Σ, we take a small disk D_x for each $x \in \Sigma \cap C'$ with boundary γ_x. Then, using the same notations, Stokes's theorem and the fact that $d\omega = 0$, implies that

$$0 = \int_{(\Sigma \smallsetminus \bigcup_{x \in \Sigma \cap C'} D_x) \times C'} d\delta^*\omega = \int_{C \times C'} \delta^*\omega - \sum_{x \in \Sigma \cap C'} \int_{\gamma_x \times C'} \delta^*\omega.$$

Therefore

$$E(C, C') = \sum_{x \in \Sigma \cap C'} E(\gamma_x, C').$$

To determine $E(\gamma_x, C')$ we proceed exactly as in Example ($*$): by homotopy, we can reduce to the case where γ_x is a circle, say horizontal, and where C' intersects the plane of the circle orthogonally at x. Then, in the notation of Definition 7.33, $F^{-1}(0,0,1)$ has only one element if the radius is sufficiently small, and one may check that the sign associated is indeed $\mathrm{or}_x(\Sigma \cap C')$. $\square$

To finish, we mention that linking arises in magnetostatics. If C is an electric circuit traversed by a uniform current of intensity I, the circulation of the magnetic field created by the circuit along C' is, up to a multiplicative constant determined by the physical units, equal to

$$I \int_{C \times C'} \delta^* \omega = IE(C', C),$$

as C.-F. Gauss knew. See also Exercise 8.

7.7. Invariance under Homotopy

We now move on to the study of cohomology in any degree. We will need properties of differential forms on manifolds of the form $M \times \mathbf{R}$. The reader is invited to consult Poincaré's lemma in Section 5.6 where we surreptitiously introduced analogous tools.

Let p denote the canonical projection $p : M \times \mathbf{R} \to M$. For each real number t, we have an injection $j_t : M \to M \times \mathbf{R}$ defined by $j_t(x) = (x, t)$. Note that

$$p \circ j_t = Id_M \quad \text{and} \quad j_u \circ p = r_u,$$

where $r_u : M \times \mathbf{R} \to M \times \mathbf{R}$ is defined by $r_u(x, t) = (x, u)$. Finally we denote the vector field $\frac{\partial}{\partial t}$ on $M \times \mathbf{R}$ by T.

Definitions 7.37. *We say a differential form $\alpha \in \Omega^p(M \times \mathbf{R})$ is* basic *if there exists a form $\beta \in \Omega^k(M)$ such that $\alpha = p^* \beta$, and* semi-basic *if $i_T \alpha = 0$.*

In other words a basic form can be expressed without t or dt, and a semi-basic form without dt, see below for more detail.

Example. A basic form is necessarily semi-basic (since $T_{(x,t)}p$ annihilates tangent vectors to the factor $\mathbf{R}$), but the converse is false. If we take for example $U = \mathbf{R}$, the form $t\,dx$ is semi-basic but not basic. Of course a semi-basic form on $M \times \mathbf{R}$ may be simply identified with a one-parameter family of differential forms on M as seen in Section 5.6.

Lemma 7.38

i) *The map $p^* : \Omega^k(M) \to \Omega^k(M \times \mathbf{R})$ is injective.*

ii) *A form $\alpha \in \Omega^k(M \times \mathbf{R})$ is basic if and only if*

$$i_T \alpha = 0 \quad \text{and} \quad L_T \alpha = 0$$

PROOF

i) For $p^* \alpha = 0$ it is necessary and sufficient that for each $x \in M$ and $t \in \mathbf{R}$,

$$^t T_{(x,t)} p(\alpha_x) = 0$$

or again, if for each k-tuple $v_1 \ldots v_k$ of tangent vectors to (x,t),

$$\alpha_x\big(T_{(x,t)}p \cdot v_1, \ldots, T_{(x,t)}p \cdot v_k\big) = 0.$$

This allows us to deduce the result since the linear tangent map $T_{(x,t)}p$ is surjective everywhere. The same reasoning proves that p^* is injective as soon as p is a surjective submersion.

ii) Being a local property (on M!) and invariant under diffeomorphism, it suffices to consider the case of open subsets of $\mathbf{R}^n$. A form $\alpha \in \Omega^k(U \times \mathbf{R})$ may be written

$$\alpha = \sum_{i_1 < i_2 < \cdots < i_k} f_{i_1 i_2 \ldots i_k}(x,t)\, dx^{i_1} \wedge \cdots \wedge dx^{i_k} +$$

$$+ \sum_{j_1 < j_2 < \cdots < j_{k-1}} g_{j_1 j_2 \ldots j_k}(x,t)\, dt \wedge dx^{j_1} \wedge \cdots \wedge dx^{j_{k-1}}.$$

We have $i_T\alpha = 0$ if and only if dt does not arise in the decomposition above, which is to say if the functions $g_{j_1 j_2 \ldots j_k}$ are zero. Then by the definition of the Lie derivative, $L_T\alpha$ is obtained by differentiating the functions $f_{i_1 i_2 \ldots i_k}$ with respect to t, thus if $L_T\alpha = 0$ the $f_{i_1 i_2 \ldots i_k}$ are independent of t. The converse is clear. $\qquad\square$

Another spin on this proof is the following property.

Lemma 7.39. *Every form $\alpha \in \Omega^k(M \times \mathbf{R})$ can be decomposed as*

$$\alpha = \beta + dt \wedge \gamma,$$

in a unique way, where $\beta \in \Omega^k(M \times \mathbf{R})$ and $\gamma \in \Omega^{k-1}(M \times \mathbf{R})$ are semi-basic.

PROOF. The result is immediate, since a form on $M \times \mathbf{R}$ is semi-basic if and only if dt does not arise in its decomposition in local coordinates. Intrinsically, we can note that $\gamma = i_T\alpha$. $\qquad\square$

The following property, which uses the notation of Section 5.6, is close to Cartan's formula.

Proposition 7.40. *Let $\alpha \in \Omega^k(M \times \mathbf{R})$ be a closed form. Then, for $a, b \in \mathbf{R}$, we have*

$$j_b^*\alpha - j_a^*\alpha = I_a^b(L_T\alpha) = d\big(I_a^b(i_T\alpha)\big) + I_a^b\big(i_T(d\alpha)\big),$$

where

$$I_a^b\alpha = \int_a^b (j_t^*\alpha)\, dt.$$

PROOF. We will first show that for $\alpha \in \Omega(M \times \mathbf{R})$, we have

$$\frac{d}{dt} j_t^* \alpha = j_t^* (L_T \alpha).$$

Note that the two sides of this equation are depend only on the β component of the preceding lemma. In other words, it suffices to consider the semi-basic case. By the arguments of Lemma 7.38, we reduce to forms on $U \times \mathbf{R}$, where U is an open subset of $\mathbf{R}^n$, of the form $f(x, t) \, dx^{i_1} \wedge \cdots \wedge dx^{i_k}$. But then both sides are clearly equal to $\partial_t f(x, t) \, dx^{i_1} \wedge \cdots \wedge dx^{i_k}$.

With this assertion proved, we have, by using Cartan's formula and the properties of one-parameter families of forms from Section 5.6:

$$\begin{aligned}
j_b^* \alpha - j_a^* \alpha &= \int_a^b \left(\frac{d}{dt} j_t^* \alpha \right) dt \\
&= \int_a^b \left(j_t^* (L_T \alpha) \right) dt \\
&= \int_a^b \left(j_t^* \big(d(i_T \alpha) + i_T (d\alpha) \big) \right) dt \\
&= d\big(I_a^b (i_T \alpha) \big) + I_a^b \big(i_T (d\alpha) \big). \qquad \square
\end{aligned}$$

An immediate consequence is the following.

Theorem 7.41. *Let f and g be two smooth maps from a manifold M to a manifold N. Suppose that there exists a smooth homotopy between f and g. Then $h^k(f) = h^k(g)$ for every integer k.*

PROOF. Let $F : M \times [0, 1] \to N$ be a homotopy, and $\alpha \in \Omega^k(N)$ a closed form. Then Proposition 7.40 says exactly that $f^* \alpha$ and $g^* \alpha$ are cohomologous. $\qquad \square$

Remark. There was a (very) slight abuse of notation in the previous proof. The considerations above do not apply directly to $M \times [0, 1]$. Instead we can deduce for example that this theory is in fact C^1, and that F can be extended to a homotopy $F_1 : M \times [a, b] \to N$, with $a < 0$, $b > 1$, such that F_1 is additionally constant in t on intervals $[a, a - \epsilon]$ and $[b - \epsilon, b]$. **We can also appeal to the theory of manifolds with boundary.**

Definition 7.42. *A (smooth) map $f : M \to N$ is a* homotopy equivalence *if there exists a (smooth) map $g : N \to M$ such that $g \circ f$ is homotopic to Id_M and $f \circ g$ is homotopic to Id_N. We then say that M and N are of the same* homotopy type.

This notion can of course be defined in a purely topological setting.

Examples

a) The map j_u of the preceding paragraph (and its "homotopy inverse" p) are homotopy equivalences.

b) If $U \subset \mathbf{R}^n$ is a star-shaped open subset with respect to $a \in U$, the inclusion $a \to U$ is a homotopy equivalence. This situation is sufficiently important to merit a definition.

Definition 7.43. *A topological space is* contractible *if the identity map is homotopic to a constant map (in other words if the constant map is a homotopy equivalence).*

The following result is then a consequence of Theorem 7.41.

Theorem 7.44. *If $f : M \to N$ is a homotopy equivalence between manifolds, then $h^k(f)$ is an isomorphism for all k. In particular, if V is a contractible manifold, $H^k(M \times V)$ and $H^k(M)$ are isomorphic.*

Examples

a) The Poincaré lemma now becomes a special case of this result.

b) The sphere with two points removed is diffeomorphic to $S^{n-1} \times (-1, 1)$ (see Figure 7.6). An explicit diffeomorphism, when these two points are the poles, is

$$(t, x) \longmapsto \left(\frac{x}{\sqrt{1 - t^2}}, t \right)$$

(we have written $x^0 = t$, $x = (x^1, \ldots, x^n)$). As a result

$$H^k\big(S^n \smallsetminus \{N, S\}\big) \simeq H^k\big(\mathbf{R}^n \smallsetminus \{0\}\big) \simeq H^k\big(S^{n-1}\big).$$

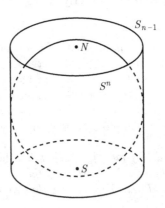

Figure 7.6: From S^{n-1} to S^n

7.8. The Mayer-Vietoris Sequence

7.8.1. Exact Sequences

This section gives a taste of algebraic topology. Ssee for example [Fulton 95] for further development. The following notion plays an important role.

Definition 7.45. *A sequence*

$$E_0 \xrightarrow{f_0} E_1 \xrightarrow{f_1} E_2 \ \ldots \ E_{n-1} \xrightarrow{f_{n-1}} E_n$$

of vector spaces and linear maps is said to be exact *if* $\operatorname{Ker} f_i = \operatorname{Im} f_{i-1}$.

In particular, $f_i \circ f_{i-1} = 0$ for all i, but this condition alone does not imply exactness.

Fundamental example

Let M be a manifold of dimension n. Denote the differential on forms of degree k by d_k. The sequence

$$\Omega^0(M) \xrightarrow{d_0} \Omega^1(M) \xrightarrow{d_1} \Omega^2(M) \ \ldots \ \Omega^{n-1}(M) \xrightarrow{d_{n-1}} \Omega^n(M) \longrightarrow 0$$

satisfies $d_k \circ d_{k-1} = 0$. It is exact if M is contractible, but it is not exact in general. De Rham cohomology exactly measures the "failure of exactness".

Exact sequences play a fundamental role in the calculation of cohomology groups. The following properties are clear and will be used systematically.

The sequence

$$E \xrightarrow{f} F \longrightarrow 0$$

is exact if and only if f is surjective. Similarly, the sequence

$$0 \longrightarrow E \xrightarrow{g} G$$

is exact if and only if g is injective. Combining these two remarks we see that the sequence

$$0 \longrightarrow E \xrightarrow{f} F \longrightarrow 0$$

is exact if and only if f is an isomorphism.

The following property is merely a translation of exactness.

Proposition 7.46. *If the sequence*

$$0 \longrightarrow E_0 \xrightarrow{f_1} E_1 \ \ldots \ E_{n-1} \xrightarrow{f_n} E_n \longrightarrow 0$$

is exact, and if the E_i are finite dimensional, we have

$$\sum_{i=0}^{n}(-1)^i \dim E_i = 0.$$

PROOF. We proceed by induction on n. For $n = 1$ this is the example above. If the sequence

$$0 \longrightarrow E_0 \xrightarrow{f_1} E_1 \xrightarrow{f_2} E_2 \longrightarrow 0$$

is exact, f_2 is surjective, then

$$E_2 \simeq E_1 / \operatorname{Ker} f_2.$$

So

$$\dim E_2 = \dim E_1 - \dim \operatorname{Ker} f_2 = \dim E_1 - \dim \operatorname{Im} f_1 = \dim E_1 - \dim E_0$$

since f_1 is injective. In the general case, we decompose the sequence

$$0 \longrightarrow E_0 \xrightarrow{f_1} E_1 \ \ldots \ E_{n-1} \xrightarrow{f_n} E_n \longrightarrow 0$$

into

$$0 \longrightarrow E_0 \xrightarrow{f_1} E_1 \ \ldots \ E_{n-2} \xrightarrow{f_{n-1}} \operatorname{Im} f_{n-1} \longrightarrow 0$$

and

$$0 \longrightarrow \operatorname{Ker} f_n \longrightarrow E_{n-1} \xrightarrow{f_n} E_n \longrightarrow 0.$$

We know that

$$\dim \operatorname{Ker} f_n - \dim E_{n-1} + \dim E_n = 0,$$

and by the induction hypothesis,

$$\sum_{i=0}^{n-2}(-1)^{n-2} \dim E_i + (-1)^{n-1} \dim \operatorname{Im} f_{n-1} = 0.$$

It suffices then to take the sum or difference of these two equalities depending on the parity in n. □

7.8.2. The Mayer-Vietoris Sequence

Now let M be a manifold, and let U, V be two open subsets such that $U \cup V = M$. Define a linear map

$$r: \ \Omega^k(M) \longrightarrow \Omega^k(U) \bigoplus \Omega^k(V) \quad by \quad r(\omega) = (\omega_{|U}, \omega_{|V})$$

and a linear map

$$s: \ \Omega^k(U) \bigoplus \Omega^k(V) \longrightarrow \Omega^k(U \cap V) \quad by \quad s(\alpha, \beta) = \alpha_{|U \cap V} - \beta_{|U \cap V}.$$

Lemma 7.47. *The sequence*

$$0 \longrightarrow \Omega^k(M) \xrightarrow{\ r\ } \Omega^k(U) \bigoplus \Omega^k(V) \xrightarrow{\ s\ } \Omega^k(U \cap V) \longrightarrow 0$$

is exact.

PROOF. The injectivity of r is clear by the equality $\operatorname{Im} r = \operatorname{Ker} s$ (two forms on two open subsets U and V glue together to a form on the union if and only if they coincide on the intersection). To see that s is surjective, introduce a partition of unity (f, g) on M subordinate to the cover (U, V). Write

$$U = (U \cap V) \cup (U \smallsetminus \operatorname{Supp} g)$$

and for $\gamma \in \Omega^k(U \cap V)$ define $\alpha \in \Omega^k(U)$ by

$$\alpha = g\gamma \ \text{on}\ U \cap V \quad \text{and}\quad 0 \ \text{on}\ U \smallsetminus \operatorname{Supp} g.$$

(since $g\gamma = 0$ on $U \cap V \smallsetminus \operatorname{Supp} g$). In the same way, define a form $\beta \in \Omega^k(V)$ starting with $-f\gamma$. By construction, $s\left((\alpha, \beta)\right) = \gamma$. □

It is clear that if $\omega \in \Omega^k(M)$ is closed (resp. exact) and if $r(\omega) = (\alpha, \beta)$, the forms α and β are closed (resp. exact). If $\alpha \in \Omega^k(U)$ and $\beta \in \Omega^k(V)$ are both closed or exact, the same is true of $s\left((\alpha, \beta)\right)$. Thus, r and s pass to the quotient and define linear maps which we denote by R and S on cohomology.

Proposition 7.48. *The sequence*

$$H^k(M) \xrightarrow{\ R\ } H^k(U) \bigoplus H^k(V) \xrightarrow{\ S\ } H^k(U \cap V)$$

is exact.

PROOF. As a consequence of the preceding discussion, $S \circ R = 0$, and so $\operatorname{Im} R \subset \operatorname{Ker} S$. To see that $\operatorname{Ker} S \subset \operatorname{Im} R$, it suffices to check that if $(\alpha, \beta) \in \Omega^k(U) \bigoplus \Omega^k(V)$ is a pair of closed forms such that $s(\alpha, \beta) = d\gamma$, then there exists $\omega \in \Omega^k(M)$, and $(\alpha_1, \beta_1) \in \Omega^{k-1}(U) \bigoplus \Omega^{k-1}(V)$ such that $r(\omega) = (\alpha - d\alpha_1, \beta - d\beta_1)$. However, by the preceding lemma, γ is of the form $s(\alpha_1, \beta_1)$. The forms $\alpha - d\alpha_1 \in \Omega^k(U)$ and $\beta - d\beta_1 \in \Omega^k(V)$ have the same restriction to $U \cap V$. Indeed, restricting to $U \cap V$, we have

$$(\alpha - d\alpha_1) - (\beta - d\beta_1) = (\alpha - \beta) - d(\alpha_1 - \beta_1) = (\alpha - \beta) - d\gamma = 0$$

by hypothesis. □

Conversely, it is not true in general that R is injective: if $\alpha = d\alpha'$ and $\beta = d\beta'$, there is no reason that the forms α' and β' have the same restriction to $U \cap V$. Similarly, S is in general not surjective: if $\gamma = s(\alpha, \beta)$, the proof of Lemma 7.47 shows that α and β are not closed in general. However this phenomena, with the aid of the differential on $\Omega^k(U \cap V)$, will allow us to map $H^k(U \cap V)$ into $H^{k+1}(M)$.

Theorem 7.49. *There exists a linear map*

$$\partial:\ H^k(U \cap V) \longrightarrow H^{k+1}(M)$$

such that the sequence

$$H^k(U) \bigoplus H^k(V) \xrightarrow{S} H^k(U \cap V) \xrightarrow{\partial} H^{k+1}(M) \xrightarrow{R} H^{k+1}(U) \bigoplus H^{k+1}(V)$$

is exact.

By applying this and the preceding result in every degree, and omitting the degree in the notation of R, S, ∂, we deduce the existence of a "long" exact sequence called the *Mayer-Vietoris sequence*.

$$0 \longrightarrow H^0(M) \xrightarrow{R} H^0(U) \bigoplus H^0(V) \xrightarrow{S} H^0(U \cap V) \xrightarrow{\partial} H^1(M) \longrightarrow \cdots$$

$$\xrightarrow{\partial} H^k(M) \xrightarrow{R} H^k(U) \bigoplus H^k(V) \xrightarrow{S} H^k(U \cap V) \xrightarrow{\partial} H^{k+1}(M) \longrightarrow \cdots$$

$$\xrightarrow{S} H^{n-1}(U \cap V) \xrightarrow{\partial} H^n(M) \xrightarrow{R} H^n(U) \bigoplus H^n(V) \xrightarrow{S} H^n(U \cap V) \longrightarrow 0$$

PROOF. Let $\gamma \in F^k(U \cap V)$. By Lemma 7.47, there exists forms $\alpha \in \Omega^k(U)$ and $\beta \in \Omega^k(V)$ such that

$$\alpha_{|U \cap V} - \beta_{|U \cap V} = \gamma.$$

Then the restrictions of $d\alpha$ and $d\beta$ to $U \cap V$ are equal, and again by the same lemma, there exists a form $\omega \in \Omega^{k+1}(M)$ such that

$$\omega_{|U} = d\alpha \quad \text{and} \quad \omega_{|V} = d\beta.$$

This shows that ω is closed.

Now let

$$(\alpha', \beta') \in \Omega^k(U) \bigoplus \Omega^k(V)$$

be a pair of forms such that $s(\alpha', \beta')$ are cohomologous to γ, and thus to the form $\gamma + d\gamma'$. Then there exists $\delta \in \Omega^k(M)$, $\alpha'' \in \Omega^k(U)$ and $\beta'' \in \Omega^k(V)$ such that

$$\alpha' = \alpha + \alpha'' + \delta_{|U}$$
$$\beta' = \beta + \beta'' + \delta_{|V}$$

with

$$d\gamma' = r(\alpha'', \beta'').$$

Thus the restrictions of α' and β' to $U \cap V$ are the same, and the pair of forms defined by (α', β') on M is $\omega + d\delta$. This shows that the correspondence between γ and ω gives a well defined map at the level of cohomology which we denote ∂.

It remains to prove that the sequence is exact. It is immediate that $\partial \circ S = 0$ and $R \circ \partial = 0$, which is to say Im $S \subset \text{Ker } \partial$ and Im $\partial \subset \text{Ker } R$. Conversely, we have $\partial[\gamma] = 0$ if and only if the construction above taken at the level of forms results in an exact form. This means that if $\gamma = s(\alpha, \beta)$, there exists a form $\delta \in \Omega^k(M)$ such that

$$d\delta_{|U} = d\alpha \quad \text{and} \quad d\delta_{|V} = d\beta.$$

But then $\alpha - \delta \in F^k(U)$, $\beta - \delta \in F^k(V)$, and consequently $[\gamma] = S([\alpha - \delta], [\beta - \delta])$.

The equality $\operatorname{Ker} R = \operatorname{Im} \partial$ is proved in the same way. If $R([\omega]) = 0$, we have $\omega_{|U} = d\alpha$ and $\omega_{|V} = d\beta$, the restriction of $\alpha - \beta$ to $U \cap V$ is closed, and by the definition of ∂, we have $[\omega] = \partial([\alpha], [\beta])$. $\qquad \square$

Remark. The map ∂ is itself "natural" in a sense that we will leave to the reader to make precise, as it takes longer to explain than to understand.

7.8.3. Application: A Few Cohomology Calculations

It is instructive to use this method to calculate the cohomology of S^1. Take two distinct points p and q, and set $U = S^1 \setminus \{p\}$, $V = S^1 \setminus \{q\}$. Then U and V are diffeomorphic to $\mathbf{R}$, thus $U \cap V$ has *two connected components*, each of which is also diffeomorphic to $\mathbf{R}$. The Mayer-Vietoris sequence becomes

$$0 \longrightarrow H^0(S^1) \longrightarrow H^0(U) \oplus H^0(V) \longrightarrow H^0(U \cap V) \longrightarrow H^1(S^1) \longrightarrow 0.$$

Thus Proposition 7.46 shows that $\dim H^1(S^1) = 1$. While we do not learn anything new, the method generalizes to S^n.

Theorem 7.50. *If $0 < k < n$, $H^k(S^n) = 0$, and $\dim H^n(S^n) = 1$.*

PROOF. We proceed by induction on n. We have just seen the result is true for $n = 1$. Choose two distinct points of S^n (the north and south poles of the embedded sphere if we want to better visualized the situation, but in reality these choices are not important), and write the same Mayer-Vietoris sequence as above. For $1 < k \leqslant n$, the part of the sequence

$$H^{k-1}(U) \oplus H^{k-1}(V) \longrightarrow H^{k-1}(U \cap V) \longrightarrow H^k(S^n) \longrightarrow H^k(U) \oplus H^k(V)$$

can be written

$$0 \longrightarrow H^{k-1}(U \cap V) \longrightarrow H^k(S^n) \longrightarrow 0.$$

By Theorem 7.44, $H^{k-1}(U \cap V)$ is isomorphic to $H^{k-1}(S^{n-1})$, and the result follows in this case by applying the induction hypothesis. For $k = 1$, we write

$$0 \longrightarrow H^0(S^n) \longrightarrow H^0(U) \oplus H^0(V) \longrightarrow H^0(U \cap V) \longrightarrow H^1(S^n) \longrightarrow 0.$$

By applying Proposition 7.46, we see that $H^1(S^n) = 0$ if $n > 1$. $\qquad \square$

Corollary 7.51. $H^k(\mathbf{R}^n \smallsetminus \{0\})$ *is one dimensional if* $k = 0$ *or* $n - 1$, *and vanishes otherwise.*

PROOF. The canonical inclusion of S^{n-1} into $\mathbf{R}^n \smallsetminus \{0\}$ is a homotopy equivalence. $\qquad\square$

The same method allow us to calculate the cohomology of projective spaces. We will treat the complex projective case, which is simpler than that of the real projective space.

Theorem 7.52. *We have*
$$H^{2k+1}(P^n\mathbf{C}) \simeq 0 \quad and \quad H^{2k}(P^n\mathbf{C}) \simeq \mathbf{R} \quad if \ \ 0 \leqslant k \leqslant n.$$

PROOF. Consider the vector space $\mathbf{C}^{n+1}$ with its canonical basis, the point p given in homogeneous coordinates by $(1, 0, \dots, 0)$, and the subset E of $P^n\mathbf{C}$ formed by the set of points whose first homogeneous coordinate vanishes. This is a compact submanifold of $P^n\mathbf{C}$ diffeomorphic to $P^{n-1}\mathbf{C}$. The chart
$$\varphi_0 : \ \left[(z^0, z^1, \dots, z^n)\right] \longmapsto \left(\frac{z^1}{z^0}, \dots, \frac{z^n}{z^0}\right)$$
is a diffeomorphism of $P^n\mathbf{C} \smallsetminus E$ to $\mathbf{C}^n \simeq \mathbf{R}^{2n}$.

On the other hand:

Lemma 7.53. *The embedding* $j : P^{n-1}\mathbf{C} \to P^n\mathbf{C} \smallsetminus \{p\}$ *given by*
$$\left[(u^1, \dots, u^n)\right] \longmapsto \left[(0, u^1, \dots, u^n)\right]$$
is a homotopy equivalence.

PROOF. Define $p : P^n\mathbf{C} \smallsetminus \{p\} \to P^{n-1}\mathbf{C}$ by
$$\left[(u^0, u^1, \dots, u^n)\right] \longmapsto \left[(u^1, \dots, u^n)\right].$$
Thus $h \circ j$ is the identity map on $P^{n-1}\mathbf{C}$, while $j \circ h$ is homotopic to the identity of $P^n\mathbf{C}$ as
$$H\big(t, [(u^0, u^1, \dots, u^n)]\big) = \left[(tu^0, u^1, \dots, u^n)\right]. \qquad\square$$

END OF THE PROOF OF THEOREM 7.52. Write $U = P^n\mathbf{C} \smallsetminus \{p\}$ and $V = P^n\mathbf{C} \smallsetminus E$. Then $U \cap V$ is diffeomorphic to $\mathbf{R}^{2n} \smallsetminus \{0\}$, and the lemma lets us proceed by induction. The result is true for $n = 1$ by the above, since $P^1\mathbf{C}$ is diffeomorphic to S^2. For $k < 2n$, the exact sequence
$$H^{k-1}(U \cap V) \longrightarrow H^k(P^n\mathbf{C}) \longrightarrow H^k(U) \bigoplus H^k(V) \longrightarrow H^k(U \cap V)$$
gives
$$0 \longrightarrow H^k(P^n\mathbf{C}) \longrightarrow H^k(P^{n-1}\mathbf{C}) \longrightarrow 0,$$

from which the result follows in this case. For $k = 2n$, consider the exact sequence

$$H^{2n-1}(U) \bigoplus H^{2n-1}(V) \longrightarrow H^{2n-1}(U \cap V)$$
$$\longrightarrow H^{2n}(P^n\mathbf{C}) \longrightarrow H^{2n}(U) \bigoplus H^{2n}(V),$$

which can also be written

$$0 \longrightarrow H^{2n-1}(\mathbf{R}^{2n} \smallsetminus \{0\}) \longrightarrow H^{2n}(P^n\mathbf{C}) \longrightarrow 0. \qquad \square$$

7.8.4. The Noncompact Case

For the most part, everything we have done so far concerns compact manifolds. However we saw a key result for $\mathbf{R}^n$, which we reformulate with the help of the following definition.

Definition 7.54. *Let M be a manifold. The quotient vector space of compactly supported closed forms of degree k by compactly supported exact forms of the same degree is called the* cohomology space with compact support *of degree k and will be denoted $H_c^k(M)$.*

Example. Theorem 7.3 says exactly that $H_c^n(\mathbf{R}^n) \simeq \mathbf{R}$, with the isomorphism given by passing the map $\alpha \mapsto \int_M \alpha$ to the quotient. One may prove this result is true for all orientable manifolds. Of course if M is compact, then $H_c^k(M) = H^k(M)$.

We can also extend degree theory to noncompact oriented manifolds, under the condition that we restrict to *proper* maps, which are maps such that the inverse image of every compact set is compact. If f is proper and smooth, it is clear that the inverse image of a compactly support form remains of compact support, and that the inverse image of a regular value is finite.

If α is a form of degree k, and β is a $n - k$ form with compact support on an oriented manifold of dimension n, the integral $\int_M \alpha \wedge \beta$ is well defined. Moreover, if we replace α by $\alpha + d\alpha'$ (with α' having compact support) and β by $\beta + d\beta'$, then by Stokes's theorem, the result will not change. Thus we have a bilinear map

$$\mathcal{PD} : \; H^k(M) \times H_c^{n-k}(M) \longrightarrow \mathbf{R}.$$

Theorem 7.55 (Poincaré duality).[1] *The bilinear form $\mathcal{PD}$ thus defined is non-degenerate when M admits a "good" finite covering.*

---

1. Proof omitted.

In particular, $\dim H^k(M) = \dim H_c^{n-k}(M)$. These good coverings are defined in Exercise 21.

7.9. Integral Methods

It is possible to calculate the cohomology of the tori T^n using an appropriate Mayer-Vietoris sequences, see Exercise 5. We will proceed in another way, using a method of interest in its own right which uses the Lie group structure of the torus. We begin with a particular case of Theorem 7.41 above.

Proposition 7.56. *Let M be a compact manifold, and let X be a vector field on M with flow φ_t. Then for every form $\alpha \in F^k(M)$, the forms $\varphi_t^*\alpha$ and α are cohomologous.*

PROOF. It suffices to use the homotopy $F(u, m) = \varphi_{tu}(m)$. We note that by the proof of Proposition 7.40, an explicit primitive of $\varphi_t^*\alpha - \alpha$ is given by

$$d\left(\int_0^t (\varphi_u^* i_X \alpha)\, du \right). \qquad \square$$

Example. Let $M = S^n$. The reader can verify, by using classical results on the structure of orthogonal matrices (see also Exercise 7 of Chapter 4) that every $g \in SO(n+1)$ is of the form $\exp X$, where $X \in \mathfrak{o}(n+1)$. As a result, if $\alpha \in F^k(S^n)$ then the forms $g^*\alpha$ and α are cohomologous. We also note, to shed further light on the result above, that this property is false if $g \in O(n+1)$: if σ is the antipodal map on S^{2p} and if ω is a volume form on S^{2p}, then $\sigma^*\omega$ and ω are not cohomologous.

We come to T^n, which is, by Section 4.5, *the* compact connected Lie group of dimension n which is isomorphic to $\mathbf{R}^n/\mathbf{Z}^n$. By Proposition 6.7 the differential forms on T^n are identified with the $\mathbf{Z}^n$-invariant differential forms on $\mathbf{R}^n$, which is to say forms that can be decomposed as

$$\sum_{i_1 < i_2 < \cdots < i_p} f_{i_1 i_2 \ldots i_p}\, dx^{i_1} \wedge \cdots \wedge dx^{i_p},$$

where the functions $f_{i_1 i_2 \ldots i_p}$ are periodic with period 1 with respect to the x^i. An important case is where the functions $f_{i_1 i_2 \ldots i_p}$ are *constants*. We again denote the p-form obtained on T^n by $dx^{i_1} \wedge \cdots \wedge dx^{i_p}$. This is hardly an abuse of notation: is the same form in local coordinates associated to the covering $p : \mathbf{R}^n \to \mathbf{R}^n/\mathbf{Z}^n$. With this notation, a p-form on T^n can be written

$$\sum_{i_1 < i_2 < \cdots < i_p} f_{i_1 i_2 \ldots i_p}\, dx^{i_1} \wedge \cdots \wedge dx^{i_p},$$

where the functions $f_{i_1 i_2 \ldots i_p}$ are now (smooth) functions *on* T^n. If the $f_{i_1 i_2 \ldots i_p}$ are constant, we say that the form has *constant coefficients*. We note that these forms have a simple intrinsic characterization: these are the forms such that $L_u^* \alpha = \alpha$ for every translation L_u (since the group is commutative, we need not distinguish between left and right translations). This is why we will hereafter denote the vector space of forms with constant coefficients by $\Omega_{\mathrm{inv}}(T^n)$. We have $\dim \Omega_{\mathrm{inv}}^p(T^n) = \binom{n}{p}$ and $\dim \Omega_{\mathrm{inv}}(T^n) = 2^n$.

In particular, $\Omega_{\mathrm{inv}}^n(T^n)$ is generated by the volume form $\omega = dx^1 \wedge \cdots \wedge dx^n$. Note that

$$\int_{T^n} dx^1 \wedge \cdots \wedge dx^n = 1.$$

Definition 7.57. *The* average *of a p-form*

$$\alpha = \sum_{i_1 < i_2 < \cdots < i_p} f_{i_1 i_2 \ldots i_p} \, dx^{i_1} \wedge \cdots \wedge dx^{i_p}$$

on T^n is the form

$$\overline{\alpha} = \sum_{i_1 < i_2 < \cdots < i_p} \left(\int_{T^n} f_{i_1 i_2 \ldots i_p} \omega \right) dx^{i_1} \wedge \cdots \wedge dx^{i_p}.$$

It is clear that $\overline{\alpha}$ has constant coefficients. We can also see this obvious property in the following way, which might seem needlessly complicated, but applies to all compact Lie groups (see Exercise 17).

We begin with the following trivial remark. If $f \in C^0(T^n)$, then the average of f is

$$\int_{T^n} f(x^1 + u^1, \ldots, x^n + u^n) \, dx^1 \wedge \cdots \wedge dx^n =$$

$$\int_{T^n} f(x^1 + u^1, \ldots, x^n + u^n) \, du^1 \wedge \cdots \wedge du^n.$$

This being said, we can consider the translations $L_u^* \alpha$ of α as a family of p-forms parametrized by $u \in T^n$. We define the integral of this family with respect to u as in 5.6.2 and integrate the coefficients with respect to the measure $d\mu$ defined by $du^1 \wedge \cdots \wedge du^n$. Then

$$\overline{\alpha} = \int_{T^n} L_u^* \alpha \, d\mu.$$

The measure $d\mu$ is clearly translation invariant, thus for $v \in T^n$ we have

$$L_v^*(\overline{\alpha}) = \int_{T^n} L_v^*(L_u^* \alpha) \, d\mu = \int_{T^n} L_{u+v}^* \alpha \, d\mu = \int_{T^n} L_u^* \alpha \, d\mu = \overline{\alpha}.$$

The main idea to calculate the cohomology of tori is to note that a closed form is cohomologous not only to all of its translates as we have just seen, but also to average of all of its translates.

Theorem 7.58. *The map* $\alpha \mapsto \overline{\alpha}$ *passes to the quotient as an isomorphism between* $H^p(T^n)$ *and* $\Omega_{\text{inv}}(T^n)$. *In particular*

$$\dim H^p(T^n) = \binom{n}{p}.$$

PROOF. The proof rests on the following key property.

Lemma 7.59. *If* $\alpha \in \Omega^p(T^n)$ *is closed,* α *and* $\overline{\alpha}$ *are cohomologous.*

PROOF. We have already seen that $L_u^* \alpha$ and α are cohomologous. Better still, if $u = \exp X$,

$$L_u^* \alpha - \alpha = d\left(\int_0^1 (i_X L_{tu}^* \alpha)\, dt \right).$$

To integrate a relation of this type with respect to u, we must control the dependence of X with respect to u. The Lie algebra of T^n is $\mathbf{R}^n$ with the zero bracket, and $\exp X = X \bmod \mathbf{Z}^n$. The exponential map is a diffeomorphism of $(-\frac{1}{2}, \frac{1}{2})^n$ to an open subset U of T^n whose complement is of measure zero. We let $\exp^{-1}$ denote the inverse diffeomorphism, and for $u \in U$ write

$$\beta(u) = \left(\int_0^1 (i_{\exp^{-1}u} L_{tu}^* \alpha)\, dt \right).$$

Now the family of forms $\beta(u)$, extended arbitrarily to all of T^n is integrable, and by integration over T^n, we obtain

$$\overline{\alpha} - \alpha = d\left(\int_{T^n} \beta(u)\, du \right). \qquad \Box$$

REMAINDER OF THE PROOF OF THEOREM 7.58. By integrating the equality $L_u^* d\alpha = dL_u^* \alpha$ over T^n, we see that

$$\overline{d\alpha} = d\overline{\alpha}.$$

Thus, the average of a closed form is closed, and the average of an exact form is zero : it is the differential of a form with constant coefficients. We deduce that the map $\alpha \mapsto \overline{\alpha}$ passes to the quotient as a map $L : H^p(T^n) \to \Omega_{\text{inv}}^p(T^n)$. As $\overline{\overline{\alpha}} = \overline{\alpha}$, L is surjective. On the other hand, the lemma tells us that if $\overline{\alpha} = 0$, then α is cohomologous to 0, in other words that L is injective. $\qquad \Box$

This method applies to spheres – where it gives nothing new – and to compact Lie groups (see Exercises 15 and 18), **and more generally to symmetric spaces** (see for example [Greub-Halperin-Van Stone 76, volume 2]).

Remark. As a bonus we have a distinguished representative of the form α in its cohomology class. More generally, it is true for compact Lie groups, where each cohomology class contains a unique bi-invariant differential form.

**Obtaining a distinguished representative of each cohomology class comes from Riemannian geometry. A Riemannian metric g on a manifold X can be extended to an inner product on each bundle $\bigwedge^p T^*X$. If X is compact, we obtain a norm from the inner product on $\Omega^p(X)$, defined by

$$\|\alpha\|_g^2 = \int_X g(\alpha_x, \alpha_x)\omega_g.$$

The *Hodge-de Rham theorem* (see [Booss-Bleecker 85]) then ensures the existence of a unique form realizing the minimum norm in each cohomology class. For example, if G is a compact Lie group equipped with a bi-invariant Riemannian metric (there are always such metrics by an integration argument analogous to that of Exercise 8 of Chapter 6) the forms which minimize the norm in their cohomology class are precisely the bi-invariant forms.**

7.10. Comments

The only compact manifolds for which there is a complete classification are manifolds of 1 and 2 dimensions. Theorem 3.45 ensures that every compact connected 1-dimensional manifold is diffeomorphic to S^1. In two dimensions, with techniques from algebraic topology [Massey 77, Chapter I] or Morse theory [Hirsch 76] one may show:

1) that every compact connected orientable manifold is diffeomorphic to S^2 or to a connected sum (see Exercise 28 of Chapter 2) of k tori;

2) that every compact connected non orientable manifold is diffeomorphic to the connected sum of k projective planes (the reader who has solved Exercise 28 of Chapter 2 can show that $P^2\mathbf{R}\sharp P^2\mathbf{R}$ is diffeomorphic to the Klein bottle).

In particular, every compact simply connected manifold of dimension 2 is diffeomorphic to S^2. The methods for calculating cohomology spaces that we have discussed allow us to see that these manifolds are pairwise non diffeomorphic: we can show using an appropriate Mayer-Vietoris sequence that

$$\dim H^1(\overbrace{T^2\sharp\cdots\sharp T^2}^{k \text{ times}}) = 2k$$

and

$$\dim H^1(\overbrace{P^2\mathbf{R}\sharp\cdots\sharp P^2\mathbf{R}}^{k \text{ times}}) = k - 1.$$

In higher dimension, even if we restrict to simply connected manifolds, the situation is much more complicated. The Poincaré conjecture (every simply connected compact 3-dimensional manifold is homeomorphic to S^3) resisted proof for more than a century, having been proved in 2003 by G. Perelman (see [Charpentier-Ghys-Lesne 10, Chapter 12] for an overview of the ideas that arise in the proof).

Starting in dimension 4, there are many compact simply connected manifolds which are non diffeomorphic. For example, S^4, $S^2 \times S^2$ and $P^2\mathbf{C}$ are not diffeomorphic: the dimensions of their second cohomology group are 0, 2 (see Exercise 10 below), and 1, respectively. For more examples, we can observe that the connected sum of two simply connected manifolds is also simply connected by Proposition 2.42. This method is far from exhausting the list of compact simply connected 4-dimensional manifolds, but it would take us too far afield to say more.

To remain in the point of view developed in this book, we mention that

$$H^*(X) \overset{\text{def}}{=} \bigoplus_{k=0}^{\dim X} H^k(X)$$

has a multiplicative structure, inherited from the multiplication of differential forms: if α, β and γ are three homogeneous *closed* forms, then

$$(\alpha + d\beta) \wedge \gamma = \alpha \wedge \gamma \pm d(\beta \wedge \gamma).$$

Thus, the exterior product passes to the quotient as a bilinear map from $H^k(X) \times H^l(X)$ to $H^{k+l}(X)$, called the *cup product*, which is of maximum rank by Poincaré duality.

The example of cohomology of degree 2 in dimension 4 is particularly interesting. Indeed, if X is a compact oriented manifold, the cup product of two classes of degree 2 is identified with a symmetric bilinear form on $H^2(X)$, which is non-degenerate by Poincaré duality. The signature of the associated quadratic form furnishes a new differential invariant, which we call the *signature* of the manifold. Taking connected sums of projective planes and products of spheres, we can see without difficulty that any real quadratic form can be realized as a cup product in dimension 4.

Alas (or perhaps happily from the mathematician's point of view) this procedure is a little primitive.

One can define, by other methods than those using differential forms, a cohomology with *integer* coefficients (see for example [Bredon 94] or [Bott-Tu 86]). We then obtain a much finer invariant by considering the cup product on integer cohomology of degree 2 as an integer coefficient quadratic form. Of course, the world of integer quadratic forms is much richer and more complicated than the real case. See [Serre 96, Chapter 5] for a glimpse of this theory. There has been an interaction between number theory, topology and hard analysis which has seen dramatic developments (see for example [Lawson 85]).

7.11. Exercises

1*. *Degree of the map* $q \mapsto q^n$ *from* S^3 *to* S^3

Here we use freely what we have already developed regarding quaternions. In particular recall that the inner product is given by

$$\langle q, q' \rangle = \Re(q\overline{q'}).$$

a) Show that if q is a pure quaternion of norm 1, $q^2 = -1$. Deduce that if q is a pure quaternion and n is an integer, we have

$$(e \cos t + q \sin t)^n = e \cos(nt) + q \sin(nt).$$

b) Let s be a nonzero quaternion. Show that for $n > 2$ the equation $q^n = s$ admits n distinct solutions if s is not real, and infinitely many solutions if s is real. More precisely, the set of solutions in this case is the disjoint union of n submanifolds diffeomorphic to S^2 (if s is negative), to $\{e\}$ and to $n-1$ submanifolds diffeomorphic to S^2 (if s is positive). What happens if $n = 2$?

c) We will show that every point in S^3 distinct from $-e$ is a regular value of the map $f : q \mapsto q^n$ from S^3 to itself (we first treat the case of values of the form $c + di$, c, d real).

c1) Calculate the differential of f.

c2) Consider $q = a + bi$ (a, b both nonzero real numbers). Then $T_q S^3 = \mathbf{R}iq \oplus E$, where E is the (real) vector space spanned by j and k. Show that

$$T_q f \cdot x = \begin{cases} nq^{n-1}x & \text{if } x \in \mathbf{R}iq \\ q^{-n}\frac{q^{2n}-1}{q-1}x & \text{if } x \in E. \end{cases}$$

(Note that $qx = x\overline{q}$ if $x \in E$.) Deduce that every quaternion of S^3 of the form $c + di$ (c, d real, $d \neq 0$) is a regular value of f.

d) Writing $q \in S^3$ in the form $ae + bs$ (a and b real, and s a pure quaternion of norm 1), show that $T_q S^3$ admits an orthogonal decomposition of the form

$$\mathbf{R}q' \bigoplus E,$$

where q' commutes with q and $uq = q\bar{u}$ for $u \in E$. Deduce a generalization of c2), and conclude that every non real quaternion of S^3 is a regular value of f.

e) Show that f has degree n.

2. *Quaternionic polynomials*

a) Imitating the constructions of the book, identify $\mathbf{H}$ with $\mathbf{R}^4$ and associate to every polynomial with coefficients in $\mathbf{H}$ a smooth map from S^4 to S^4.

b) Show that every nonconstant polynomial with coefficients in $\mathbf{H}$ has a zero.

3*. *A useful lemma... used in Section 7.4.4*

a) Let f be a diffeomorphism of $\mathbf{R}^n$ such that $f(0) = 0$. Find a homotopy H between f and its differential at 0, such that the H_t are again diffeomorphisms (such a homotopy is called an *isotopy*).

b) Show that every diffeomorphism of $\mathbf{R}^n$ preserving the orientation is isotopic to the identity.

4**. *Examples of links*

a) Use the preceding exercise to make sense of the notion of linking of curves in S^3.

b) Let H be the Hopf fibration of S^3 to S^2. Show that for all a, $b \in S^2$, the curves $H^{-1}(a)$ and $H^{-1}(b)$ are linked, and their linking number is 1.

c) Consider the Möbius strip with boundary, which is the image of $[0, 4\pi] \times [0, 3/4]$ under the map

$$(\theta, r) \longmapsto \left(\cos\theta \left(1 + r \cos\frac{\theta}{2}\right), \sin\theta \left(1 + r \cos\frac{\theta}{2}\right), r \sin\frac{\theta}{2} \right).$$

Calculate the linking number of the closed curves obtained by taking $r = 1/4$ and $r = 3/4$.

5*. *Existence of maps from S^n to S^n with arbitrary degree*

a) Let f be a strictly increasing smooth map from $[0, 1)$ to $[a, +\infty)$, where $a > 0$. Show that the map

$$x \longmapsto f(\|x\|^2)x$$

defines a smooth diffeomorphism from the open ball $B(0, 1)$ to $\mathbf{R}^n$.

b) With an appropriate choice of f, and using stereographic projection, show that there exists a smooth map g from $\mathbf{R}$ to S^n such that

b1) g is a diffeomorphism from $B(0,1)$ to $S^n \smallsetminus \{p\}$;

b2) $g(x) = p$ if $\|x\| \geqslant 1$ (here $p \in S^n$ is fixed once and for all).

c) Let X be a manifold of dimension n. Show that for all $m \in X$ there exists an open subset U containing m and a smooth map h from X to S^n such that

c1) h is a diffeomorphism from U to $S^n \smallsetminus \{p\}$;

c2) $h(x) = p$ if $x \in X \smallsetminus U$.

d) Deduce from c) that if X is compact connected and oriented, there exists smooth maps from X to S^n of every degree.

6. Let G be a compact and connected Lie group, and f_k the map $x \mapsto x^k$ from G to itself.

a) For all $g \in G$, write

$$\varphi_{g,k} = L_{g^{-k}} \circ f_k \circ L_g.$$

Show that $T_g f_k$ and $T_e \varphi_{g,k}$ have the same rank, and that

$$T_e \varphi_{g,k} = \sum_{r=0}^{k-1} \operatorname{Ad} g^{-r}.$$

b**) Show that at every regular point $T_g f_k$ preserves the orientation. Deduce that f_k is surjective.

c) Show that for $G = SO(2n)$ or $SO(2n+1)$, the degree of f is equal to kn.

7*. Show that every map from S^n to S^n with degree not equal to $(-1)^{n+1}$ has a fixed point.

8. *Ampère's theorem*

Let C and C' be two disjoint simple closed curves in oriented three dimensional Euclidean space, given by arc-length parametrizations $f : \mathbf{R}/L\mathbf{Z} \to E$ and $g : \mathbf{R}/L'\mathbf{Z} \to E$. The magnetic field created by an electrical current of intensity I traversing C is given by

$$H_x = \frac{1}{4\pi} \int_0^L I f'(s) \wedge \frac{\overrightarrow{f(s)x}}{\left\|\overrightarrow{f(s)x}\right\|^3}$$

where $\wedge$ denotes the cross product (this is the Biot-Savart law, see for example [Feynman-Leighton-Sands 63, volume II, Chapter 14]).

a) Show that

$$\oint_{C'} H = I \int_0^{L'} \int_0^L \frac{\det\left(f'(s), \overrightarrow{f(s)g(t)}, g'(t)\right)}{\left\|\overrightarrow{f(s)g(t)}\right\|^3} \, ds \, dt.$$

b) Deduce that

$$\oint_{C'} H = IE(C', C).$$

9. Summarize the different proofs of the fact that $H^1(S^1) \simeq \mathbf{R}$.

10. *Cohomology of a finite quotient*

Let Γ be a finite group acting freely on a manifold X, let $Y = X/\Gamma$ be the quotient manifold, and $p : X \to Y$ the corresponding covering map.

a) Show that

$$h^k(p) : \ H^k(Y) \longrightarrow H^k(X)$$

  is injective.

b) Show that the image of $h^k(p)$ from $H^k(Y)$ to $H^k(X)$ is formed by γ-invariant cohomology classes for all $\gamma \in \Gamma$.

c) *Application.* Calculate the cohomology of real projective space and the Klein bottle (defined in Exercise 3 of Chapter 6).

11. *Cohomology of a product*

a) Let X and Y be two compact connected orientable manifolds of dimensions p and q respectively. Show that the groups $H^p(X \times Y)$ and $H^q(X \times Y)$ are nonzero.

b) **More generally, the "Künneth formula" ensures that

$$H^r(X \times Y) \simeq \bigoplus_{p+q=r} H^p(X) \bigotimes H^q(Y).**$$

  Check this formula in the case of a product of spheres.

12*. Show that every smooth map from S^n to T^n (for $n > 1$!) has degree zero. Try to generalize this result by changing either the domain or target manifold.

13. Calculate the cohomology of T^2, and then the cohomology of T^n using an appropriate Mayer-Vietoris sequence.

14*. *Hopf invariant*

a) Let f be a smooth map from S^3 to S^2, and $\alpha \in \Omega^2(S^2)$. Show that the form $f^*\alpha$ is exact.

b) Suppose we fix orientations on S^2 and S^3. Let $\alpha \in \Omega^2(S^2)$, and let β be a primitive of $f^*\alpha$. Show that the integral

$$\int_{S^3} \beta \wedge f^*\alpha$$

is independent of the choice of β.

c) Show that this integral is zero if α is exact.

d) Let $\alpha \in \Omega^2(S^2)$ be such that $\int_{S^2} \alpha = 1$. Show that

$$\int_{S^3} \beta \wedge f^*\alpha$$

is also independent of the choice of α satisfying this condition.

e) The above shows that this integral depends only on f. We call this the *Hopf invariant of f* and we denote it by $H(f)$. If φ is a smooth map from S^3 to S^3, show that

$$H(f \circ \varphi) = \deg(\varphi)H(f).$$

f) If g is a smooth map from S^2 to S^2, show that

$$H(g \circ f) = \deg(g)^2 H(f).$$

g) Show that $H(f) = 0$ if f is not surjective. Calculate $H(f)$ when f is the Hopf fibration. Show that $H(f_1) = H(f_2)$ if f_1 and f_2 are homotopic, and deduce that the Hopf fibration is not homotopic to a constant map.

15*. *Invariant forms and cohomology*

a) By imitating the method of Section 7.9, show that every closed form on S^n is cohomologous to an $SO(n+1)$-invariant form.

b) Show that if $0 < k < n$, every $SO(n+1)$-invariant form $\alpha \in \Omega^k(S^n)$ is zero. Deduce another method to calculate the cohomology of spheres.

16. *Quaternionic projective space*

a) Repeating the constructions of Section 2.5, define the projective space $P^n\mathbf{H}$ as an appropriate quotient of $\mathbf{H}^{n+1} \setminus \{0\}$, and equip it with a smooth manifold structure. Warning: there are two ways to proceed with either left or right multiplication, but these give diffeomorphic manifolds.

b) In the same way, show the existence of a fibration

$$p: S^{4n+3} \longrightarrow P^n \mathbf{H}$$

whose fibers are diffeomorphic to S^3, and deduce that $P^1\mathbf{H}$ is diffeomorphic to S^4.

c) Repeating the arguments of Theorem 7.52, show that the cohomology of $P^n\mathbf{H}$ is zero in every degree that is not a multiple of 4, and that $H^{4k}(P^n\mathbf{H}) \simeq \mathbf{R}$ for $0 \leqslant k \leqslant n$.

17. *Bi-invariant forms on a Lie group*

a) Let G be a Lie group. Denote by $\mathcal{I}$ the map $x \mapsto x^{-1}$. What is the differential of $\mathcal{I}$ at the identity element?

b) A differential form α on G is *bi-invariant* if

$$L_g^*\alpha = \alpha \quad \text{and} \quad R_g^*\alpha = \alpha$$

for any $g \in G$.

Show that if α is bi-invariant, then the same is true for $d\alpha$ and $\mathcal{I}^*\alpha$.

c) Show that if α is bi-invariant (or even only left or right invariant) and $\alpha_e = 0$, then $\alpha = 0$.

d) Show that for every form $\alpha \in \Omega^k(G)$, we have

$$(\mathcal{I}^*\alpha)_e = (-1)^k \alpha_e.$$

Deduce that if α is bi-invariant and of degree k, we have

$$\mathcal{I}^*\alpha = (-1)^k \alpha.$$

Deduce that every bi-invariant form on a Lie group is closed.

e) Using the techniques of Section 7.9, show that for a compact connected Lie group G,

$$H^p(G) \simeq \Omega_{\text{inv}}^p(G),$$

where we have denoted the vector space of bi-invariant forms of degree p by $\Omega_{\text{inv}}^p(G)$ (use the fact that the exponential map of a compact Lie group is surjective). Show that the algebra $\text{Inv}_G \bigwedge \mathfrak{G}^*$ of alternating $\text{Ad}(G)$-invariant forms on $\mathfrak{G}$ is isomorphic to $\Omega_{\text{inv}}(G)$.

f) *Example.* Take $G = SO(n)$. Show that

$$(X, Y, Z) \longmapsto \text{tr}(XYZ - XZY)$$

is an alternating trilinear form on $\mathfrak{so}(n)$, which is $\text{Ad}(G)$-invariant, and nontrivial if $n \geqslant 3$. Deduce that $H^3(SO(n)) \neq 0$ for $n \geqslant 3$.

18. *Cohomology of the sphere with two punctures*

a) Show that the connected sum of two orientable manifolds is orientable.

b) Using an appropriate Mayer-Vietoris sequence, show that

$$\dim H^1(T^2 \sharp T^2) = 4.$$

19*. *Moser's theorem*

Let ω_0 and ω_1 be two volume forms on a compact n-dimensional manifold M, such that $\int_M \omega_0 = \int_M \omega_1$. We will show that there exists a diffeomorphism ϕ of M such that $\phi^*\omega_1 = \omega_0$. This results in the existence of a family of diffeomorphisms $t \mapsto \phi_t$ $(0 \leqslant t \leqslant 1)$ such that $\phi_0 = Id$ and

$$\phi_t^*\omega_t = \omega_0, \quad \text{where } \omega_t = (1-t)\omega_0 + \omega_1.$$

To show the existence of such a family, we use Moser's trick (compare to the proof of Theorem 3.44).

a) Show that ω_t is a volume form for all $t \in [0,1]$.

b) Deduce that if $\alpha \in \Omega^{n-1}(M)$ there exists a unique time-dependent vector field X_t such that $i_{X_t}\omega_t = \alpha$.

c) Using Theorem 5.30, show that there exists a time-dependent vector field X_t which generates a family of diffeomorphisms ϕ_t having the desired properties.

20. Let M be a compact orientable $n + 1$-dimensional manifold with connected boundary ∂M, and let f be a smooth map from M to a compact orientable manifold of dimension n. Show that the degree of $f_{|\partial M}$ is zero.

21*. *Good coverings*

We say an finite open covering $(U_i)_{1 \leqslant i \leqslant r}$ of a n-dimensional manifold M is a good covering if every nonempty intersection of open subsets of the covering is diffeomorphic to $\mathbf{R}^n$.

a) Show that if M admits a good covering, the de Rham cohomology spaces are finite dimensional. (Use induction on r and a Mayer-Vietoris sequence.)

b) Give examples of good coverings for S^n, T^n, $P^2\mathbf{R}$.

Note. One may prove with techniques of Riemannian geometry (see [Do Carmo 92]) that every compact manifold admits a good covering.

Chapter 8

The Euler-Poincaré Characteristic and the Gauss-Bonnet Theorem

8.1. Introduction

The Gauss-Bonnet theorem is at the heart of the geometry of manifolds. It mixes topology (triangulations, cohomology spaces), differential geometry (index of singular points of vector fields) and Riemannian geometry. We do not have the space to illustrate all of these ideas in detail. To keep with the spirit of the book, the proofs we give will use differential geometry to the greatest extent possible. We nonetheless believe it would be interesting to sketch a purely Riemannian proof in this introduction. The price we pay is using certain notions that have not been introduced (geodesics, geodesic curvature), of which we give the idea.

8.1.1. From Euclid to Carl-Friedrich Gauss and Pierre-Ossian Bonnet

It was known since at least the time of Euclid that the sum of the angles of a triangle was π. This result is equivalent to the parallel postulate in Euclidean geometry.

It is possible to extend Euclid's results to curved triangles, by introducing a correcting terms that takes into account the failure of the sides to be straight lines. We introduce, for an arc of a C^2 curve parametrized by arc-length, the geodesic curvature $k(s)$ defined by $\tau'(s) = k(s)n(s)$ (see for example [Do Carmo 76, 4.4]). Here, s is the arc-length parameter, $\tau(s)$ is the oriented unit tangent vector, and $n(s)$ is the vector such that the frame $(\tau(s), n(s))$ is positively oriented.

For a triangle formed by three C^2 arcs with angles $(\beta_i)_{1 \leqslant i \leqslant 3}$, we then have

$$\beta_1 + \beta_2 + \beta_3 = \pi + \int_T k(s)\, ds$$

or, by introducing the exterior angles α_i,

$$\alpha_1 + \alpha_2 + \alpha_3 + \int_T k(s)\, ds = 2\pi.$$

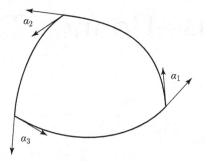

Figure 8.1: Gauss-Bonnet for a triangle

In case there is no angle at a point, $\alpha_i = 0$ and $\beta_i = \pi$, we obtain

$$\int_T k(s)\, ds = 2\pi.$$

This is the *Umlaufsatz* or theorem of turning tangents, see [Berger-Gostiaux 88, 9.5] or [Chavel 83, 4.6].

These results are natural: knowing that $k(s) = \varphi'(s)$, where φ is the angle $\tau(s)$ forms with a fixed vector, this simply says that the unit tangent vector turns exactly 2π. Natural does not mean easy to prove however, as in the example of Jordan's theorem which ensures that the complement of a simple closed curve has two connected components.

This formula was generalized by C.-F. Gauss (who did not publish it) and by P.-O. Bonnet to triangles constrained to a surface. The curves which replace straight lines are geodesics, which is to say curves that minimize length, and the function $k(s)$, whose vanishing characterizes geodesics, is the algebraic measure of the orthogonal projection of the acceleration onto the tangent plane. (See [Do Carmo 76, p. 248].)

However, there is now an additional correction term which involves the Gaussian curvature K of the surface (defined below, Definition 8.11). As we will see, K measures the failure of the surface to be locally isometric to the Euclidean plane.

The formula becomes

$$\beta_1 + \beta_2 + \beta_3 = \pi + \int_T k(s)\,ds + \iint_T K\,dA, \tag{8.1}$$

where dA is a measure on the surface defined in 6.29, which comes from the ambient Euclidean structure.

In fact, this formula remains true for any Riemannian surface (not necessarily embedded in $\mathbf{R}^3$), and dA is again the natural measure associated to the metric (see Definition 8.8). See [Spivak 79, volume 3, p. 396] or [Do Carmo 76, Chapter 2] for detailed explanation.

A particularly important case is *Girard's formula*, which ensures that any geodesic triangle (which is to say a triangle whose sides are arcs of great circles) on the sphere of radius 1 satisfies

$$\text{Area}(T) = \beta_1 + \beta_2 + \beta_3 - \pi.$$

(See [Berger 87, 18.3.8.4].)

8.1.2. Sketch of a Proof of the Gauss-Bonnet Theorem

Now let S be a compact surface equipped with a Riemannian metric g and a triangulation (for this notion, which we think of intuitively here, see Definition 8.4). The sum of the left hand side of equation (8.1) over all of the triangles equals $2\pi v$, where v is the number of vertices. The sum of the right hand side is $\pi f + \iint_T K\,dA$, where f is the number of faces, because in the integrals over the boundary of the triangles, each edge arises twice with opposite orientation. We thus obtain

$$\iint_S K\,dA = \pi(2s - f).$$

The number e of edges satisfies $3f = 2e$ (if we count 3 edges per face, each is counted twice), so that $2v - f = 2v - 3f + 2f = 2(v - e + f)$. Finally

$$\frac{1}{2\pi} \iint_S K\,dA = v - e + f \quad \text{(Gauss-Bonnet theorem)}.$$

The interest in $v - e + f$ comes from the fact that this number, taken for any triangulation is unchanged if we cut each face into further triangles. This formula at once shows that $\frac{1}{2\pi}\iint_S K\,dA$ is an *integer* which is independent of the Riemannian metric on S, and that $e - v + f$ is independent of the triangulation of S chosen. This integer is the Euler-Poincaré characteristic, denoted $\chi(S)$.

8.1.3. Abstract

We will not adopt this point of view, which is amply developed in the references we have cited. To remain in the spirit of the book, we will prove the Gauss-Bonnet theorem with less Riemannian geometry and more differential geometry, without passing through the Gauss-Bonnet formula for triangles.

We first show that for a compact surface equipped with a triangulation or a tiling, $v - e + f$ is equal to the alternating sum of its cohomology spaces.

We then give the rudiments of Riemannian geometry by the moving frame method, where a Riemannian metric g is given by local g-orthonormal frames.

Then, if X is a vector field on a compact orientable Riemannian surface, the idea is to introduce the vector field of orthonormal frames $(\frac{X}{\|X\|}, Y)$. This is only defined on $S \setminus \{p_1, \ldots, p_r\}$ (we have supposed that X has only a finite number of zeros, denoted $(p_i)_{1 \leqslant i \leqslant r}$). An argument using Stokes's theorem shows that the integral of curvature is equal to the sum of the indices of X at the singular points, multiplied by 2π. We are then in the same situation as we were at the end of 8.1.2: having obtained an equality between two mathematical objects with no relation between them, a metric and vector field, we play one off the other. There are some technicalities: the method of moving frames requires that we use C^2 vector fields. However we will associate to every triangulation a piecewise C^1 vector field whose sum of indices is equal to $\chi(S)$ (See Section 8.5.2). Working around these delicate points is handled by Lemmas 8.19 and 8.20.

8.2. Euler-Poincaré Characteristic

8.2.1. Definition; Additivity

For a manifold whose cohomology spaces are finite dimensional, for example a compact manifold, it is very convenient to encode the dimensions of these spaces in the *Poincaré polynomial*, defined by

$$P_M(t) = \sum_{i=0}^{\infty} \dim\big(H^k(M)\big) t^k.$$

Definition 8.1. *The* Euler-Poincaré characteristic *of a manifold M whose cohomology groups are finite dimensional is the integer*

$$\chi(M) = \sum_{i=0}^{\infty} (-1)^k \dim H^k(M),$$

in other words, the value of the Poincaré polynomial at $t = -1$.

We give a few examples.

Manifold M	$P_M(t)$	$\chi(M)$
Contractible	1	1
S^n	$1 + t^n$	$1 + (-1)^n$
$P^n \mathbf{C}$	$\sum_{k=0}^n t^{2k}$	$n + 1$
T^n	$(1 + t)^n$	0
$P^{2n} \mathbf{R}$	1	1
Compact orientable	reciprocal polynomial	0 in odd dimensions
Product $M_1 \times M_2$	$P_{M_1}(t) P_{M_2}(t)$	$\chi(M_1)\chi(M_2)$

We remind the reader that if P is a polynomial, then P is said to be a *reciprocal polynomial* if

$$P(t) = t^{\deg(P)} P\left(\frac{1}{t}\right).$$

The next to last line in the table is a reformulation of Poincaré duality. The last line gives a concise and easily remembered way to express the cohomology of a product of two manifolds. This is the *Künneth formula*, which will not be proved here. See [Bott-Tu 86, I.5] or [Karoubi-Leruste 87, II.5] for a proof).

The Euler-Poincaré characteristic clearly gives less information than the cohomology spaces, but it is easier to calculate, and it admits many pleasant geometric interpretations.

Theorem 8.2 ("Additivity" of the Euler-Poincaré characteristic). *Let M be a manifold. Suppose that $M = U \cup V$, where U and V are open subsets, and that the cohomology spaces of at least three of the following four manifolds M, U, V and $U \cap V$ are finite dimensional. Then the same is true for the fourth space, and we have*

$$\chi(U \cup V) + \chi(U \cap V) = \chi(U) + \chi(V).$$

PROOF. Everything comes from the Mayer-Vietoris sequence. The first part is a consequence of the rank theorem. For the second part, it suffices to notice that in the long exact sequence of Theorem 7.49, there is a gap of 2 between the cohomology of $U \cup V$ and that of $U \cap V$, and then to apply Proposition 7.46. □

Corollary 8.3. *If M and N are two compact manifolds of the same even dimension,*

$$\chi(M \sharp N) = \chi(M) + \chi(N) - 2.$$

PROOF. By Exercise 28 of Chapter 2, $M \sharp N = M' \cup N'$, where M' and N' are respectively diffeomorphic to M and N with a closed ball removed, and $M' \cap N'$ is diffeomorphic to $I \times S^{n-1}$. By the preceding theorem, we have

$$\chi(M) = \chi(M') + 1$$
$$\chi(N) = \chi(N') + 1$$
$$\chi(M \sharp N) = \chi(M') + \chi(N')$$

from which the result follows. $\qquad\qquad\qquad\qquad\qquad\qquad\qquad\qquad\square$

8.2.2. Tilings

A result of Euler (which was suspected by Descartes) ensures that for every convex polyhedron in $\mathbf{R}^3$ having f faces, v vertices and e edges, $f + v - e = 2$ (see [Hopf 83, p. 3] for a pleasant argument). The fact that $2 = \chi(S^2)$ is more than a coincidence: this fact is true for any polyhedron drawn on a surface homeomorphic to S^2 and can be seen in a more general setting.

Definitions 8.4

a) *A polygon in a surface S is the image of a polygon P of the plane under a diffeomorphism defined on an open subset containing P. The* vertices *of the polygon are the images under ϕ of the vertices of P. The* edges *are the $\phi((a, b))$, where a and b denote two consecutive vertices in P.*

b) *A* tiling *of S is a finite number of polygons $P_1, \ldots, P_F$ whose interiors (called the* faces *of the tiling), edges and vertices form a partition of S. When the polygons are triangles, we say that the tiling is a* triangulation.

We note that S is necessarily compact, and that an edge is shared between two polygons. Tilings always exist (Theorem 8.18 gives a more precise statement). We can use this result to obtain a classification of surfaces, see [Fulton 95, Chapter 17] or [Massey 77, Chapter I].

Theorem 8.5. *For every tiling of a compact surface S having f faces, v vertices and e edges,*

$$f + v - e = \chi(S).$$

PROOF. By decomposing the polygons into triangles, we reduce to the case of a triangulation (each polygon with n sides is replaced by $n - 2$ triangles: we have added $n - 3$ edges and $n - 3$ faces, so that $f + s - a$ is unchanged, see Figure 8.2).

Let $\sigma_1, \ldots, \sigma_f$ be triangles of the new tiling. For each of these, we have a 2-simplex (a triangle) σ_i' of $\mathbf{R}^2$, an open subset W_i containing σ_i' and a diffeomorphism $\phi_i : W_i \to \phi_i(W_i)$ such that $\sigma_i = \phi_i(\sigma_i')$. It will be convenient

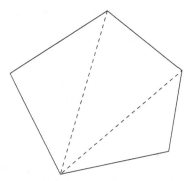

Figure 8.2: Decomposition of a polygon into triangles

to equip S with a distance function defining its topology (this is only a convenience which allows us to shorten what follows).

First step. We chose a point q_i in the interior of each 2-simplex σ_i (for example the image under ϕ_i of the barycenter of the vertices of σ_i') and a $r > 0$ such that the closed ball $\overline{B}(q_i, 2r)$ is included in the interior of σ_i for all i. We write

$$M_1 = M \smallsetminus \bigcup_{i=0}^{f} \overline{B}(q_i, r)$$

$$U_1 = \bigcup_{i=0}^{f} B(q_i, 2r).$$

By Theorem 8.2,

$$\chi(M) + \chi\left(\bigcup_{i=0}^{f} B(q_i, 2r) \smallsetminus \overline{B}(q_i, r) \right) = \chi(M_1) + \chi(U_1).$$

On the left hand side, one has disjoint annuli which retract onto S^1 and are therefore of zero characteristic. We thus have

$$\chi(M) = \chi(M_1) + \chi(U_1) = \chi(M_1) + f.$$

Second step. Using the parametrizations ϕ_i, join the pair of points q_i, q_j corresponding to adjacent simplexes by a piecewise C^1 arc that meets the common edge at only a single point. Choose these arcs so that their only point in common are the endpoints, see Figure 8.3.

Let Δ be the union of these arcs. We write

$$M_2 = M_1 \smallsetminus \overline{V}_\rho(\Delta)$$

$$U_2 = V_{2\rho}(\Delta) \smallsetminus \bigcup_{i=0}^{f} B(q_i, r/2) \quad \text{with } \rho << r.$$

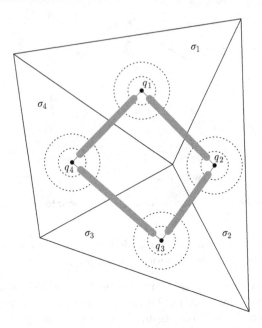

Figure 8.3: Mayer-Vietoris decomposition of M

We have again

$$\chi(M_1) + \chi(M_2 \cap U_2) = \chi(M_2) + \chi(U_2).$$

Now, U_2 has the homotopy type of e disjoint segments, and $M_2 \cap U_2$ of $2e$ such segments. On the other hand, M_2 is homeomorphic to the union of v disjoint open discs, so that

$$\chi(M_2) = v - e$$

and $\chi(M) = v - e + f$ as stated.   □

Remark. This method generalizes to higher dimensions at the price of a few more technical results on triangulations. The method is the same: **from the manifold we remove appropriate neighborhoods of k-skeletons of the dual triangulation.**

Corollary 8.6. *Let M be a compact manifold, and let $p : M' \to M$ be a covering map of order d. Then $\chi(M') = d\chi(M)$.*

PROOF. Fix a triangulation of M. It then suffices to remark that every simplex is contained in a trivializing simply connected open subset. As a consequence, the preimage under p of these simplexes gives a triangulation of M', which contains d times the simplexes in each dimension.   □

From additivity property of Theorem 8.2 and the multiplicativity with respect to finite coverings that we have just seen, we may suspect that the Euler-Poincaré characteristic localizes and can be interpreted in terms of measure. This is indeed the case, as we will see, so long as we have a Riemannian metric.

8.3. Invitation to Riemannian Geometry

As we saw at the end of Chapter 5, a Riemannian metric g is a symmetric bilinear form such that for every $m \in M$, the associated quadratic form is positive definite on $T_m M$. In local coordinates, g can be written

$$\sum_{1 \leqslant i,j \leqslant n} g_{ij} \, dx^i \, dx^j,$$

with

$$g_{ij}(m) = g_m(\partial_i, \partial_j).$$

Instead of working with coordinates, we can instead specify vector fields forming an orthonormal frame (it suffices to apply the Gram-Schmidt orthonormalization procedure to the frame field $(\partial_i)_{1 \leqslant i \leqslant n}$). If $(X_1, \ldots, X_n)$ is such a vector field on an open subset U, the metric on U may be written as

$$\sum_{i=1}^{n} (\theta^i)^2,$$

where the 1-forms θ^i are the coordinates with respect to the X_i. This viewpoint is known as the *method of moving frames*. The field of orthonormal frames $(X_1, \ldots, X_n)$ and the dual system of forms $(\theta_1, \ldots, \theta_n)$ (the *coframe*) mutually determine each other.

Warning. These two points of view are in a certain manner contradictory: there is no reason that there exist local coordinates such that the vector fields $(\partial_i)_{1 \leqslant i \leqslant n}$ form an orthonormal frame. Indeed if such is the case, the metric can be written in these coordinates as $\sum_{i=1}^{n} (dx^i)^2$, while the fundamental theory of Riemannian geometry (Theorem 8.12) says precisely that this is impossible in general.

We begin with a simple remark.

Lemma 8.7. *If on an open subset U a Riemannian metric may be written*

$$\sum_{i=1}^{n} (\theta^i)^2$$

with closed *forms, the metric is locally isometric to a Euclidean metric.*

PROOF. Let $a \in U$ be fixed. By the Poincaré lemma, (Theorem 5.44), there exists an open subset V containing a and smooth functions f_i on V such that $\theta_i = df^i$. As the quadratic form $\sum_{i=1}^{n}(\theta^i)^2$ is non-degenerate, $df_1 \wedge \cdots \wedge df_n \neq 0$ at each point of V, and the map $x \mapsto F(x) = (f^1(x), \ldots, f^n(x))$ from V to $\mathbf{R}^n$ is of maximum rank. Let $W \subset V$ be an open subset containing a for which F is a diffeomorphism to its image. Then on W we have

$$g = \sum_{i=1}^{n}(df^i)^2 = F^* \left(\sum_{i=1}^{n}(dy^i)^2 \right). \qquad \square$$

Unfortunately, this result only scratches the surface on the question of when a Riemannian metric is locally Euclidean: a metric g can be locally decomposed in many ways as the sum of squares of linear forms, but there is no simple way to know if these forms are closed.

From this point on, we will consider only dimension 2. The method of moving frames is easy to describe in this case. Indeed if (X_1, X_2) and (X'_1, X'_2) are two orthonormal frames for the same metric on an open subset U, and if these vector fields define the same local orientation on U, there exists a map R from U to $SO(2)$ such that

$$\begin{pmatrix} X'_1 \\ X'_2 \end{pmatrix} = R \begin{pmatrix} X_1 \\ X_2 \end{pmatrix} \quad \text{and} \quad \begin{pmatrix} \theta'^1 \\ \theta'^2 \end{pmatrix} = R \begin{pmatrix} \theta^1 \\ \theta^2 \end{pmatrix}. \qquad (8.2)$$

If we introduce complex-valued forms $\Theta = \theta^1 + i\theta^2$ and $\Theta' = \theta'^1 + i\theta'^2$, (8.2) may be written

$$\Theta' = e^{-i\phi}\Theta \quad \text{with } R = \begin{pmatrix} \cos\phi & \sin\phi \\ -\sin\phi & \cos\phi \end{pmatrix}. \qquad (8.3)$$

We remark in passing that $\theta^1 \wedge \theta^2$ is independent of the orthonormal (co)frame chosen, and gives a globally defined volume form.

Definition 8.8. *The form $\theta^1 \wedge \theta^2$ is the canonical volume form of the oriented Riemannian manifold.*

One may easily check that for an embedded surface this volume form coincides with that of Definition 6.29.

The real point of departure of the method of moving frames is the following result:

Lemma 8.9. *Let g be a Riemannian metric on a surface S which may be written in an open subset U in the form $g = (\theta^1)^2 + (\theta^2)^2$. Then there exists* *a unique form* $\omega \in \Omega^1(U)$ *such that*

$$d\theta^1 = -\omega \wedge \theta^2$$
$$d\theta^2 = \omega \wedge \theta^1.$$

PROOF. Immediate: if $d\theta^1 = a\theta^1 \wedge \theta^2$ and $d\theta^2 = b\theta^1 \wedge \theta^2$, then $\omega = -a\theta^1 - b\theta^2$. $\qquad\qquad\qquad\qquad\qquad\qquad\qquad\qquad\qquad\qquad\qquad$ □

Definition 8.10. *The form ω is called the* connection form *associated to the frame field* (X_1, X_2).

Remark. In a smooth manifold, we do not have a means to compare tangent vectors at different points. However, additionally specifying a Riemannian metric allows such a comparison. This is what justifies the name for ω. **The Levi-Civita connection of the metric g is given here by

$$\nabla_{X_1} X_1 = \omega(X_1)X_2 \qquad \nabla_{X_2} X_1 = \omega(X_2)X_2$$
$$\nabla_{X_1} X_2 = -\omega(X_1)X_1 \qquad \nabla_{X_2} X_2 = -\omega(X_2)X_1.$$

We will not use this result.**

With the 1-form $\Theta = \theta^1 + i\theta^2$ introduced above, the equations of Lemma 8.9 can be written

$$d\Theta = i\omega \wedge \Theta.$$

Let U' be another open subset on which the metric can be written $g = (\theta'^1)^2 + (\theta'^2)^2$. If $U \cap U'$ is nonempty and the two frames define the same orientation, then by (8.2)

$$\Theta' = (e^{-i\phi})\Theta.$$

Computing differentials, we obtain

$$\begin{aligned} d\Theta' &= (e^{-i\phi})\,d\Theta - i(e^{-i\phi})\,d\phi \wedge \Theta \\ &= (e^{-i\phi})i\omega \wedge \Theta - i(e^{-i\phi})\,d\phi \wedge \Theta \\ &= i(\omega - d\phi) \wedge (e^{-i\phi})\Theta \\ &= i(\omega - d\phi) \wedge \Theta'. \end{aligned}$$

Thus, the form ω' associated to the coframe (θ'^1, θ'^2) can be written

$$\omega' = \omega - d\phi. \qquad\qquad\qquad\qquad\qquad (8.4)$$

Warning. We have used the usual abuse of notation with the angular form: if the function $e^{-i\phi}$ is well defined on $U \cap U'$, the function ϕ itself only admits local determinations up to $2k\pi$, so that $d\phi$ is nonetheless well defined. It is closed, but not necessarily exact. Later it will prove to be useful to work with open subsets that need not be star shaped.

As a result of (8.4) $d\omega = d\omega'$ on $U \cap U'$. Thus there exists a closed 2-form Ω on the manifold, equal to $d\omega$ on every open subset equipped with a local coframe. On such an open subset Ω may be written $K\theta^1 \wedge \theta^2$, where K is a smooth function. If we change the (co)frame while keeping the same orientation, Ω and $\theta^1 \wedge \theta^2$ will not change, therefore neither will K. Finally, if we interchange θ_1 and θ_2, the forms ω, $d\omega$ and $\theta^1 \wedge \theta^2$ are changed to their opposite, thus K is unchanged. It is thus a smooth function on the manifold.

Definitions 8.11. *If (M, g) is a Riemannian manifold, then K is the* Gaussian curvature. *Further, if the manifold is oriented, Ω is* curvature form *or* Euler form.

We use an index g if we want to emphasize the dependence with respect to the metric.

We are now ready to formulate the fundamental theorem of Riemannian geometry (in dimension 2).

Theorem 8.12. *The Gaussian curvature is a local Riemannian invariant. More precisely:*

i) *if $f : (M, g) \to (M', g')$ is an isometry between Riemannian manifolds of dimension 2 whose Gaussian curvatures are K_g and $K_{g'}$, then $K_g = K_{g'} \circ f$;*

ii) *(M, g) is locally isometric to the Euclidean plane if and only if K_g is identically zero.*

PROOF

i) If (θ^1, θ^2) is a local coframe for (M', g') on an open subset U, then $(f^*\theta^1, f^*\theta^2)$ is a local coframe for (M, g) on the open subset $f^{-1}(U)$, and by construction f preserves the local orientations defined by these coframes. If ω is the connection form associated to the coframe (θ^1, θ^2), as

$$df^*\Theta = f^*(d\Theta) = if^*\omega \wedge f^*\Theta,$$

the connection form of g for the coframe $(f^*\theta^1, f^*\theta^2)$ is $f^*\omega$. The curvature form is therefore

$$\Omega_g = d(f^*\omega) = f^*d\omega = f^*\Omega_{g'},$$

from which i) follows.

ii) For the Euclidean plane, we can take $\theta^1 = dx$ and $\theta^2 = dy$, and the curvature is zero. Conversely, let (θ^1, θ^2) be an orthonormal coframe of a Riemannian manifold with zero curvature. Then $d\omega = 0$. By shrinking the open subset of the definition, we can write ω in the form $d\phi$. We then introduce the coframe (θ'^1, θ'^2) defined by

$$\theta'^1 + i\theta'^2 = (e^{-i\phi})(\theta^1 + i\theta^2).$$

For this new coframe, by equation (8.4) we have $\omega' = 0$, while the forms θ'^1 and θ'^2 are closed. We therefore have reduced to the case of Lemma 8.7.   □

Remark. We had to apply the Poincaré lemma *twice*. In fact curvature is an invariant of order two. It turns out that there are no Riemannian invariants of order 1. Indeed, if (M, g) is a Riemannian manifold, one can show that for all $m \in M$, there exists a chart (U, ϕ) (called an exponential chart, see [Do Carmo 92] or [Gallot-Hulin-Lafontaine 05]) such that $\phi(m) = 0$ and

$$\phi^{-1*}(g)_x = \sum_{i=1}^{n} dx^{i2} + O(\|x\|^2).$$

Unfortunately, this result is difficult to see with the method of moving frames.

Two curvature calculations

On the unit sphere of three dimensional Euclidean space, in "latitude-longitude" coordinates

$$(u, v) \longmapsto (\cos u \cos v, \cos u \sin v, \sin u)$$

the metric induced by the Euclidean metric $dx^2 + dy^2 + dz^2$ can be written

$$du^2 + \cos^2 u \, dv^2.$$

The singularity observed at $u = \pm\frac{\pi}{2}$, which is to say at the poles, is in fact a false singularity, its corresponds to values of the parameters for which the coordinates do not give an immersion. On $S^2 \smallsetminus \{N, S\}$, the forms du and $(\cos u)dv$ form an orthonormal coframe. We note that

$$d(du) = -(\sin u) \, dv \wedge dv \quad (= 0!)$$
$$d((\cos u)dv) = (\sin u) \, dv \wedge du$$

The connection form is thus $(\sin u)dv$, the curvature form is $(\cos u)du \wedge dv$ and the Gaussian curvature is equal to 1 on $S^2 \smallsetminus \{N, S\}$, and thus on all of S^2 by continuity.

Another example where the curvature is easily calculated is the *Poincaré half plane*, which is to say the upper half plane in $\mathbf{R}^2$ of points with positive y coordinate, equipped with the metric $\frac{dx^2 + dy^2}{y^2}$. We will calculate curvature using the coframe $(\frac{dx}{y}, \frac{dy}{y})$. We obtain

$$d\left(\frac{dx}{y}\right) = \frac{dx}{y} \wedge \frac{dy}{y} \quad \text{and} \quad d\left(\frac{dy}{y}\right) = -\frac{dx}{y} \wedge \frac{dx}{y}.$$

The curvature form is therefore $-d(\frac{dx}{y}) = -\frac{dx \wedge dy}{y^2}$ and the curvature equals -1.

8.4. Poincaré-Hopf Theorem

8.4.1. Index of a Vector Field: Revisited

In two dimensions, the index of a vector field is the degree of a map from S^1 to S^1. We may therefore adopt the point of view of Section 7.4.1. Let X be a vector field on an open subset U of the oriented Euclidean plane having an isolated zero at a. Let $r > 0$ be such that $\overline{D}(a,r) \subset U$. We may suppose that r is so small that there are no other zeros in this disk.

Let u be a fixed unit vector and $R(\phi_p)$ the rotation from u to $\frac{X_p}{\|X_p\|}$. This defines a function on $D(a,r) \smallsetminus \{0\}$ having the same regularity as X, and by Section 7.4.1,

$$\mathrm{ind}_a X = \frac{1}{2\pi} \int_{C(a,r)} d\phi$$

(intuitively, the degree is the number of turns that X winds around a).

If we change the reference vector, $d\phi$ is unchanged. However two generalizations will be useful.

1. We can replace u by a vector field that is nonvanishing on the disk.

2. We can replace the Euclidean angle by the angle defined by any Riemannian metric.

Indeed, let U be a nonvanishing vector field on $D(a,r)$. By shrinking r, we can suppose that at every point $p \in C(a,r)$ the inner product $\langle u, U_p \rangle$ is positive. We can then replace $R(\phi_p)$ by the rotation $R(\phi_p^1)$ from $\frac{U_p}{\|U_p\|}$ to $\frac{X_p}{\|X_p\|}$, as the two maps form $C(a,r)$ to S^1 are homotopic.

Now let g be any Riemannian metric on the disk. We can again replace $R(\phi_p^1)$ by the rotation $R(\phi_p^2)$ of the Euclidean plane $(T_p\mathbf{R}^2 \simeq \mathbf{R}^2)$ equipped with the inner product g_p from $\frac{U_p}{\|U_p\|}$ to $\frac{X_p}{\|X_p\|}$ (note that the norm is also defined by g).

In fact, if we do this for the family of metrics $tg + (1-t)(dx^2 + dy^2)$, for $t \in [0,1]$, we obtain a homotopy between $R(\phi^1)$ and $R(\phi^2)$.

8.4.2. A Residue Theorem

Theorem 8.13. *Let (S,g) be a compact oriented Riemannian surface, let Ω_g be its curvature form, and let X be a vector field on S having r zeros $p_1, \ldots, p_r$. Then*

$$\frac{1}{2\pi} \int_S \Omega_g = \sum_{i=1}^{r} \mathrm{ind}_{p_i} X.$$

PROOF. On $S \setminus \{p_1, \ldots, p_r\}$ we introduce the frame field formed by $\widetilde{X} = \frac{X}{\|X\|}$ and the field $\widetilde{Y}$ such that at every point $(\widetilde{X}_m, \widetilde{Y}_m)$ is an direct orthonormal frame. Let $\omega \in \Omega^1(S \setminus \{p_1, \ldots, p_r\})$ be the associated connection form. If X has no zeros, ω is defined on all of S, and

$$\int_S \Omega = \int_S d\omega = 0$$

which proves the result in this case.

Otherwise for each zero p_i, obtain a chart (U_i, f_i), where $p_i \in U_i$ and $f_i(p_i) = 0$. If $r > 0$ is sufficiently small, the closed disk $D(0, r)$ and its boundary $C(0, r)$ are contained in the $f_i(U_i)$. Then

$$\int_S \Omega = \lim_{r \to 0} \int_{S \setminus \bigcup_{i=1}^k f_i^{-1}(D(0,r))} \Omega$$

$$= -\lim_{r \to 0} \sum_{i=1}^r \int_{f_i^{-1}(C(0,r))} \omega$$

by Stokes's theorem.

We can already see that everything significant happens in the neighborhood of these zeros. However note that the form ω is not defined at the p_i.

We now study each integral

$$\int_{f_i^{-1}(C(0,r))} \omega$$

separately. On U_i, the Riemannian metric can also be given by a positively-oriented orthonormal frame (Z_1^i, Z_2^i). Let ω_0 be the corresponding connection form. By equation (8.4),

$$\omega = \omega_0 - d\phi,$$

where ϕ denotes the angle between the vectors Z_1^i and $\widetilde{X}$. Then

$$\int_{f_i^{-1}(C(0,r))} \omega = \int_{f_i^{-1}(C(0,r))} \omega_0 - \int_{f_i^{-1}(C(0,r))} d\phi.$$

The form ω_0 is well defined and smooth on all of U_i, thus the first term of the right hand side tends to 0 as r tends to 0. The second term is constant and equal to the index of U at p_i and therefore to the index the vector field X (up to multiplication by 2π), by the preceding section. Finally

$$-\lim_{r \to 0} \sum_{i=1}^r \int_{f_i^{-1}(C(0,r))} \omega = \sum_{i=1}^r \operatorname{ind}_{p_i} X. \qquad \square$$

We thus have an equality between two quantities which do not *a priori* have anything to do with each other. The first applies to any Riemannian metric (and note there is no reason *a priori* to obtain an integer), and the second to any vector field. This result simultaneously shows that $\sum_{a \in Z(X)} \mathrm{ind}_a X$ is independent of the vector field X (with isolated zeros) on S and that $\int_S \Omega_g$ is independent of the metric.

A particular choice of a metric or of a vector field gives a result valid for *every* metric and *every* vector field.

Corollary 8.14

i) *For every Riemannian metric g on S^2,*

$$\frac{1}{2\pi} \int_{S^2} \Omega_g = 2,$$

  as before, Ω_g denotes the curvature form of the metric.

ii) *For every vector field X on S^2 having finitely many zeros,*

$$\sum_{x \in Z(X)} \mathrm{ind}_x(X) = 2.$$

To prove this result we have a choice: we can take the metric with constant curvature on S^2, or we can take the vector field which is the infinitesimal generator of the group of rotations about the north-south axis. The two poles are zeros of index 1. Finally we could also take the gradient of a coordinate function of the round sphere embedded in $\mathbf{R}^3$. Such a function has two non-degenerate critical points, a maximum and minimum, and its gradient has index 1 at these two points.

Corollary 8.15

i) *For every metric g on T^2,*

$$\int_{T^2} \Omega_g = 0.$$

ii) *For every vector field X on T^2 having finitely many zeros,*

$$\sum_{x \in Z(X)} \mathrm{ind}_x(X) = 0.$$

Again there is a choice: we can take a nonvanishing vector field or a metric with zero curvature. It is easy to see that there exists such fields and such metrics: if Γ is a lattice of $\mathbf{R}^2$ then constant vector fields and the metric $dx^2 + dy^2$ are Γ-invariant and pass to the quotient.

Warning. Two locally Euclidean tori R^2/Γ and R^2/Γ' are not isometric in general. Such tori are isometric if and only if the lattices Γ and Γ' are isometric. See for example [Gallot-Hulin-Lafontaine 05, 2.24].

8.5. From Poincaré-Hopf to Gauss-Bonnet

8.5.1. Proof Using the Classification Theorem for Surfaces

We have just checked the common value of both sides of the equation in Theorem 8.13 for the sphere S^2 and the torus T^2 is the Euler-Poincaré characteristic.

The result remains true for other compact surfaces. We could prove this using the same method (with a metric or a well chosen vector field), but this it not so easy.

To pass to the general case, there exists an instructive but pedestrian method. We can invoke the theorem of classification of surfaces stated in Section 7.10. We know that a compact orientable surface S is a connected sum of p tori. By Corollary 8.3, we have $\chi(S) = 2 - 2p$.

A fallout of the proof of the classification theorem is the existence of a function $f \in C^\infty(S)$ having $2p + 2$ non-degenerate critical points, a maximum, a minimum, and p "saddle points" of index 1. This allows us to extend the results of Corollaries 8.14 and 8.15 to every compact surface.

Theorem 8.16. *Let S be a compact surface, g a Riemannian metric on S, and X a vector field with finitely many zeros. Then*

$$\chi(S) = \frac{1}{2\pi} \int_S \Omega_g = \sum_{x \in Z(X)} \mathrm{ind}_x X.$$

PROOF. If S is orientable and a connected sum of p tori, it suffices to apply Theorem 8.13 to the gradient of a function f of the type we have discussed.

If S is not orientable, let $p : \widetilde{S} \to S$ be its orientation covering. Under these conditions:

1. By Corollary 8.6, $\chi(\widetilde{S}) = 2\chi(S)$.

2. Let X be a vector field on S, and $\widetilde{X}$ its lift under p. Above each zero of X there are two zeros of $\widetilde{X}$, which have the same index.

3. If g is a Riemannian metric on S, then the curvature form Ω_g is not defined *a priori*. However its integral is well defined nonetheless! To see this we

proceed as follows. If U is a connected orientable open subset of S, the choice of an orientation allows us to define Ω_g and $\int_U \Omega_g$. If we change the orientation to its opposite, Ω_g is changed to $-\Omega_g$, and by the definition of the integral of forms, the integration operator

$$\int : \ \Omega^2(U) \longrightarrow \mathbf{R}$$

is also changed by a sign, so that $\int_U \Omega_g$ is independent of the orientation. Now introduce a tiling of S by domains $(D_i)_{1 \geqslant i \geqslant N}$ homeomorphic to disks, and set

$$\int_S \Omega_g = \sum_{i=1}^{n} \int_{D_i} \Omega_g.$$

The above shows that the right hand side is independent of the choice of tiling. Moreover, the covering p is trivial above each domain, *i.e.*, for each i, the preimage $p^{-1}(D_i)$ is the union of two disjoint domains D_i' and D_i''; if we equip $\widetilde{S}$ with the metric $\widetilde{g} = p^*g$, then above each of them p is a local isometry. Thus

$$\int_{\widetilde{S}} \Omega_{\widetilde{g}} = \sum_{i=1}^{n} \int_{D_i'} \Omega_{\widetilde{g}} + \sum_{i=1}^{n} \int_{D_i''} \Omega_{\widetilde{g}}$$

$$= 2 \left(\sum_{i=1}^{n} \int_{D_i} \Omega_g \right)$$

$$= 2 \int_S \Omega_g.$$

Putting these remarks together gives the result in the non orientable case. To summarize the proof tersely: we pass to the orientation covering and everything gets multiplied by two.   □

Remark. Incidentally this proves that $\int_S \Omega_g$ can be defined even though S is not orientable. **Instead, we could have introduced the *density* dA_g defined by the metric, and considered $\int_S K_g \, dA_g$.**

8.5.2. Proof Using Tilings: Sketch

A more direct method consists in associating to every tiling of a surface a vector field having a zero of index 1 at each vertex, a zero of index 1 at the interior of each face, and a zero of index -1 in the interior of each edge.

We begin with a triangle of the plane. We draw the segments connecting the midpoints, and orient the twelve segments thus obtained in a way that each vertex is always a source, and the center of mass a sink (this procedure is known as the *barycentric subdivision* by topologists).

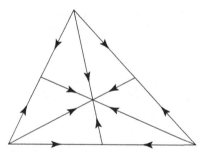

Figure 8.4: Barycentric subdivision

This construction when, transported to simplexes of a triangulation, suggests the existence of a vector field having the following zeros:

1) the vertices of the triangulation, of index $+1$ (sources);

2) a zero exactly at the interior of each simplex of dimension 2, of index 1 (sink);

3) a zero exactly at the interior of each simplex of dimension 1, of index -1 (saddle points).

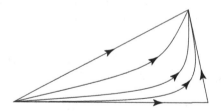

Figure 8.5: Zooming in on a triangle of the subdivision

The sum of the indices of such a vector field equals $f+v-e$. Using everything above as well as Theorem 8.13, we see that

1) the quantity $\int_S \Omega_g$ is independent of the Riemannian metric on S;

2) the quantity $\sum_{x\in Z(X)} \mathrm{ind}_x X$ is independent of the vector field X (assumed to have only finitely many zeros);

3) the quantity $f + v - e$ is independent of the triangulation.

Furthermore, each of these three quantities is equal to $\chi(S)$ by 8.5.

8.5.3. Putting the Preceding Arguments Together

We can pose two objections to the above arguments. The existence of a vector field having the "right" indices was mostly suggested rather than proved. In the best case this vector field is only piecewise C^1, but Theorem 8.13 assumes the vector field is C^2. To make things rigorous, we need to be little more precise with our triangulations.

Definitions 8.17. *A* finite simplicial complex *is a finite set S and a family Φ of subsets of S satisfying the following properties:*

i) *the singleton elements of S belong to Φ;*

ii) *if $F \in \Phi$, then so is every finite subset of F.*

The singleton elements are called the vertices *of the complex, the subsets with $k + 1$ elements are the* faces *of dimension k. The maximum value of k is the* dimension *of the complex.*

We associate to each simplicial complex its *geometric realization* defined in the following manner. We index the elements of the canonical basis of $\mathbf{R}^S$ by elements of S. Let $(e_s)_{s \in S}$ be this basis. To each face F we associate the convex hull $|F|$ of the e_s for $s \in F$. The geometric realization is then the union of the $|F|$ for F taken over Φ, equipped with the topology induced by $\mathbf{R}^S$. This is a compact topological space.

Examples

a) If Φ consists of every subset of a set with $n + 1$ elements, the geometric realization is the *standard simplex* of dimension n (a segment for $n = 1$, a triangle for $n = 2$).

b) Let S be a set with six elements labeled n, s, a, b, c, d, and consider the simplicial complex whose faces of dimension 2 are

$$\{n, a, b\}, \quad \{n, b, c\}, \quad \{n, c, d\}, \quad \{n, d, a\},$$
$$\{s, a, b\}, \quad \{s, b, c\}, \quad \{s, c, d\}, \quad \{s, d, a\}.$$

This is called the octahedral complex. The geometric realization of S is homeomorphic to S^2.

In fact, the geometric realization of a simplicial complex is a polyhedron, and simplicial complex itself codes the incidence relations of the polyhedron.

Warning. Comparing to the notion of polyhedron from elementary geometry (see [Berger 87]) this situation here is simultaneously more general (the dimension of maximal faces for the inclusion can vary) and less general (the faces are simplexes).

The existence theorem for tilings invoked in Section 8.2.2 can now be refined. The following result can be found in [Massey 77, Chapter I].

Theorem 8.18.[1] *Let M be a compact C^1 manifold of dimension n. There exists a simplicial complex and a homeomorphism f from its geometric realization X to M that has the following property: for every face F of dimension n, there exists a chart (U, ϕ) such that $f(|F|) \subset U$ and such that the restriction of $\phi \circ f$ to $|F|$ is affine.*

Warning. This result, whose proof uses the embedding theorem and a transversality argument (see [Whitney 57, IV.B]), is false for topological manifolds as soon as $n \geqslant 3$. This is beyond the scope of the book.

We are now able to make the informal arguments of Section 8.5.2 more precise. First, on a triangle (before subdivision), we can give an explicit formula for the vector field we suggested earlier. Denoting the vertices by a, b, c, we take the vector field defined by

$$X(m) = u^2\, \overrightarrow{ma} + v^2\, \overrightarrow{mb} + w^2\, \overrightarrow{mc},$$

where we have denoted the barycentric coordinates of the point m by u, v, w. Since this construction is equivariant under affine transformations, Theorem 8.18 allows us to obtain a piecewise C^1 vector field having a zero of index 1 on each face and vertex and a zero of index -1 on each edge on every compact triangulated surface.

The existence of a C^2 vector field having the same properties is a consequence of the following lemmas.

Lemma 8.19. *Let X be a continuous vector field on a compact manifold M, let $D \subset M$ be a domain with boundary and let g be a Riemannian metric. For every $\alpha > 0$, there exists a vector field Y which is smooth on D such that*

$$\forall m \in D, \quad \|Y_m - X_m\|_{g_m} < \alpha.$$

PROOF. It suffices to complete the proof for *a* Riemannian metric.

Indeed, by compactness, there exists for every pair (g_1, g_2) of such metrics, strictly positive constants A and B such that

$$A(g_1)_m(v, v) \leqslant (g_2)_m(v, v) \leqslant B(g_1)_m(v, v)$$

for all $m \in M$ and $v \in T_m M$.

We thus embed M in $\mathbf{R}^N$ and we work with the Riemannian metric induced from the Euclidean metric. The restriction of X to D can be seen as a continuous function on D with values in $\mathbf{R}^N$, which we can approximate uniformly to within α by polynomials $Z(m)$ restricted to D by the Stone-Weierstrass theorem. The orthogonal projection of $Z(m)$ to the tangent space to M at m gives the desired vector field. $\square$

---

1. Proof omitted.

Lemma 8.20. *Let X be a continuous vector field on a compact manifold M having a finite number of zeros. There exists a C^2 vector field on M having the same zeros and indices.*

PROOF. Let $a_1, \ldots, a_r$ be the zeros of X. For each of these, choose a chart (U_i, ϕ_i) such that $\phi_i(a_i) = 0$. We can always suppose that the U_i are mutually disjoint, and we choose an $r > 0$ such that $\phi_i^{-1}(\overline{B}(0,r)) \subset U_i$ for all i. We take

$$D = M \setminus \bigcup_{1 \leqslant i \leqslant r} \phi_i^{-1}(\overline{B}(0,r)).$$

For a Riemannian metric g, introduce the vector field $X' = \frac{X}{\|X\|}$ on D. By Lemma 8.19 there exists a vector field X'' on D such that $\|X'_m - X''_m\| < 1/2$ for all $m \in D$.

Now we must fill in the holes, for which it suffices to work chart by chart. Take $Y_i = (\phi_i)_* X''$ on $S(0,r)$, and extend this to the ball by writing $Y_i(\rho x) = \frac{\rho^2}{r^2} f(\rho) U_i(x)$ for $x \in S(0,r)$, where $f : [0,r] \to \mathbf{R}^{+*}$ is a C^2 function such that $f(r) = 1$, $f'(r) = 1$, $f''(r) = 0$. The desired vector field may be taken to be equal to X'' on D and to $(\phi_i^{-1})_* Y_i$ on $\phi_i^{-1}(\overline{B}(0,r))$. $\qquad\square$

These two lemmas now complete the argument.

Corollary 8.21. *Let X be a continuous vector field on a compact surface S having finitely many zeros $a_1, \ldots, a_p$. Then*

$$\sum_{k=1}^{p} \mathrm{ind}_{a_k} X = \chi(S).$$

8.6. Comments

The case of embedded surfaces

A two-dimensional submanifold S of three-dimensional Euclidean space is naturally equipped with the induced Riemannian metric, obtained by restricting the ambient inner product to each tangent space. On every open subset $U \subset S$ equipped with a local orientation, we can define the normal vector field n_x. If (X_1, X_2) is a orthonormal frame on U compatible with the orientation, n_x will be the unit normal such that the frame $((X_1)_x, (X_2)_x, n_x)$ is positively oriented. The *Gauss map* is the map from U to S^2 which associates to x the point $\Gamma(x)$ defined by $\overrightarrow{0\Gamma(x)} = n_x$. The *shape operator* (or Weingarten map) $\mathcal{S}$ is the differential of $x \mapsto n_x$. One verifies that this is a *symmetric* endomorphism of (T_x, g_x). Either one of these takes into account the shape of S in the ambient space. See [Do Carmo 76, Chapter 3] for more details. We also use the *second fundamental form* (in this terminology, the first fundamental form is the induced metric) which is the symmetric bilinear form on $T_x S$ defined by

$$II_x(u, v) = g_x(u, \mathcal{S} \cdot v).$$

The interpretation of this form is as follows: if u is a unit tangent vector to S at x, $II_x(u, u)$ is the curvature of the plane curve formed by intersecting S with the plane spanned by n_x and u.

We then have the following result (see [Do Carmo 76, p. 234]).

Theorem 8.22 (Gauss's *theorema egregium*)

i) *If ϖ is the canonical volume form of S^2, and Ω_g is the curvature form,*
   $\Gamma^* \varpi = \Omega_g$.

ii) *If K_g is the curvature*, $\det(\mathcal{S}) = K_g$.

If S is orientable, which is the case if it is a compact submanifold, the Gauss map can be defined globally. Then we have

Corollary 8.23. *Let S be a compact oriented submanifold of $\mathbf{R}^3$ seen as oriented Euclidean space, and let Γ be the Gauss map defined by this orientation. Then*

$$\deg(\Gamma) = \frac{1}{2}\chi(S).$$

PROOF. By definition of the degree, $\int_S \Gamma^* \varpi = \deg(\Gamma) \int_{S^2} \varpi$. However, by Proposition 6.31, $\int_{S^2} \varpi = 4\pi$, while $\Gamma^* \varpi = \Omega_g$. Thus it suffices to apply Theorem 8.16. □

From these two results we obtain the following expression for the volume of tubular neighborhoods of a compact surface.

Theorem 8.24. *With the same hypothesis as Corollary 8.23, let $V_r(S)$ be a tubular neighborhood of S. Then*

$$\mathrm{vol}\big(V_r(S)\big) = 2r\big(\mathrm{area}(S)\big) + \frac{4\pi r^3}{3}\chi(S).$$

The volume of tubular neighborhoods of submanifolds was first studied by Hermann Weyl in a prophetic article (cf. [Weyl 39]). For a systematic study, see [Gray 04].

Canonical Riemannian metrics on a surface

We saw that on S^2 there is a metric of constant curvature $+1$ (this metric is essentially unique by [Gallot-Hulin-Lafontaine 05, 3.F]), and on T^2 there are metrics of zero curvature. These results extend to $P^2\mathbf{R}$ and the Klein bottle by passing to the quotient.

The other surfaces have negative Euler-Poincaré characteristic. If these surfaces admit a metric with constant curvature, the curvature must be negative by the Gauss-Bonnet theorem. Such metrics do exist (this is one of the major discoveries of H. Poincaré, see [Charpentier-Ghys-Lesne 10] and also, for a very detailed account, [DE Saint-Gervais 10]). These can be obtained by a quotient of the Poincaré half-plane by a discrete and cocompact group of isometries. The study of these metrics has remained an active area of research. For an introduction, see [Buser 92].

A few words on higher dimensions

In higher dimensions, everything that we have said about vector fields and triangulations generalizes without difficulty. However curvature becomes, in the words of Misha Gromov, a "monster of linear algebra", starting from the fact that one must discover the correct integrand. This is very nicely explained in the final chapter of the work by M. Spivak, which has an evocative title: "The Gauss-Bonnet theorem and what it means for mankind". We know of no better reference to invoke the ubiquity of the Gauss-Bonnet theorem (which due to the decisive contribution of S.S. Chern in higher dimensions is also called the Chern-Gauss-Bonnet theorem).

8.7. Exercises

1. Find a free action of $\mathbf{Z}/2\mathbf{Z}$ on $P^{2n+1}\mathbf{C}$. Show that there does not exist such an action on $P^{2n}\mathbf{C}$.

2. Write an S^1-invariant metric in polar coordinates as $dr^2 + f^2(r)\,d\theta^2$ (away from the origin). Calculate the curvature.

3. Calculate the curvature of a torus of revolution equipped with the metric induced by Euclidean space.

4. Give an example of a vector field on S^2 having a zero of index 2.

5**. Write the underlying simplicial complex of a regular icosahedron.

6*. *Ramified coverings*

Let S and S' be two surfaces. A *ramified covering* is a map from S to S' such that for all $x \in S$ there exists charts (U, φ) $(x \in U)$ and (U', φ') $(f(x) \in U')$ such that

$$\varphi(x) = \varphi'\big(f(x)\big) = 0 \quad \text{and} \quad \varphi' \circ f \circ \varphi^{-1}(z) = z^{k_x}$$

where k is a positive integer, and where we identify $\mathbf{R}^2$ with $\mathbf{C}$.

a) Show that the integer k is independent of the charts having the property above. We call this integer the *index of ramification* of x.

b) If S and S' are holomorphic manifolds of dimension 1, every holomorphic map is a ramified covering.

In the sequel, and in the following exercises, we suppose that S is *compact* and *connected*.

c) Show that f is surjective.

d) Show that the set Σ of points where $k_x > 1$ is finite.

e) Show that the restriction of f to $S \setminus \Sigma$ is a covering of $S' \setminus f(\Sigma)$. Its degree (which is necessarily finite under our hypothesis) is also the degree of f as a map from S to S'.

7*. *Riemann-Hurwitz formula*

Let $f : S \to S'$ be a ramified covering of degree d. Show that

$$\chi(S) = d\chi(S') - \sum_{x \in \Sigma}(k_x - 1)$$

Hint. Starting with a triangulation of S', show that there exists another triangulation such that the points of $f(\Sigma)$ are vertices, and imitate the proof of Corollary 8.6.

8*. *A smooth projective algebraic curve of degree 3 is a torus*

Let $P(x) = x^3 + px + q$ be a polynomial of degree 3 with complex coefficients, with three distinct zeros.

a) Supposing that P is real with real zeros, sketch the curve with equation $y^2 = P(x)$ in $\mathbf{R}^2$. How many connected components does it have?

b) Show that this equation defines a complex submanifold of $\mathbf{C} \times \mathbf{C}$ of dimension 1, and that this submanifold is connected.

c) Equip $P^2\mathbf{C}$ with a system of homogeneous coordinates x, y, t. Show that the equation $y^2 t = x^3 + pxt^2 + qt^3$ defines a compact and connected submanifold of $P^2\mathbf{C}$. Denote this submanifold by E.

d) Show that the map $(x, y) \mapsto x$ of the curve $y^2 = P(x)$ to $\mathbf{C}$ extends by continuity to a unique holomorphic map from E to $P^1\mathbf{C}$.

e) Find the ramification points of this map. Deduce that $\chi(E) = 0$ and that E is diffeomorphic to a torus.

Appendix:
The Fundamental Theorem
of Differential Topology

Manifolds with boundary (of dimension n) are defined in the same way as manifolds in Section 2.2, by taking charts with values in the half-space

$$\left\{(x^1, \ldots, x^n): \ x^1 \leqslant 0\right\}$$

The only additional problem is how to define smooth functions on a half-space. It suffices to replace partial derivatives with respect to x^1 at $(0, x^2, \ldots, x^n)$ by left derivatives. The *interior* $\mathrm{int}(M)$ of a manifold with boundary M is the set of points which are diffeomorphic to an open subset of $\mathbf{R}^n$. This is clearly a smooth manifold of dimension n. The reader is invited to check that $M \smallsetminus \mathrm{int}(M)$ is a smooth manifold of dimension $n - 1$, denoted ∂M.

Examples of manifolds with boundary include the half-space itself, closed balls, and also regular domains. Theorem 6.24 extends to manifolds with boundary: the boundary of an oriented manifold with boundary is orientable and inherits a natural orientation. Stokes's theorem extends to oriented manifolds with boundary.

The notion of a manifold with boundary does not appear to play an important role in this book. However this notion is unavoidable, as it is really this which arises in Stokes's theorem. To illustrate we will sketch, in the spirit of [Guillemin-Pollack 74] (even though this result is not in this book) the proof what we could call with good reason the "fundamental theorem of differential topology".

Theorem. *Let X be a compact connected manifold (without boundary) of dimension at least 1. Then the identity map from X to X is not homotopic to a constant map.*

PROOF. Let H be such a homotopy, which is to say a smooth map from $X \times [0,1]$ to X such that for all $x \in X$, $H(x,0) = x$ and $H(x,1) = a$, where $a \in X$ is fixed. By Sard's theorem, the function H admits at least one regular values y. If $\dim(X) \geqslant 1$, y is distinct from a.[1]

Now $H^{-1}(y)$ is a compact submanifold of $X \times [0,1]$ of dimension 1, possibly with boundary, and

$$\partial H^{-1}(y) = \partial(X \times [0,1]) \cap H^{-1}(y) = (X \times \{0\} \cup X \times \{1\}) \cap H^{-1}(y).$$

By the definition of H, we have

$$H^{-1}(y) \cap (X \times \{0\}) = (y,0) \quad \text{and} \quad H^{-1}(y) \cap (X \times \{1\}) = \emptyset,$$

therefore $\partial H^{-1}(y) = (y,0)$. But this is impossible, since by a refinement of Theorem 3.45 (which is easy to show using the same methods, or see [Guillemin-Pollack 74]) every compact connected manifold of dimension 1 with nonempty boundary is a compact interval of $\mathbf{R}$: in any case, $\partial H^{-1}(y)$ must have an even number of elements. $\square$

Remarks

a) Besides the classification of manifolds with boundary of dimension 1, this result uses an extension of the notions of regular point and regular value to manifolds with boundary. If $f : M \to M'$ is a smooth map from one manifold with boundary to another, a point x of M is a regular point if either it belongs to the interior $M \smallsetminus \partial M$ and if it is regular for the restriction of f to the interior of M, or if it is regular for the restriction of f to the boundary. A point y of the target is a regular value if $f^{-1}(y)$ consists of regular points in the sense above. We also prove the result used above without difficulty (*ibidem*): $f^{-1}(y)$ is a submanifold with boundary of M and $\partial f^{-1}(y) = (\partial M) \cap f^{-1}(y)$.

b) In this proof the "difficult" Sard theorem (the case where the dimension of the domain manifold is strictly greater than the dimension of the target, see [Golubitsky-Guillemin 73, Chapter 2, § 1] or [Hirsch 76, Chapter 3, Theorem 1.3]) plays a key role. One must assume from the start that the homotopy is C^2. Then an approximation argument (see Lemma 7.24) allows us to pass to the general case.

---

1. This theorem is clearly false in dimension 0. Still it is not useless to indicate where the hypothesis on dimension arises. I thank Antoine Chambert-Loir for this remark, which is deeper than it first seems.

Solutions to the Exercises

Chapter 1

2. *Laplacian and isometries*

b) Testing the property on functions of the form $f(x) = x^k x^l$, we see that A must be orthogonal. A direct calculation then shows that this condition is sufficient.

c) Note that

$$\Delta(f \circ T) = \sum_{i,k,l} \partial^2_{kl}(f \circ T)\partial_i T^k \partial_i T^l + \sum_k \partial_k (f \circ T)\Delta T^k,$$

where T^k denotes the k-th component of T. Testing the property on linear forms and functions of the form $x^k x^l$, we check that the Jacobian of T at each point is an orthogonal matrix.

It follows that T preserves the length of curves, and consequently the Euclidean distance. It is thus an affine isometry.

3. *Differentiation and integration*
We find

$$F'(x) = h\big(b(x), x\big) - h\big(a(x), x\big) + \int_{a(x)}^{b(x)} h(t, x)\, dt$$

5. We reduce to the case where the center of the ball is the origin and work in polar coordinates.

Warning. $\mathbf{C} \smallsetminus \{0\}$ *is not* biholomorphically equivalent to $\mathbf{C}$ with a closed disk removed.

7. We again obtain a simple closed curve for a), a curve with a double point at the origin for b), and a curve with a cusp at the origin for c).

8*. *Cartan decomposition of the linear group*

a) If $(e_i)_{1 \leqslant i \leqslant n}$ is an orthonormal basis of eigenvectors of S, the associated eigenvalues, which are equal to $\langle S(e_i), e_i \rangle$ are strictly positive. It follows that

$$\forall x \in \mathbf{R}^n, \quad \langle Sx, x \rangle \leqslant k\|x\|^2, \quad \text{with } k = \inf_{1 \leqslant i \leqslant n} \lambda_i.$$

b) Uniqueness of T comes when $S = T^2$, then S and T commute. Denoting the open subset of $\mathrm{Sym}(n)$ consisting of strictly positive endomorphisms by $\mathrm{Sym}^+(n)$, check that the map $S \mapsto S^2$ is a bijection on $\mathrm{Sym}(n)$ with invertible differential. If $S \in \mathrm{Sym}^+(n)$ and if $ST + TS = 0$, we see that $T = 0$ by taking a basis which diagonalizes S.

9*. The differential of f_S is surjective at $A = I$. It then suffices to apply Theorem 1.18. In other words, if T is sufficiently close to S, the quadratic form q_T is equivalent to q_S modulo a change of basis which depends differentiably on T.

It is elementary to show that a quadratic form associated to a matrix T sufficiently close to S (which we suppose to be non-degenerate) has the same signature as the form associated to S, by Sylvester's law of inertia (cf. [Berger 87, Chapter 13]). But the calculation of the differential allows us to make this result more precise. Under this form, it plays a key role in proof of the Morse lemma.

10*. *Constant rank theorem*

See [Demazure 00, Chapter IV].

11*. *Morse lemma*

The first part is elementary. Using the map g of Exercise 9, we see that

$$h(x) = {}^t g(h(x)) S g(h(x)),$$

and therefore

$$f(x) = f(0) + {}^t u(x) S u(x) \quad \text{where } u(x) = g(h(x)) \cdot x.$$

The differential of u at 0 is the identity map, and the inverse function theorem applies. If $y \mapsto v(y)$ is the local inverse to u, we have

$$f(v(y)) = f(0) + {}^t y S y.$$

13. We find

$$(x^2 + y^2 + z^2 + a^2 - r^2)^2 - 4a^2(x^2 + y^2) = 0;$$

a convenient parametrization is given by

$$(\theta, \varphi) \longmapsto \big((a + r \cos \theta) \cos \varphi, (a + r \cos \theta) \sin \varphi, r \sin \theta\big).$$

15*. We check that the differential of the map

$$(x, v) \longmapsto \big(f(x), T_x f \cdot v\big)$$

at the point (x, v) is surjective when $T_x f$ is surjective.

16. The set of lines is the given by the surface

$$x \cos \frac{z}{b} - y \sin \frac{z}{b} = 0,$$

called the *helicoid*. **It is the only ruled minimal surface in $\mathbf{R}^3$, see for example [Spivak 79, volume 3].**

18. Let C be the intersection. Considering the rank of the matrix

$$\begin{pmatrix} x & y & z \\ 2x - 1 & y & 0 \end{pmatrix}$$

we see that $C \smallsetminus (1, 0, 0)$ is a submanifold of dimension 1. If C is itself a submanifold of dimension 1, then the tangent at $(1, 0, 0)$ will be parallel to the plane $x = 0$, and thus the projection of C onto this plane parallel to the line $y = z = 0$ will be a submanifold in a neighborhood of $(0, 0)$. The equation of this projection, obtained by eliminating x is

$$y^2 - z^2 + z^4 = 0.$$

This is not a submanifold, since the tangent vectors at 0 are multiples of $(1, 1)$ and $(1, -1)$.

19. *Pseudo-orthogonal group*

Let $J_{p,q}$ be the matrix

$$\begin{pmatrix} I_p & 0 \\ 0 & -I_q \end{pmatrix}, \quad \text{where } p + q = n.$$

The condition of the statement is equivalent to

$${}^t A J_{p,q} A = J_{p,q},$$

and we proceed as in the case of the orthogonal group by using the map $A \mapsto {}^t A J_{p,q} A$, from $M_n(\mathbf{R})$ to $\mathrm{Sym}(n)$.

22. *Position of a hypersurface with respect to the tangent plane*

b) It suffices to apply Taylor's formula.

c) The first assertion also comes from Taylor's formula (take x in a space where Q is positive definite, and in a space where Q is negative definite).

Using Exercise 11 allows us to be more precise: there exist smooth functions $\varphi^1, \ldots, \varphi^n$, whose differentials at 0 are independent, defined on a neighborhood of 0 and such that for x in this neighborhood,

$$f(x) = \left(\varphi^1(x)\right)^2 + \cdots + \left(\varphi^p(x)\right)^2 - \left(\varphi^{p+1}(x)\right)^2 - \cdots - \left(\varphi^n(x)\right)^2.$$

This allows us to see the intersection $S \cap T_0 S \smallsetminus \{0\}$ with an appropriate neighborhood of 0 is mapped by a diffeomorphism to an open subset of the $n-1$ dimensional cone given by the equation

$$\sum_{i=1}^{p} y_i^2 - \sum_{i=p+1}^{n} y_i^2 = 0,$$

with the origin removed.

23. *The exponential is not a group morphism*

b) It suffices to identify the coefficients of t^2 in the series expansion of both sides.

25. The function f can be expanded as an entire series about 0 in $D(0, r)$, and can be written

$$\sum_{n=k}^{\infty} a_n z^n \quad \text{with } a_k \neq 0$$

where again

$$a_k z^k \left(1 + \sum_{n=1}^{\infty} c_n z^n\right) = z^k g(z).$$

Here, k denotes the smallest integer such that $a_k \neq 0$. The inverse function theorem, applied to the function $u \mapsto u^k$ at 1, shows that g can be written h^k, where h is holomorphic on a disk $D(0, r')$, of radius possibly less than r. Then, since $h(0) = 1$, the inverse function theorem also applies to $z \mapsto zh(z)$ at 0, whose inverse is the **C**-diffeomorphism required.

Chapter 2

2. If a has coordinates (x_0, y_0), and d has equation $ux + vy + w = 0$, check that

$$(\operatorname{dist}(a, d))^2 = \frac{(ux_0 + vy_0 + w)^2}{u^2 + v^2}.$$

It then suffices to use the charts for the manifold of lines seen in the introduction.

4. *The unitary and special unitary groups*

a) We use the same method as in Section 1.5.2 for the orthogonal group by
showing that the map $A \mapsto {}^t\overline{A}A$ from $M_n(\mathbf{C})$ to the set of Hermitian
matrices is a submersion for A such that ${}^t\overline{A}A = I$. Warning: the set of
Hermitian matrices of order n is a *real* vector space of dimension n^2.

The exponential map gives, by imitating the proof of Proposition 1.32,
a parametrization defined on a neighborhood of 0 to the vector space of
antihermitian matrices, also of dimension n^2.

b) We use the map

$$A \longmapsto \det A \quad \text{from } U(n) \text{ to } S^1 \simeq \{z \in \mathbf{C} : |z| = 1\},$$

for example, after noting that a unitary matrix has a determinant of unit
modulus.

c) A direct calculation shows that the special unitary matrices of order 2 can
be written

$$\begin{pmatrix} a & -\overline{b} \\ b & \overline{a} \end{pmatrix},$$

where of course $|a|^2 + |b|^2 = 1$.

5. *Projective group*

a) We can either use the *real* Hopf fibration or conjugate the transformation
$t \mapsto \frac{at+b}{ct+d}$ by stereographic projection i_N. We find the map from S^1 to S^1
defined by

$$(x, y) \longmapsto \left(\frac{2(ac + bd) + 2(ac - bd)x + 2(ad + bc)y}{a^2 + b^2 + c^2 + d^2 + (a^2 - b^2 + c^2 - d^2)x + 2(ab + cd)y} \atop \frac{a^2 + b^2 - c^2 - d^2 + (a^2 - b^2 - c^2 + d^2)x + 2(ab - cd)y}{a^2 + b^2 + c^2 + d^2 + (a^2 - b^2 + c^2 - d^2)x + 2(ab + cd)y} \right)$$

It is clear that it is preferable (and more natural!) in this case to consider
S^1 as the projective line.

6. *Projective quadrics*

b) For an appropriate basis of $\mathbf{R}^4$, the quadric Q is the set of points whose
homogeneous coordinates satisfy

$$x^2 + y^2 + z^2 - t^2 = 0.$$

The last homogeneous coordinate from a point of Q is thus nonzero, and we finish by sending Q to $\mathbf{R}^3$ by the chart

$$[x, y, z, t] \longmapsto \left(\frac{x}{t}, \frac{y}{t}, \frac{z}{t}\right).$$

c) The equation of Q in homogeneous coordinates can be written as

$$x^2 + y^2 - z^2 - t^2 = 0, \quad \text{and also} \quad XY - ZT = 0.$$

Note that $\frac{X}{Z} = \frac{T}{Y}$ and $\frac{X}{T} = \frac{Z}{Y}$, and define a smooth map $f = (f_1, f_2)$ from Q to $P^1\mathbf{R} \times P^1\mathbf{R}$ by setting

$$f_1([X, Y, Z, T]) = \begin{cases} [X, Z] & \text{for } (X, Z) \neq (0, 0) \\ [T, Y] & \text{for } (T, Y) \neq (0, 0) \end{cases}$$

and

$$f_2([X, Y, Z, T]) = \begin{cases} [X, T] & \text{for } (X, T) \neq (0, 0) \\ [Z, Y] & \text{for } (Z, Y) \neq (0, 0). \end{cases}$$

d) If the quadratic form is of type (r, s) (with $rs \neq 0$ if we require that Q be nonempty), we check that Q is diffeomorphic to the quotient of $S^{r-1} \times S^{s-1}$ by the "double antipodal map" $(x, y) \mapsto (-x, -y)$.

8*. To embed T^3 in $\mathbf{R}^4$, the idea is to "thicken" the embedding $(\theta, \varphi) \mapsto (e^{i\theta}, e^{i\varphi})$ from T^2 into $\mathbf{C}^2 \simeq \mathbf{R}^4$. Take for example

$$(\theta, \varphi, \psi) \longmapsto \left(e^{i\theta}\left(1 - \frac{\cos \psi}{2}\right), e^{i\varphi}\left(1 - \frac{\sin \psi}{2}\right)\right).$$

The product of the standard embedding with itself gives an embedding of $S^2 \times S^2$ into $\mathbf{R}^6$, whose image is included in a sphere S^5. It then suffices to use an appropriate stereographic projection.

11. *A little more on submersions*

b) Let X be the hyperbola with equation $xy = 1$ in $\mathbf{R}^2$, and f the restriction of the first projection to X. This is a submersion (even a local diffeomorphism) at every point which it is not surjective.

c) It suffices to remark that the Hopf fibration is trivializable above S^2 with one point removed.

13. *Veronese surface*

d) We remark that $V(P^2\mathbf{R})$ is contained in a sphere S^4, and we take a stereographic projection with respect to a point not in the image.

The explanation of this mysterious formula is the following. We associate to each line v the operator p_v of orthogonal projection onto v. In an orthonormal basis where a unit vector in the v direction has coordinates (x, y, z), the matrix of p_v is

$$\begin{pmatrix} x^2 & xy & xz \\ xy & y^2 & yz \\ xz & yz & z^2 \end{pmatrix}.$$

In the same way we can associate to a k-plane P of $\mathbf{R}^n$ equipped with its Euclidean structure the orthogonal projection operator onto P, thus obtaining an embedding of the Grassmannian $G_{k,n}$ of k-planes of $\mathbf{R}^n$ into Euclidean space.

14. *A useful fibration*

b) Every unit vector v can be completed to an orthonormal basis depending differentiable on v, provided that v is not collinear to a given vector v_0. This comes from the Gram-Schmidt orthonormalization procedure.

16. *Conformal compactification of $\mathbf{R}^n$; Möbius group*

The following three transformations: isometry $x \mapsto Ax$, homothety $x \mapsto \lambda x$ and translation $x \mapsto x + a$ are projective maps defined by the matrices

$$\begin{pmatrix} 1 & 0 & \cdots & 0 & 0 \\ 0 & & & & 0 \\ \vdots & & A & & \vdots \\ 0 & & & & 0 \\ 0 & 0 & \cdots & 0 & 1 \end{pmatrix}, \quad \begin{pmatrix} \frac{1}{\lambda} & 0 & \cdots & 0 & 0 \\ 0 & & & & 0 \\ \vdots & & I_n & & \vdots \\ 0 & & & & 0 \\ 0 & 0 & \cdots & 0 & \lambda \end{pmatrix},$$

and

$$\begin{pmatrix} 1 & & 2a & & 0 \\ 0 & & & & \\ \vdots & & I_n & & {}^t a \\ 0 & & & & \\ 0 & 0 & \cdots & 0 & 1 \end{pmatrix}.$$

Similarly, we associate to inversion $x \mapsto \frac{x}{|x|^2}$ the projective map defined by

$$\begin{pmatrix} 0 & 0 & \cdots & 0 & 1 \\ 0 & & & & 0 \\ \vdots & & I_n & & \vdots \\ 0 & & & & 0 \\ 1 & 0 & \cdots & 0 & 0 \end{pmatrix}.$$

17*. *Blowup*

c) The inverse diffeomorphism is given by $r(x,y) = ([x,y],(x,y))$.

d) It suffices to apply the Hadamard lemma to c.

e) One obtains a diffeomorphism from M to E by proceeding as follows: to a line d in $\mathbf{R}^2$ we associate the orthogonal line that passes through the origin, and the intersection of this line with d. Explicitly this map is given by

$$[(u,v,w)] \longmapsto ([u,v], -wuu^2 + v^2, -wvu^2 + v^2).$$

f) The map $\widehat{\varphi}$ is necessarily given by

$$\widehat{\varphi}([x,y],x,y) = ([\varphi(x,y)], \varphi(x,y)) \ \text{ if } \ (x,y) \neq (0,0)$$
$$\widehat{\varphi}([X,Y],0,0) = ([d\varphi_0 \cdot (X,Y)],0,0).$$

One must check this map is smooth. Denote the coordinates of φ by f and g, and start with the parametrization

$$\psi : \ (x,t) \longmapsto ([1,t],(x,xt)).$$

For $x \neq 0$, we have

$$(\varphi \circ \psi)(x,t) = ([f(x,xt), g(x,xt)], f(x,xt), g(x,xt)).$$

By the Hadamard lemma, there exist smooth functions f_1, f_2, g_1, g_2, equal at 0 to the partial derivatives of f and g such that

$$f(x,xt) = xf_1(x,xt) + xtf_2(x,xt) \quad \text{and} \quad g(x,xt) = xg_1(x,xt) + xtg_2(x,xt).$$

Then

$$(\varphi \circ \psi)(x,t) = ([f_1(x,xt) + tf_2(x,xt), g_1(x,xt) + tg_2(x,xt)], f(x,xt), g(x,xt)).$$

However, by the hypotheses on φ,

$$\begin{pmatrix} f_1 + tf_2 \\ g_1 + tg_2 \end{pmatrix} = \begin{pmatrix} f_1 & f_2 \\ g_1 & g_2 \end{pmatrix} \begin{pmatrix} 1 \\ t \end{pmatrix}$$

is nonzero for all t if x is sufficiently close to 0. The remainder is left to the reader.

Note. This construction works in exactly the same way in complex geometry, where it is especially important. See for example [Audin 04, Chapter VIII].

18. In Exercise 5, $\Gamma = Gl(n+1,\mathbf{R})$ and $\Gamma_o = \mathbf{R}^*I$. The natural action of $SO(n+1)$ on $X = P^n\mathbf{R}$ is effective if and only if n is even.

19. The map $z \mapsto z^n$ descends to the quotient and gives a continuous and bijective map from $\mathbf{C}/\Gamma$ to $\mathbf{C}$; the inverse bijection is also continuous and is the map which sends z to any of its n-th roots, identified modulo Γ. This lets us transport the $\mathbf{C}$-differential structure to $\mathbf{C}/\Gamma$. The quotient map is then smooth, but its differential at 0 vanishes.

22. *Lens spaces*

a) This comes from the fact that 1 is an eigenvalue of every matrix in $A \in SO(2n+1)$.

b) More generally, we can consider the action of $\mathbf{Z}/p\mathbf{Z}$ on S^3 given by

$$k \cdot (z, z') = (u^k z, u^{kr} s'),$$

where r is a number that is relatively prime with p. Let the manifold thus obtained be denoted $L_{r,p}$. One proves, and this is difficult, that $L_{1,5}$ and $L_{2,5}$ are not homeomorphic. For more details on these manifolds, which have subtle topological properties, see [Wolf 84] and [Milnor 66].

24. *Coverings and local diffeomorphisms*

a) It is clear that $f(X)$ is open (since f is a local diffeomorphism) and closed (it is a compact subset of Y), which shows that f is onto. From there it suffices to use Theorem 2.14.

b) For example, consider the covering of the projective space by the sphere and restrict it to the sphere with one point removed.

25. S^1 with a finite set of points removed is not connected as soon as there are more than two points!

26. No. For example, we can restrict $\varphi_i \times \psi_j$ to an open subset of $U_i \times V_j$ which is not a product of open subsets.

28*. *Connected sum*

b) I was very discrete on the issue of the uniqueness of the connected sum. The only reference I know of on this subtle issue is a series of exercises in [Dieudonné 72, XVI.26, problems 12 to 15].

d**) It suffices to note that $\mathbf{R}^2 \sharp P^2 \mathbf{R}$ is diffeomorphic to $P^2 \mathbf{R}$ with a point removed, and to use Exercise 17 e).

30. Let $f : M \smallsetminus F \to \{0,1\}$ be a continuous function. We must show that f is constant. If $a \in F$ there exists an open subset U containing a but none of the other points of F.

The function f is constant on $U \smallsetminus \{a\}$, as it is a connected subset by the hypothesis on dimension. It extends by continuity to U, and to a continuous map $\tilde{f} : M \to \{0,1\}$, which is constant since M is connected.

Chapter 3

1. The Stone-Weierstrass theorem applies since by Corollary 3.5 the algebra of smooth functions separates points.

3. We argue by contradiction using Taylor's formula.

4. Let f be a continuous function which vanishes at 0, and is positive or zero in a neighborhood of 0, and let $g = \sqrt{f}$. Then

$$\delta \dot{f} = \delta \dot{g}^2 = g(0)\delta \dot{g} = 0,$$

so $\delta \dot{f} = 0$. From linearity we deduce that δ vanishes for all germs that vanish at zero (by writing $f = \sup(f, 0) + \inf(f, 0)$), and then for all germs.

5. Let f be the function of a real variable defined by $f(x) = \exp(-1/x)$ if $x > 0$ and $f(x) = 0$ if $x \leqslant 0$, and let g be the function $x \mapsto f(-x)$. Now f and g are smooth, and have non-vanishing germs at 0, while $fg = 0$. On the other hand, the ring of germs of analytic functions is a subring of the ring $\mathbf{R}[\![X]\!]$ of formal power series, which is an integral domain.

7. *North-South and North-North dynamics on the sphere*

a) It suffices, as in Section 2.3, to write $i_N^{-1} \circ h_t \circ i_N$ in the chart $(S^2 \setminus S, i_S)$. Then

$$(i_S \circ i_N^{-1} \circ h_t \circ i_N \circ i_S^{-1})(y) = e^{-t}y.$$

d*) It suffices to start with a one-parameter group of translations.

Remark. Checking the Poincaré-Hopf theorem for these examples (Corollary 8.14) is easy but instructive.

8. Start with the fact, which is clear but fundamental, that a vector field is invariant under its own flow. If $[X, Y] = 0$, by Theorem 3.38

$$\frac{d}{ds}\varphi_{s*}Y_{|s=0} = 0.$$

But we also have

$$\frac{d}{ds}\varphi_{s*}Y_{|s=s_o} = \frac{d}{ds}\varphi_{s*}(\varphi_{s_o*})Y_{|s=0} = -[X, \varphi_{s_o*}Y],$$

and

$$[X, \varphi_{s_o*}Y] = \varphi_{s_o*}[\varphi_{-s_o*}X, Y] = \varphi_{s_o*}[X, Y] = 0.$$

Thus $\varphi_{s*}Y = Y$ for all s. However the flow of $\varphi_{s*}Y$ is $\varphi_s \circ \psi_t \circ \varphi_{-s}$ by Proposition 3.37. The result is elementary in the other direction.

9. It suffices to prove the result for open subsets of $\mathbf{R}^n$. Taking the vector fields ∂_i for Y, we see that the coefficients of X are (locally) constant. We can then take $x^i \partial_i$ as the test vector field.

12*. *Examples of parallelizable manifolds*

a) We use Exercise 10 by remarking that there is more than one way to view the unit sphere of $\mathbf{R}^4$ as

$$\{(z, z') \in \mathbf{C}^2 : |z|^2 + |z'|^2 = 1\}.$$

b**) Let T be the vector field $z \mapsto iz$ on S^1 seen as the set of complex numbers z of modulus 1. With S^n embedded in the usual way in $\mathbf{R}^{n+1}$, introduce $(n+1)$ vector fields X^i whose value at x is the orthogonal projection onto $T_x S^n$ of the i-th basis vector e_i. Explicitly,

$$X_x^i = e_i - x^i x.$$

Now introduce $n+1$ vector fields on $S^1 \times S^n$ defined by

$$x^i T + X^i,$$

and check that their values at every point (z, x) of $S^1 \times S^n$ are independent vectors.

15*. *Transitivity of the group of diffeomorphisms*

a) It suffices to use the flow of a vector field of the form $f(b - a)$, where f is a bump function with support contained in $B(0, r')$ and equal to 1 on $B(0, r)$ (of course $r < r'!$).

c) Say that two points of M are equivalent if they correspond under a diffeomorphism. We obtain an equivalence relation whose equivalence classes are open by b).

d) An argument analogous to a) lets us argue by induction on k Note that the result is grossly false in dimension 1, as soon as $k \leqslant 3$.

16*. *Normal form of a non-vanishing vector field*

a) The Jacobian matrix of F at 0 is

$$\begin{pmatrix} X^1(0) & 0 & \cdots & 0 \\ X^2(0) & & & \\ \vdots & & I_{n-1} & \\ X^n(0) & & & \end{pmatrix}$$

b) We have

$$(F^{-1} \circ \varphi_t^X \circ F)(x^1, \ldots, x^n) = (x^1 + t, x^2, \ldots, x^n),$$

and we apply Proposition 3.37.

20. *Normal bundle*

c) Let $f_1, \ldots, f_p$ be the components of f. By hypothesis the differentials df_i are everywhere linearly independent. Thus the normal bundle admits p everywhere linearly independent sections, given by

$$N_i = \sum_{k=1}^{n} \partial_k f_i \partial_k.$$

21*. *Tautological bundle*

a) Above the open subset $U_i = \{[x] \in P^n \mathbf{R} : x_i \neq 0\}$, take the trivialization

$$([x], v) \longmapsto ([x], v^i).$$

b) See Exercise 17 e) of Chapter 2.

c**) The vector bundle γ_n is of rank 1. It is thus a trivialization if and only if there exists an everywhere non-vanishing section. We can then suppose this section has norm 1 everywhere (for the norm on the fibers induced by the Euclidean norm of $\mathbf{R}^{n+1}$). We thus obtain a right inverse s of the canonical projection $p : S^n \mapsto P^n \mathbf{R}$. This is impossible: the map s will be surjective as it is open and S^n is compact. We deduce that p is injective, a contradiction.

22*. *Constructions of some vector bundles*

See for example [Hirsch 76, Chapter 4]. In fact, every "functorial" construction on vector spaces (dual, product, tensor, etc.) has a meaning for vector bundles.

23**. *Tangent bundle to $P^n \mathbf{R}$*

b) To the equivalence class of (x, v) we associate the linear map from $(\gamma_n)_{[x]}$ to $(\gamma_n^{\perp})_{[x]}$ which maps λx to λv. We thus obtain a morphism of vector bundles from $TP^n \mathbf{R}$ to $\mathrm{Hom}(\gamma_n, \gamma_n^{\perp})$ which is injective on the fibers. This is an isomorphism since two fibers have the same rank.

We can also show that the Whitney sum of $TP^n \mathbf{R}$ and the trivial bundle of rank 1 is isomorphic to the Whitney sum of $n+1$ copies of γ_n (see [Milnor-Stasheff 74, p. 45]), which requires a little more agility in handling vector bundles, without being really difficult.

24*. *Tubular neighborhood of a submanifold*

Take inspiration from the proof of Lemma 6.18, or see [Hirsch 76, 4.5], [Berger-Gostiaux 88, 2.7].

Chapter 4

3. *The multiplicative group of quaternions*

b) Imitate the proofs of Section 1.6.

d) It suffices to redo the presentation of Exercise 2 a). We note in passing that the Lie algebra of the group of quaternions of norm 1 consists of *pure* quaternions.

4. Let $\rho : G \rightarrow H$ be a local diffeomorphism. By the inverse function theorem, there exists an open subset U of G containing the identity such that $U \cap \mathrm{Ker}\,\rho = \{e\}$, thus $\mathrm{Ker}\,\rho$ is a discrete subgroup of G. We then easily see that ρ descends to the quotient as a Lie group isomorphism from $G/\mathrm{Ker}\,\rho$ to H.

5*. *Quaternions and rotations*

a) Note that
$$\rho(s) \cdot h + \overline{\rho(s) \cdot h} = s(h + \overline{h})\overline{s}.$$

On the other hand,
$$\|\rho(s) \cdot h\|^2 = sh\overline{s}\overline{sh\overline{s}} = s\|h\|^2\overline{s} = \|h\|^2 s\overline{s} = \|h\|^2.$$

b) By a), $\det\big(\rho(s)\big) = \pm 1$. However since s runs over a connected space, the values taken by the determinant form a connected subset of $\mathbf{R}$.

c) The linear tangent map to ρ at e associates to the pure quaternion σ the linear map $T_e\rho(\sigma)$ defined by
$$T_e\rho(\sigma) \cdot h = \sigma h + h\overline{\sigma} = [\sigma, h].$$

We deduce that $T_e\rho$ is injective by Exercise 2 d).

d) The image of ρ is open and closed. The fact that $\mathrm{Ker}(\rho) = \pm e$ again comes from Exercise 2.

e) Write $s = \mathrm{Re}(s) + \sigma$, where σ is pure. Then $\rho(s) \cdot \sigma = \sigma$.

e2) This comes from the transitivity of the action of $SO(3)$ on the unit sphere and the surjectivity of ρ.

e3) The quaternion $s = \alpha + \beta t$ is thus conjugate in $\mathbf{H}$ to $\alpha + i\beta$. To calculate the angle we can reduce to the previous case. The axis of rotation of $\rho(s)$ is then i. To find the angle, it suffices to calculate the transformation of a vector orthogonal to i, for example j. We obtain
$$\rho(s) \cdot j = (\alpha + i\beta)j(\alpha - i\beta) = (\alpha^2 - \beta^2)j + 2\alpha\beta k.$$

If the axis of rotation is oriented by i, the angle θ satisfies
$$\cos\theta = \alpha^2 - \beta^2 \quad \text{and} \quad \sin\theta = 2\alpha\beta.$$

7. *Examples of exponentials*

c) The image of the exponential of $\mathfrak{Gl}(2, \mathbf{R})$ is the set of matrices of $Sl(2, \mathbf{R})$ whose eigenvalues are positive.

8. Let h be such a morphism. Then $u \mapsto h(e^{iu})$ is a morphism from $\mathbf{R}$ to S^1, which must be of the form $u \mapsto e^{iau}$ by Proposition 4.20. As this morphism must be periodic with period 2π, a is an integer.

9. *Comparison between $SL(2, \mathbf{R})$ and the Lorentz group of dimension 3*

For questions b) and c), the arguments are essentially the same as those of Exercise 5.

For d), look in Exercise 16 of Chapter 2 what happens in dimension 1.

11*. Four.

13. *Some homogeneous spaces*

a) Once more it suffices to note that

$$S^{2n+1} = \left\{ (z^0, \ldots, z^n) \in \mathbf{C}^{n+1} : \sum_{i=0}^{n} |z^i|^2 = 1 \right\}.$$

We thus check that $U(n+1)$ and $SU(n+1)$ act transitively on S^{2n+1}, with the stabilizer of $(1, 0, \cdots, 0)$ consisting of matrices of the form

$$\begin{pmatrix} 1 & 0 & \ldots & 0 \\ 0 & & & \\ \vdots & & A & \\ 0 & & & \end{pmatrix},$$

where A belongs to $U(n)$ or $SU(n)$ in each case.

b) *Remark.* **In the same way, the group

$$Sp(n+1) = Gl(n+1, \mathbf{H}) \cap O(4n+4)$$

acts transitively on

$$S^{4n+3} = \left\{ (q^0, \ldots, q^n) \in \mathbf{H}^{n+1} : \sum_{i=0}^{n} |q^i|^2 = 1 \right\},$$

and there is a diffeomorphism $S^{4n+3} \simeq Sp(n+1)/Sp(n)$.**

It was shown in the 1950s by Armand Borel that we thereby obtain all compact groups that act transitively on the spheres, after we add three exceptional cases.

c) The analogous embedding of $U(n)$ into $SU(n+1)$ is given by

$$A \longmapsto \begin{pmatrix} \overline{\det(A)} & 0 & \dots & 0 \\ 0 & & & \\ \vdots & & A & \\ 0 & & & \end{pmatrix}.$$

14*. *Orbits of a compact group action*

In the case of $P^n\mathbf{R}$, the orbits of $G \simeq O(n)$ other than the fixed point are diffeomorphic to S^{n-1}, with the exception of one of them which is diffeomorphic to $P^{n-1}\mathbf{R}$.

In the case of $P^n\mathbf{C}$, the orbits of $G \simeq U(n)$ other than the fixed point are diffeomorphic to S^{2n-1}, with the exceptions of one of them which is diffeomorphic to $P^{n-1}\mathbf{C}$.

The structure of orbits of a compact Lie group acting differentiably on a manifold is well understood. See for example [Audin 04, Chapter I] or [Duistermaat-Kolk 99].

16*. *Manifold of matrices of a given rank*

The matrices M and PMQ^{-1} have the same rank. Conversely, if M has rank r, there exists matrices P and Q in $Gl(p,\mathbf{R})$ and $Gl(q,\mathbf{R})$ respectively such that

$$M = P \begin{pmatrix} I_r & 0 \\ 0 & 0 \end{pmatrix} Q^{-1}.$$

The orbits of the action of $Gl(p,\mathbf{R}) \times Gl(q,\mathbf{R})$ on $M_{p,q}(\mathbf{R})$ are matrices of rank r $(0 \leqslant r \leqslant \inf(p,q))$. From this we deduce the set of matrices (p,q) of rank r is a homogeneous space of dimension $r(p+q-r)$.

17*. **The adjoint representation is an analytic function on G with values in $Gl(\dim \mathfrak{G}, \mathbf{C})$. It is thus constant by the maximum principle, since G is compact and connected.**

Chapter 5

2*. *Forms of degree 2; symplectic group*

a) If $\omega \neq 0$, there exists necessarily independent vectors x and y such that $\omega(x,y) \neq 0$, hence two independent vectors a and b such that $\omega(a,b) = 1$. The kernel of the linear forms $x \mapsto \omega(a,x)$ and $x \mapsto \omega(b,y)$ are distinct hyperplanes, whose intersection is of codimension 2, and furnishes the subspace E' required. This lets us proceed by induction on the dimension of E.

b) By the definition of the exterior product, F contains $\bigcap_{1\leqslant i\leqslant 2p}\operatorname{Ker}\theta^i$. To check the opposite inclusion, we introduce the basis $(e_i)_{1\leqslant i\leqslant n}$ whose dual is $(\theta^i)_{1\leqslant i\leqslant n}$, and note that

$$w(x, e_i) = \begin{cases} -\theta^{i+p}(x) & \text{for } 1 \leqslant i \leqslant p \\ \theta^{i-p}(x) & \text{for } p+1 \leqslant i \leqslant 2p \end{cases}$$

c) Let $(\alpha^i)_{1\leqslant i\leqslant 2p}$ be a family of $2p$ linear forms on a vector space. Noting that the multilinear alternating forms of even degree generate a *commutative* subalgebra of $\bigwedge E^*$, we prove by induction on k that

$$\left(\sum_{i=1}^{p} \theta^{2i-1} \wedge \theta^{2i}\right)^p = p!\, \theta^1 \wedge \cdots \wedge \theta^{2p}.$$

d) Let

$$J_n = \begin{pmatrix} 0 & I_n \\ -I_n & 0 \end{pmatrix}.$$

The matrices of endomorphisms of $\mathbf{R}^{2n}$ leaving ω invariant are characterized by the relation

$${}^tM J_n M = J_n.$$

From here, to show that $Sp(n,\mathbf{R})$ is a Lie group, we proceed as we did for the orthogonal group, except that the map $M \mapsto {}^tM J_n M$ sends $M_{2n}\mathbf{R}$ to the antisymmetric matrices. The Lie algebra of $Sp(n,\mathbf{R})$ consists of marries such that

$${}^tM J_n + J_n M = 0,$$

which is to say block matrices

$$\begin{pmatrix} A & C \\ B & D \end{pmatrix} \text{ such that } \begin{cases} {}^tB = B \\ {}^tC = C \\ {}^tA = -D \end{cases}$$

3*. *An application of exterior algebra to Lie groups*

a) and b) To $g \in Gl(4,\mathbf{R})$, associate the map $\rho(g)$ from E to E defined by

$$\rho(g)(\alpha \wedge \beta) = {}^tg^{-1}(\alpha) \wedge {}^tg^{-1}(\beta)$$

if α and β are linear forms. The calculations from the end of Section 5.2 show that ρ is a group morphism, and sends $Sl(4,\mathbf{R})$ to $O(3,3)$. Check that the differential of ρ at the identity is injective. It is then surjective for dimensional reasons. Thus $\operatorname{Im}(\rho)$ is an open subgroup (and thus closed) of $O(3,3)$ which is connected as it is the continuous image of a connected set. It is thus the connected component of the identity of $O(3,3)$.

6*. *Forms invariant under a group*

a) Use the fact that

$$\omega = i_X(dx^0 \wedge \cdots \wedge dx^n), \text{ where } X \text{ is the radial vector field.}$$

To see that Ω is the only form of degree n which is invariant under $Sl(n+1, \mathbf{R})$, note first that $Sl(n+1, \mathbf{R})$ acts transitively on $\mathbf{R}^{n+1} \smallsetminus \{0\}$, and such a form is determined on $\mathbf{R}^{n+1} \smallsetminus \{0\}$ by its value at $e_0 = (1, 0, \ldots, 0)$ for example. We thus reduce to showing that $e^{1*} \wedge \cdots \wedge e^{n*}$ is the only n-linear alternating form (up to a factor) which is invariant under the subgroup of $Sl(n+1, \mathbf{R})$ which fixes e_0 (here we denoted the basis dual to the canonical basis of $\mathbf{R}^{n+1}$ by $(e^{i*})_{0 \leqslant i \leqslant n}$).

b) Take inspiration from 12, c4) further below.

7. The primitive of

$$\alpha = \sum_{1 \leqslant i < j \leqslant n} \alpha_{ij} dx^i \wedge dx^j$$

thus obtained is

$$\beta = \sum_{1 \leqslant i < j \leqslant n} \left(\int_0^1 \alpha_{ij}(ux)\, du \right) (x^i\, dx^j - x^j\, dx^i).$$

8. *Forms invariant under a Lie group*

b) It suffices to calculate $d\omega(V_0, \ldots, V_p)$ for left invariant vector fields by applying Theorem 5.24.

c) We have $dX^{-1} = -X^{-1}dX X^{-1}$ (compare to the case of the linear group seen in Section 1.2). If $\Omega = X^{-1}dX$, we have

$$d\Omega + \Omega \wedge \Omega = 0,$$

where the matrix $\Omega \wedge \Omega$ is defined by

$$(\Omega \wedge \Omega)_i^j = \sum_k (\Omega)_i^k \wedge (\Omega)_k^j.$$

If U and V are two left invariant vector fields, we deduce that

$$d\Omega([U, V]) = (\Omega \wedge \Omega)(U, V).$$

We discover the expression for the bracket by evaluating each side at the identity element.

d) Restrict the matrices $X^{-1}dX$ and $(dX)X^{-1}$ to G. The vector space of left invariant forms of degree 1 is generated by $a^{-1}da$ and $a^{-1}db$, and that of the right invariant forms by $a^{-1}da$ and $-a^{-1}bda + db$. The left invariant (resp. right invariant) forms of degree 2 are proportional to $a^{-2}db \wedge da$ (resp. to $a^{-1}db \wedge da$): a very simple example of a noncommutative group makes property b) false.

9. *Interior product and Lie derivative*

We remark that the linear operator $Q = L_X \circ i_Y - i_Y \circ L_X$ decreases the degree of a homogeneous form by 1, and that if α is homogeneous of degree p, we have

$$Q(\alpha \wedge \beta) = Q(\alpha) \wedge \beta + (-1)^p \alpha \wedge Q(\beta).$$

From here, one imitates the proof of the Cartan formula.

Note. A little terminology to help explain this property: d is an antiderivation of degree 1, and i_X and Q are antiderivations of degree -1.

12. *Forms invariant under rotation*

a) Set $r = \sqrt{x^2 + y^2}$. One has $\alpha = d(r^2/2)$, by the invariance of α under the action of $SO(2)$, since the differential commutes with pullback. On the other hand, $\beta = i_X \omega$, where X is the radial vector field defined by $X_p = p$ if $p \in \mathbf{R}^2$, and $\omega = dx \wedge dy$. For $g \in SO(2)$ we have

$$(g^*\beta)_p(v) = \beta_{g(p)}(g(v)) \text{ since } g \text{ is linear}$$
$$= \omega_{g(p)}(g(p), g(v)) = \det(g(p), g(v))$$
$$= \det(p, v) \text{ since } \det(g) = 1$$
$$= \beta_p(v).$$

b) If a form $\gamma \in \Omega^1(\mathbf{R}^2)$ is $SO(2)$-invariant, it is determined once we know the linear forms γ_p as p runs over a half-line.

Let the rotation of angle θ be denoted R_θ. If $v = (v_1, v_2)$ is a vector in $\mathbf{R}^2$, then *a priori*

$$\gamma_{(r,0)}(v) = a(r)v_1 + b(r)v_2, \quad \text{where } a \text{ and } b \text{ are smooth functions on } \mathbf{R}^+.$$

Now

$$\gamma_{(r\cos\theta, r\sin\theta)}(v) = \gamma_{(r,0)}(R_\theta^{-1} \cdot v)$$
$$= a(r)(v_1 \cos\theta + v_2 \sin\theta) + b(r)(-v_1 \sin\theta + v_2 \cos\theta).$$

In other words,

$$\gamma_{(x,y)}(v) = \frac{a(r)}{r}(xv_1 + yv_2) + \frac{b(r)}{r}(-yv_1 + xv_2)$$

is

$$\gamma = \frac{a(r)}{r}\alpha + \frac{b(r)}{r}\beta.$$

Now if γ is defined on all of $\mathbf{R}^2$, we have

$$\gamma_{(r,0)}(v) = a(r)v_1 + b(r)v_2,$$

where a and b are smooth odd functions on $\mathbf{R}$. By Hadamard's lemma, we have $a(r) = ra_1(r)$ and $b(r) = rb_1(r)$, where a_1 and b_1 are smooth and additionally even. Applying Taylor's formula, we see that if f is an even function of class C^k ($1 \leqslant k \leqslant \infty$) on $\mathbf{R}$, then $t \mapsto f(\sqrt{t})$ is also C^k on $[0, \infty)$[1]. Reusing the preceding reasoning, we see that a $SO(2)$-invariant form on $\mathbf{R}^2$ can be written

$$f(r^2)\alpha + g(r^2)\beta,$$

where f and g are smooth functions on $[0, \infty)$.

c1) This was treated in a) (the dimension is not important here), c2) is classical, and c3) is a consequence of the decomposition

$$\mathbf{R}^3 = \mathbf{R}p \bigoplus T_p S^2(\|p\|),$$

where $S^2(r)$ denotes the sphere with center 0 and radius r.

c4) For $p \in S^2$, we define $s_p \in SO(3)$ by

$$s_p(p) = p; \quad s_p(v) = -v \text{ if } \langle v, p \rangle = 0.$$

Now, if $\gamma \in \Omega^1(S^2)$ is $SO(3)$-invariant, we have

$$\gamma_p(v) = (s_p^*\gamma)_p(v) = \gamma_p(s_p \cdot v) = -\gamma_p(v),$$

where $\gamma = 0$.

d) Using a generalization of c4) that we leave to the reader, the same result is true for $SO(n)$-invariant 1-forms on $\mathbf{R}^n \setminus \{0\}$. If $1 < k < n - 1$, we can use the same ideas to prove that there are no $SO(n)$-invariant k-forms on $\mathbf{R}^n \setminus \{0\}$ other than 0. Writing $\omega = dx^1 \wedge \cdots \wedge dx^n$, we see that the $SO(n)$-invariant n-forms are of the form $f(r)\omega$, and the $SO(n)$-invariant $(n-1)$-forms are of the form $f(r)i_X\omega$.

13. We find

$$(I^{-1})^*\alpha = \frac{4\,du \wedge dv}{(1 + u^2 + v^2)^2}.$$

14*. *Darboux's theorem*

See [Arnold 78, Chapter 8, § 43].

---

1. A function g is said to be of class C^1 on $[0, \infty)$ if it is continuously differentiable at 0 and of $\lim_{t \to 0} g'(t) = g'_d(0)$. Denote the function $[0, \infty)$ thus obtained by g'. We say that g is C^k on $[0, \infty)$ if g' is C^{k-1}, etc.

17*. *Darboux's theorem by Moser's trick*

a) See Exercise 2.

b) We write $\omega_t = \widetilde{\omega} + t(\omega - \widetilde{\omega})$, and we remark that the coefficients of $(\omega - \widetilde{\omega})$ can be made less than a given $\epsilon > 0$ for an appropriate choice of r. Then

$$\omega_t^n = \widetilde{\omega}^n(1 + f),$$

where f is bounded in absolute value by $n!\epsilon$.

c) Let $X(t)$ be the infinitesimal generator of φ_t. By Theorem 5.30,

$$\frac{d}{dt}\varphi_t^*\omega_t = \varphi_t^*(\omega - \widetilde{\omega} + d i_{X(t)}\omega_t).$$

Furthermore, by the Poincaré lemma, there exists a form α of degree 1 on $B(a, r)$ such that $d\alpha = \omega - \widetilde{\omega}$. To check that $\frac{d}{dt}\varphi_t^*\omega_t = 0$, it suffices to choose $X(t)$ such that

$$\alpha + i_{X(t)}\omega_t = 0.$$

Since the form ω_t has rank n everywhere, we see by Exercise 2 that this condition determines a unique vector field $X(t)$. We thus have

$$\varphi_t^*\omega_t = \varphi_0^*\widetilde{\omega} = \omega = \varphi_1^*\omega.$$

Chapter 6

1. *Orientability and oriented atlases*

a) Let $(U_i, \varphi_i)_{i \in I}$ be an atlas of M. Write $\Phi_i = T\varphi_i$. The transition maps from the atlas $(TU_i, \Phi_i)_{i \in I}$ of TM are

$$(\varphi_j \circ \varphi_i^{-1})(x, u) = \left((\varphi_j \circ \varphi_i^{-1})(x), T_x(\varphi_j \circ \varphi_i^{-1}) \cdot u\right).$$

The Jacobian matrices are of the form

$$\begin{pmatrix} T_x(\varphi_j \circ \varphi_i^{-1}) & 0 \\ A & T_x(\varphi_j \circ \varphi_i^{-1}) \end{pmatrix}$$

and their determinant equals $\left(\det\left(T_x(\varphi_j \circ \varphi_i^{-1})\right)\right)^2$.

2. *Orientability and volume forms*

a) If $\omega = dx^1 \wedge \cdots \wedge dx^n$, check that $i_{\nabla f}\omega$ induces a volume form on $f^{-1}(0)$.

b) Similarly, the form α defined by

$$\alpha_x(v_1, \ldots, v_{n-p}) = \omega(\nabla f_x^1, \ldots, \nabla f_x^p, v_1, \ldots, v_{n-p})$$

induces a volume form on $f^{-1}(0)$.

c) No, since every compact or second countable manifold, whether orientable or not embeds into an appropriate $\mathbf{R}^n$. Conclusion: if $X \subset \mathbf{R}^n$ is a nonorientable submanifold, it is impossible to realize X as the set of zeros of a submersion defined on a whole neighborhood of X.

4. If one of the p_k is odd, $I(p_0, \ldots, p_n) = 0$. If they are all even,

$$I(p_0, \ldots, p_n) = \frac{\prod\limits_{k=0}^{n} \Gamma\left(\frac{1+p_k}{2}\right)}{\Gamma\left(\frac{1+n}{2}\right)}.$$

6. We find $\dfrac{4\pi a^2 c}{3}$.

7. *Archimedes's formula*

a) Let ω be the volume form $x\, dy \wedge dz + y\, dz \wedge dx + z\, dx \wedge dy$ on the sphere $S^2 \subset \mathbf{R}^3$.

It will be convenient to consider

$$f : \; (\varphi, \theta) \longmapsto (\cos\theta \cos\varphi, \cos\theta \sin\varphi, \sin\theta)$$

as a diffeomorphism from $S^1 \times (-\pi 2, \pi 2)$ to $S^2 \smallsetminus \{N, S\}$. For $z \neq 0$, we have

$$f^*\omega = f^* \left(\frac{dx \wedge dy}{z} \right) = \cos\theta \, d\varphi \wedge d\theta,$$

thus $f^*\omega = \cos\theta \, d\varphi \wedge d\theta$ on all of $S^2 \smallsetminus \{N, S\}$ by continuity. As $\cos\theta \, d\varphi \wedge d\theta = -d(\sin\theta \, d\varphi)$, we can take the following form as a primitive of ω on $S^2 \smallsetminus \{N, S\}$:

$$(f^{-1})^*(-\sin\theta \, d\varphi) = \frac{z(y\, dx - x\, dy)}{x^2 + y^2} = \frac{z(y\, dx - x\, dy)}{1 - z^2}.$$

Another possible primitive of $\cos\theta d\varphi \wedge d\theta$ is $(1 - \sin\theta)d\varphi$. Thus

$$(f^{-1})^*((1 - \sin\theta)\, d\varphi) = \frac{(1-z)(x\, dy - y\, dx)}{1 - z^2} = \frac{x\, dy - y\, dx}{1 + z}.$$

The form above, which is defined on $S^2 \smallsetminus \{S\}$, is again a primitive of ω.

b) We have

$$\int_{\Sigma_{h,k}} \omega = \int_{f^{-1}(\Sigma_{h,k})} \cos\theta \, d\varphi \wedge d\theta = \int_{\partial f^{-1}(\Sigma_{h,k})} -\sin\theta \, d\varphi = 2\pi(k - h).$$

By continuity (or by using the other primitive) this formula is again valid if $h = -1$ or $k = 1$.

8*. *Haar measure of a Lie group*

a) This is a particular case of what we saw in Exercise 8 of Chapter 5 on invariant forms on a Lie group.

b) Of course we find respectively dx, $x^{-1}dx$, $dx \wedge dy$ (if $z = x + iy$), $dx^1 \wedge \cdots \wedge dx^n$.

c) If Y and $X = (x_i^j)_{1 \leqslant i, \, j \leqslant n}$ are two (n, n) matrices, the matrix YX can be obtained by juxtaposing n column matrices YX^j, where $X^j = (x_i^j)_{1 \leqslant i \leqslant n}$. For fixed j, by the results of Section 5.2.2

$$Y^* \left(\bigwedge_{1 \leqslant i \leqslant n} dx_i^j \right) = \det(Y) \bigwedge_{1 \leqslant i \leqslant n} dx_i^j,$$

thus

$$L_Y^* \left(\bigwedge_{1 \leqslant i, \, j \leqslant n} dx_i^j \right) = \left(\det(Y) \right)^n \bigwedge_{1 \leqslant i, \, j \leqslant n} dx_i^j,$$

hence the invariance under left translations of

$$\det(X)^{-n} \bigwedge_{1 \leqslant i, \, j \leqslant n} dx_i^j.$$

The invariance under right translations is seen in the same way.

9*. *Compact subgroups of the linear group*

The group G is closed in $Gl(n, \mathbf{R})$, and is thus a Lie subgroup. Abusing notation, denote the right Haar measure of G by dg. Let φ be any positive definite quadratic form on $\mathbf{R}^n$. Write

$$q(x) = \int_G \varphi(gx) \, dg.$$

By linearity of the integral, we obtain a quadratic form. This form is positive definite, as $q(x)$ is obtained by integrating the strictly positive function (for $x \neq 0$) $g \mapsto \varphi(gx)$ against the volume. Finally, since the diffeomorphism R_h of G preserves the orientation, we have

$$q(x) = \int_G R_h^*(\varphi(gx) \, dg) = \int_G \varphi(ghx) R_h^* \, dg$$

$$= \int_G \varphi(ghx) \, dg = q(hx).$$

In other words, G is included in the orthogonal group of the quadratic form q. This quadratic form which is positive definite is equivalent to the standard form $\langle x, x \rangle$, which can be seen by noting there exists a $g \in Gl(n, \mathbf{R})$ such that $q(x) = \langle gx, gx \rangle$. As a result

$$gGg^{-1} \subset O(\langle \, , \, \rangle) = O(n).$$

In particular, this result implies that $O(n)$ is a *maximal* compact subgroup of $Gl(n, \mathbf{R})$ (which is to say if $H \supset O(n)$ is compact, then $H = O(n)$, and that every maximal compact subgroup of $Gl(n, \mathbf{R})$ is conjugate to $O(n)$).

More generally, one shows that all maximal compact subgroups of a Lie group are conjugate. See for example [Helgason 78].

10. *Modulus of a Lie group*

a) If ω is left invariant, then $R_g^* \omega$ is also, since right and left translations commute. This is also a volume form since R_g is a diffeomorphism. The remainder is left to the reader.

b) The only compact subgroup of $\mathbf{R}_+^*$ is 1.

c) If g is the affine transformation $x \mapsto ax + b$, $\mathrm{mod}(g) = a^{-1}$.

d) See the reference given in the statement.

11.** *Cauchy-Crofton formula*

a) It suffices to check the invariance under rotations about the origin and translations. Denoting an oriented line by (p, θ), a rotation of angle α transforms (p, θ) to $(p, \theta + \alpha)$. Translation by the vector (a, b) transforms (p, θ) to $(p + a \cos \theta + b \sin \theta, \theta)$.

b) Introduce the angle α of the oriented tangent to the curve with arc-length parameter s with the axis Ox. Now, $(x'(s), y'(s)) = (\cos \alpha, \sin \alpha)$, and we see that F is smooth since

$$F(s, \varphi) = \left(x(s) \sin(\alpha + \varphi) - y(s) \cos(\alpha + \varphi), \varphi + \alpha - \frac{\pi}{2} \right).$$

We note that p is also equal to $x(s) \cos \theta + y(s) \sin \theta$, thus (with a little abuse of notation)

$$dp = (x'(s) \cos \theta + y'(s) \sin \theta) \, ds + (\text{something}) \, d\theta$$
$$= \cos(\alpha - \theta) \, ds + (\text{something}) \, d\theta = \sin \varphi \, ds + (\text{something}) \, d\theta.$$

However $F^* d\theta = d\varphi + (\text{something}) \, ds$, where

$$F^*(dp \wedge d\theta) = \sin \varphi \, ds \wedge F^* d\theta = \sin \varphi \, ds \wedge d\varphi.$$

c) The formula shows that (s, φ) is a critical point if and only if $\varphi = k\pi$: the critical values are the tangent lines to the curve. They form a set of measure zero. It is not necessary to invoke Sard's theorem to see this, since the set of critical points is already of measure zero. Now, if a line D is not tangent to C at any point, $D \cap C$ is finite: otherwise, being a compact set, it will have an accumulation point which would necessarily be a point of tangency of D with C.

As we will not use Stokes's theorem, we can consider $dp \wedge d\theta$ and $\sin \varphi \, ds \wedge d\varphi$ as positive measures ($\varphi \in [0, \pi]!$). Let M_k be the set of lines such that $\operatorname{card}(D \cap C) = k$, and M_k' its inverse image under F. Then

$$\int_{M_k'} \sin \varphi \, ds \wedge d\varphi = k \int_{M_k} dp \wedge d\theta,$$

hence

$$\sum_k \int_{M_k'} \sin \varphi \, ds \wedge d\varphi = \int_M \operatorname{card}(D \cap C) \, dp \wedge d\theta.$$

The left hand side equals

$$\int_{[0,L] \times [0,\pi]} \sin \varphi \, ds \wedge d\varphi = L \int_0^\pi \sin \varphi \, d\varphi = 2L.$$

13. *Archimedes's theorem*

See [Berger-Gostiaux 88, 6.5].

14. *Laplacian*

a) We have

$$f \Delta g - g \Delta f = \operatorname{div}(f \nabla g - g \nabla f).$$

b) Start with the identity

$$f \Delta f = \operatorname{div}(f \nabla f) + \langle \nabla f, \nabla f \rangle.$$

15*. *Tubular neighborhood of a curve*

a) Let $n(s)$ be the unit normal to c at a point corresponding to parameter s, chosen so that the orthonormal frame $(c'(s), n(s))$ is positively oriented. The tubular neighborhood is given by the parametrization

$$F : \ (s, t) \longmapsto c(s) + t n(s) \quad \text{with } s \in [0, L] \text{ and } t \in [-r, r].$$

If we denote the curvature by $k(s)$, we have

$$\frac{\partial F}{\partial s} = (1 - t k(s)) c'(s)$$

$$\frac{\partial F}{\partial t} = n(s)$$

and therefore

$$F^*(dx \wedge dy) = \det \left(\frac{\partial F}{\partial s}, \frac{\partial F}{\partial t} \right) ds \wedge dt = \big(1 - tk(s)\big)\, ds \wedge dt.$$

Then

$$\text{area}\big(V_r(c)\big) = \int_{V_r(c)} F^*(dx \wedge dy) = \int_{[0,L] \times [-r,r]} \big(1 - tk(s)\big)\, ds\, dt.$$

The last integral equals

$$\int_0^L \left(\int_{-r}^r \big(1 - tk(s)\big)\, dt \right) ds = \int_0^L 2r\, ds = 2rL.$$

Chapter 7

3*. *A useful lemma... used in Section 7.4.4*

a) Such a homotopy is given by

$$H(t,x) = \frac{f(tx)}{t} \quad \text{if } t \neq 0,$$
$$H(0,x) = T_0 f \cdot x.$$

(everything works thanks to Hadamard's lemma).

b) Replacing f by $f - f(0)$, we can reduce to the case where $f(0) = 0$. If f preserves orientation, $T_0 f \in Gl^+(n, \mathbf{R})$, which is connected (indeed $Gl^+(n, \mathbf{R})$ is diffeomorphic to $\mathbf{R}^{n(n+1)/2} \times SO(n)$ by Exercise 8 of Chapter 1) and $SO(n)$ is connected by Theorem 4.28.

Note. These arguments are known as "Alexander's trick".

4*. *Examples of linkings*

a**) If (c,d) is a pair of smooth curves in S^3, there exists a point p which does not belong to $\text{Im}(c) \cup \text{Im}(d)$ by the easy part of Sard's theorem. Let h by a diffeomorphism from $S^3 \setminus \{p\}$ to $\mathbf{R}^3$ preserving orientation. Then by the preceding exercise and Proposition 7.34, $E(h \circ c, h \circ d)$ is independent of h. The result obtained no longer depends on the choice of p, by c) of the preceding exercise (we let the reader fill the holes).

b) We use the fact that S^2 with a point removed is an open subset trivializing a diffeomorphism to $\mathbf{R}^2$. Then if a, b, c, d are four distinct points of S^2, we can always suppose these belong to such an open subset U. Further, we can join a to c, and b to d by smooth *disjoint* paths contained in U. Then, using the trivialization of H, we easily see that the pair of curves $\big(H^{-1}(a), H^{-1}(b)\big)$ and $\big(H^{-1}(c), H^{-1}(d)\big)$ are homotopic. For an explicit calculation, take for example $a = N$ and $b = S$: after a suitable stereographic projection, the curves $H^{-1}(b)$ and $H^{-1}(a)$ become a circle in the $x0y$ plane and the axis $0z$ in $\mathbf{R}^3$.

c) Take as a target the calculation of the degree of $(1, 0, 0)$. We find the linking number equals 2.

5*. *Existence of maps from S^n to S^n of every degree*

b) Let i_p denote stereographic projection with respect to p. The function g defined by

$$g(x) = \begin{cases} i_p^{-1}\big(f(\|x\|^2)x\big) & \text{if } \|x\| < 1 \\ p & \text{if } \|x\| \geqslant 1. \end{cases}$$

is then smooth as soon as f tends to infinity "sufficiently fast", for example if $f(t) = e^t$.

c) Let $\varphi : \mathbf{R}^n \to V$ be a diffeomorphism to an open subset V of X containing m such that $\varphi(0) = m$, and let $U = \varphi\big(B(0, 1)\big)$. Then the map $h : X \to S^n$ defined by

$$h(x) = \begin{cases} g\big(\varphi^{-1}(x)\big) & \text{if } x \in V \\ p & \text{if } x \notin V \end{cases}$$

works.

d) It is clear that $\deg(h) = \pm 1$. The degree equals 1 if we oriented S^n in a way that $g_{|B(0,1)}$ preserves orientation, and if the chart φ is compatible with the orientation of X. To find a map of degree k we similarly proceed by starting with k diffeomorphisms of $\mathbf{R}^n$ on disjoint open subsets $V_1, \ldots, V_k$, and preserving the orientation (which is possible since X is orientable). Proceeding as in c) we see there exists open subsets $U_i \subset V_i$ and a smooth map $h : X \mapsto S^n$ which for all i is a orientation preserving diffeomorphism from U_i to $S^n \smallsetminus \{p\}$ and which sends $S^n \smallsetminus \bigcup U_i$ to p.

6. *Degree of the k-th power function*

a) Proceed by induction on k, by deriving the identity:

$$\psi_{k,g}(\exp tX) = (g \exp tX)^k g^{-k} = (g \exp tX)^{k-1} g^{1-k} g^k (\exp tX) g^{-k}.$$

b**) By compactness, the eigenvalues of $\mathrm{Ad}\, g$ have modulus 1, and by connectedness $\det(\mathrm{Ad}\, g) = 1$. We deduce the product of eigenvalues of $T_e \psi_{k,g}$ is nonnegative. Thus $\deg(f_k) > 0$ since there exist regular values.

c) If g is the product of n different plane rotations of $\frac{2p\pi}{k}$, it is a regular value of $f_k : SO(2n) \to SO(2n)$. The case of $SO(2n + 1)$ can be done in the same way.

7*. If f has no fixed point, the points $f(x)$ and $-x$ are never diametrically opposite, thus f is homotopic to the antipodal.

10. *Cohomology of a finite quotient*

a) Let $\alpha \in F^k(Y)$. We must show that if $p^*\alpha$ is exact, then so is α. Since for all $\gamma \in \Gamma$ we have $\gamma^* \circ p^* = p^*$, from the equation $p^*\alpha = d\beta$ we deduce

$$p^*\alpha = \frac{1}{\text{card}(\Gamma)} \sum_\gamma \gamma^* d\beta = d\left(\frac{1}{\text{card}(\Gamma)} \sum_\gamma \gamma^* \beta \right).$$

The form

$$\bar\beta = \frac{1}{\text{card}(\Gamma)} \sum_\gamma \gamma^* \beta$$

is Γ-invariant, thus of the form $p^*\beta'$. We have

$$p^*\alpha = d(p^*\beta') = p^*(d\beta')$$

thus $\alpha = d\beta'$ since p^* is injective.

b) This is proved in the same way.

c) If K is the Klein bottle, then $H^1(K) \simeq \mathbf{R}$ and $H^2(K) = 0$.

For all $k > 0$, $H^k(P^n\mathbf{R}) = 0$, with one exception: $H^{2m+1}(P^{2m+1}\mathbf{R}) \simeq \mathbf{R}$.

11. *Cohomology of a product*

a) For example let $\pi : X \times Y \to X$ be the projection onto the first factor, and let α be a volume form on X. The form $\pi^*\alpha$ is closed, while

$$\int_{X \times \{y\}} \pi^*\alpha = \int_X \alpha \neq 0.$$

b) Mayer-Vietoris once more.

12*. We begin with the volume form $\omega = dx^1 \wedge \cdots \wedge dx^n$ on T^n. Then $f^*(dx^1)$ is a closed form of degree 1 on S^n. It is thus exact, and the same is true for $f^*\omega = f^*(dx^1) \wedge f^*(dx^2 \wedge \cdots \wedge dx^n)$.

13. The decomposition $T^n = (T^{n-1} \times S^1 \smallsetminus \{p\}) \cup (T^{n-1} \times S^1 \smallsetminus \{q\})$ allows us to proceed by induction on n.

14*. *Hopf Invariant*

g) If f is not surjective, we imitate the argument which shows that a non-surjective map is of degree 0 (the end of the proof of Theorem 7.18).

The invariance under homotopy is a consequence of Theorem 7.41. We find $H(f) = 1$ for the Hopf fibration.

**One can show that $H(f)$ is an integer. This can be seen by showing that if a and b are two regular values of f, then

$$H(f) = E\big(f^{-1}(a), f^{-1}(b)\big)$$

(compare to Exercise 4 and see [Hopf 35]).**

The Hopf fibration is the first example of a map of one manifold to another of lower dimension which is not homotopic to a constant.

15. *Invariant form and cohomology*

a) We already know if $g \in SO(n+1)$ and $\alpha \in F^k(S^n)$, the forms $g^*\alpha$ and α are cohomologous. Using the same argument as in the case of the torus (compare to Theorem 7.58), we will show that α is cohomologous to

$$\bar{\alpha} = \int_{SO(n+1)} g^*\alpha \, dg,$$

where dg denotes the Haar measure normalized by the condition $\int dg = 1$. To see this, it suffices to note, by imitating the argument of Lemma 7.59, that the exponential of $SO(n+1)$ (which is surjective) admits a measurable right inverse; it suffices to find this inverse on the set (of full measure) of g whose eigenvalues are distinct.

b) This is Exercise 6 of Chapter 5.

17. *Bi-invariant forms on a Lie group*

b) Use the fact that $\mathcal{I} \circ L_g = R_{g^{-1}} \circ \mathcal{I}$.

d) If α is bi-invariant of order k, then $d\alpha$ is bi-invariant of order $k + 1$. We have

$$\mathcal{I}^*(d\alpha) = (-1)^{k+1} d\alpha = d(\mathcal{I}^*\alpha) = (-1)^k d\alpha.$$

e) If G is compact and connected and if $\alpha \in F^p(G)$, the forms $L_g^*\alpha$ and $R_g^*\alpha$ are cohomologous to α by Proposition 7.56. Using the fact that right and left translations commute, and that the Haar measure of a compact group is bi-invariant, we see that the form

$$\bar{\alpha} = \iint_{G \times G} R_g^*(L_h^*\alpha) \, dg \, dh$$

is bi-invariant. We show that this form is cohomologous to α as in Lemma 7.59 with the help of a measurable right inverse of exp. From there, knowing that every bi-invariant form on G is closed, the proof of the fact that $\alpha \mapsto \bar{\alpha}$ descends to the quotient as a isomorphism from $H^p(G)$ to $\Omega^p_{\text{inv}}(G)$ is the same as for $G = T^n$.

f) The form in question is $Ad(G)$-invariant since the trace is invariant under conjugation. We check that it is nontrivial if $n \geqslant 3$ by evaluating on the standard basis of $\mathfrak{so}(3)$. **Calculating the cohomology of compact Lie groups in principle reduces to a purely algebraic problem by the preceding problem. For more details, see volume IX of the work of Dieudonné, and [Greub-Halperin-Van Stone 76, volume II].**

18*. *Cohomology of the sphere with two punctures*

b) We write $T^2 \sharp T^2 = U \cup V$ where U and V are the complements of a closed ball in T^2, and $U \cap V$ is an annulus. We then need to calculate the cohomology of U. This can be done by using a Mayer-Vietoris sequence on a decomposition of the form $T^2 = U \cup B$, where B is an open ball.

Chapter 8

2. We find

$$-\frac{f''}{f}.$$

3. We use the notation of Exercise 13 of Chapter 1. The parametrization obtained can be written

$$F: \ (\phi, \theta) \longmapsto (a + r\cos\theta)\, u(\phi) + r\sin\theta\, k,$$

with $u(\phi) = (\cos\phi, \sin\phi, 0)$ and $k = (0, 0, 1)$. We compute

$$\partial_\phi F = (a + r\cos\theta)\, u\left(\phi + \frac{\pi}{2}\right) \quad \text{and} \quad \partial_\theta F = -r\sin\theta u(\phi) + r\cos\theta\, k.$$

In this parametrization the metric induced from the Euclidean metric can be written

$$(a + r\cos\theta)^2\, d\phi^2 + r^2\, d\theta^2.$$

We can thus take the orthonormal coframe given by $\theta^1 = (a + r\cos\theta)d\phi$ and $\theta^2 = rd\theta$. Taking the equalities $d\theta^1 = -(r\sin\theta)d\theta \wedge d\phi$ and $d\theta^2 = 0$, in the notation of Section 8.3, we deduce

$$\omega = (\sin\theta)\, d\phi, \quad \Omega = (\cos\theta)\, d\phi \wedge d\theta, \quad K = \frac{\cos\theta}{r(a + r\cos\theta)}.$$

We note, unsurprisingly, that the integral of Ω over the torus vanishes.

4. See Exercise 7 of Chapter 3.

5**. See [Berger 87, 12.4].

6*. *Ramified coverings*

a) We can give a geometric characterization of k_x: for every sufficiently small open subset U containing x, there are k_x points of $U \smallsetminus \{x\}$ where f takes the same value.

b) See Exercise 25 of Chapter 1 (and its solution).

c) As a result of the hypothesis made on f, f is an open map. Thus, $f(S)$ is an open and closed subset of S'.

d) It follows from the definition that Σ consists of isolated points, and its complement is the set of points where f is a local diffeomorphism. Thus Σ is a discrete and closed subset of S, and thus finite since S is compact.

e) Σ and $f(\Sigma)$ are discrete and closed. On $S' \smallsetminus f(\Sigma)$, which is connected by Exercise 30 of Chapter 2, the cardinality of $f^{-1}(y)$ is finite, locally constant by Theorem 2.14 and therefore constant.

For (much) more details on ramified coverings, see [Fulton 95] or [Farkas-Kra 91].

7**. *Riemann-Hurwitz formula*

If p is ramification point interior to a face, we obtain a new triangulation where p is a vertex by joining this point to three vertices. We proceed in this way for each of the ramification points interior to the faces. If p is a ramification point interior to an edge, we arbitrarily choose one of the 2-simplexes for which this edge is a side, and we obtain a new triangulation where p is a vertex by joining to the vertex opposite in these 2-simplexes. Iterating this procedure, we obtain a triangulation where the set of vertices contains all of the points of $f(\Sigma)$. If this triangulation has f faces, e edges and v vertices, then $\chi(S') = f + v - e$.

Suppose that there are v_1 ramification points. By properties of coverings, since f is trivial above each face (with the vertices that are points of ramification removed if necessary), we can lift our triangulation by f to a triangulation of S having df faces, de edges, and $v_1 + d(v - v_1)$ vertices. Thus,

$$\chi(S) = d\chi(S') - (d-1)v_1.$$

For more details, see [Fulton 95, 19.c].

8*. *A smooth projective algebraic curve of degree 3 is a torus*

a) It suffixes to apply Theorem 1.21. The submanifold obtained has two connected components, one diffeomorphic to S^1, and the other to $\mathbf{R}$.

b) We have a complex submanifold of dimension 1 of $\mathbf{C} \times \mathbf{C}$ for the same reasons.

Denote this submanifold by E_1. Let a_1, a_2, a_3 be the three zeros of P. We will show that $E_1 \setminus \{(a_2, 0), (a_3, 0)\}$ is path connected. Let $p = (b, c) \in E_1$, with $c \neq 0$, and $q = (a_1, 0)$. Let $\gamma : [0, 1] \to \mathbf{C}$ be a continuous path connecting b to a_1 and such that $\gamma((0, 1])$ avoids a_1, a_2 and a_3. By imitating the argument of Theorem 7.12, we see that there exists a path $\gamma_1 : [0, 1] \to \mathbf{C}$ connecting c and 0 and such that $(\gamma_1(t))^2 = P(\gamma(t))$.

c) This is an example of what we saw in Section 2.6.2 (a projective hypersurface). Note that E_1 is identified with the set of points of E such that $t \neq 0$, which is to say to $E \setminus [(0, 1, 0)]$. In particular, E is connected.

d) As x tends to infinity, the point (x, y) of E_1 tends to $[(0, 1, 0)]$. In a neighborhood of this point, we take the chart $[(x, y, t)] \mapsto (X = \frac{x}{y}, T = \frac{t}{y})$, in which the equation for E can be written

$$X^3 + pT^2 X + qT^3 - T = 0.$$

Finally, if we take the chart given by X for E in a neighborhood of $[(0, 1, 0)]$, and in the range the chart $(x, u) \mapsto \frac{u}{x}$ of $P^1\mathbf{C}$, the map r reads

$$X \longmapsto \frac{T}{X} = X^2 + pT^2 + q\frac{T^3}{X}.$$

A direct calculation shows that $T'(0) = 0$ and $T''(0) \neq 0$. This shows that this last map, which is a priori not defined for $X = 0$, extends by continuity to 0, and the extension map is holomorphic and of the form

$$X \longmapsto X^2 + O(X^3)$$

in the neighborhood of 0.

e) We have just seen that the point at infinity of E (in other words the point $[(0, 1, 0)]$) is a ramification point of index 2. An analogous argument shows that the zeros of P also give ramification points of order 2. The map is clearly a local diffeomorphism at the points (x, y) such that $P(x) \neq 0$.

The Riemann-Hurwitz formula gives

$$\chi(E) = 2\chi(P^1\mathbf{C}) - 4 = 0.$$

Knowing that E is orientable (since it is a complex manifold), we know that E is diffeomorphic to a torus by the classification theorem for surfaces.

One can refine this result: with complete different methods of complex analysis (modular forms) one may show there exists a ring Λ of $\mathbf{C}$ such that E is $\mathbf{C}$-diffeomorphic to $\mathbf{C}/\Lambda$. See [Hellegouarch 01, Chapter V].

Bibliography

General Overview

This bibliography contains mainly books and monographs: I wanted to give a short list to better encourage consulting the references given. The references for the various results cited in the end of chapter comments have inevitably elongated the list however. In suggesting references to learn more about these subjects, I don't pretend to give the most recent reference, just a typical reference and one accessible to a reader of the present book.

In what follows, I have made a few comments without systematically repeating the remarks from the end of chapter comments. The reader should also be aware of the extensive bibliography at the end of volume 5 of [Spivak 79], as well as that of [Guillemin-Pollack 74].

Prerequisites

For differential calculus, see for example [Lang 86], and the first chapter of [Hörmander 90]. For point set topology [Dugundji 65] contains more than is necessary (notably the basics of homotopy theory).

The local theory of curves and surfaces in $\mathbf{R}^3$ is not logically necessary, but helps enormously in understanding. See for example [Audin 03] and [Do Carmo 76].

Other points of view

The first part of [Berger-Gostiaux 88] is close to this book (however it contains no material on Lie groups), and contains global results on plane curves. The second part is devoted to the local theory of surfaces, and has a fairly complete exposition of Riemannian geometry without proofs. It possesses an abundant bibliography.

[Lee 03] is also similar to this book with the inconvenience (or advantage!) of being much longer.

[Guillemin-Pollack 74] studies manifolds through the thread of transversality.

We end this review with two much larger works. [Doubrovine-Novikov-Fomenko 85; 92] (eight hundred pages) covers practically everything in this book and the books we cite, and addresses virtually everything mentioned in the chapter comments. However one must often complete proofs for oneself or consult another book.

Finally [Spivak 79]: five volumes, more than two thousand pages! The first volume more or less treats the themes of this book (with a further introduction to Riemannian geometry). The second volume takes up the basics of Riemannian geometry from an historical point of view, and notably contains an annotated translation of Riemann's celebrated inaugural lecture. The final three volumes are devoted to Riemannian geometry.

Higher level books on the topology of manifolds

Naturally following this book we mention for example [Hirsch 76], [Bott-Tu 86]. These references are quite readable for someone who has mastered the content of this book. [Doubrovine-Novikov-Fomenko 90] is also very interesting, but more difficult.

Related subjects

Singularities, dynamical systems

[Demazure 00] are course notes from l'École polytechnique that are relatively simple and spares no explanations, examples and in particular physical examples.

In a very different style (70 page pocketbook versus 300, with numerous exercises), we cite [Arnold 92].

We also mention [Hirsch-Smale-Devaney 03], which is very pleasant and of a more classical style.

Finally, one must not be afraid of the 800 pages of [Katok-Hasselblatt 95], from which one can even benefit by browsing.

Symplectic geometry and topology

The "why" of symplectic geometry is not readily clear. In addition to an introduction, one finds answers to this question in [Audin 04] and [Arnold 78]. A more recent reference is [McDuff-Salamon 98], which takes into account the upheavals that occurred starting in 1985.

Lie groups

We only mention books which focus on "groups" with respect to "Lie algebras".

[Stillwell 08], falsely naive despite its title, treats basic questions with elegance and economy.

[Duistermaat-Kolk 99] has a very geometric point of view. Notably it treats actions of compact Lie groups on manifolds, which is rarely done in other books.

[Onishchik-Vinberg 90], very complete, is more difficult.

Riemannian geometry

[Milnor 63] has a rapid and effective exposition. [Do Carmo 92] considers mostly the "metric" aspects of the subject, while [Chavel 83] is concerned primarily with "analysis on manifolds". [Gallot-Hulin-Lafontaine 05] is a compromise between these points of view.

For a comprehensive survey, see the monumental [Berger 03].

Complex geometry

This subject was introduced (a little) through the exercises. For a very geometric introduction after this book, see [Debarre 05], [Griffiths-Harris 94]. We mention also that [Fulton 95] contains an introduction to Riemann surfaces, and [Farkas-Kra 91] provides very complete exposure to the subject.

References

M. ADACHI, *Embeddings and Immersions*. Transl. Math. Monogr., vol. 124 (Amer. Math. Soc., 1993).

C. ADAMS, *The Knot Book* (W.H. Freeman & Co., 1994). This subject is experiencing rapid growth, and involves a lot of algebra. However, this book, which is neither the most recent nor the most up to date, has a pleasant geometric point of view (AN).

F. APÉRY, *Models of the Real Projective Plane* (Friedr. Vieweg & Sohn, 1987).

V.I. ARNOLD
▸ *Mathematical Methods of Classical Mechanics*. Grad. Texts in Math., vol. 60 (Springer, 1978). Translated from the Russian.
▸ *Catastrophe Theory* (Springer, 1992).

M. ATIYAH, *Geometry of Yang-Mills Fields* (Scuola Normale Superiore di Pisa, 1979).

H. ATTOUCH, G. BUTTAZZO and G. MICHAILLE, *Variational Analysis in Sobolev and BV Spaces; Applications to PDEs and Optimization.* MPS/SIAM Ser. Optim., vol. 6 (Math. Program. Soc. & Soc. Ind. Appl. Math., 2006).

M. AUDIN
▶ *Geometry.* Universitext (Springer, 2003). Translated from the French.
▶ *The Topology of Torus Actions on Symplectic Manifolds,* 2nd ed., Progress in Math., vol. 93 (Birkhaüser, 2004). Contains, amongst other things, an introduction to symplectic geometry and basic results on the action of compact groups on manifolds (AN).

M. BERGER
▶ *Geometry I & II.* Universitext (Springer, 1987). Translated from the French.
▶ *A Panoramic View of Riemannian Geometry* (Springer, 2003).

M. BERGER and R. GOSTIAUX, *Differential Geometry: Manifolds, Curves and Surfaces.* Grad. Texts in Math., vol. 115 (Springer, 1988). Translated from the French.

B. BOOSS and D. BLEECKER, *Topology and Analysis.* Universitext (Springer, 1985). For "analysis on manifolds" (AN).

R. BOTT and J. MILNOR, "On the Parallelizability of Spheres", Bull. Amer. Math. Soc. **64**, pp. 87–89 (1958).

R. BOTT and L. TU, *Differential Forms in Algebraic Topology,* 2nd ed. Grad. Texts in Math., vol. 82 (Springer, 1986).

G. BREDON, *Topology and Geometry* (Springer, 1994). A very complete reference work on the topology of manifolds: homotopy groups, homology and cohomology groups with real an integer coefficients, multiplicative structures, duality, etc. (AN).

H. BREZIS, *Functional Analysis, Sobolev Spaces and Partial Differential Equations* (Masson, 1983).

P. BUSER, *Geometry and Spectra of Compact Riemann Surfaces* (Birkhaüser, 1992).

É. CARTAN, *La géométrie des espaces de Riemann,* 2nd ed. (Gautiers-Villars, 1951). The books of Élie Cartan have remained stimulating, but prior exposure to the subjects they treat is desirable (AN).

M. CHAPERON, *Calcul différentiel et calcul intégral 3ème année* (Dunod, 2008). Also treats the real analytic case. An introduction to singularities and normal forms (AN).

E. CHARPENTIER, E. GHYS and A. LESNE (eds.), *The Scientific Legacy of Poincaré*. Hist. Math., vol. 36 (Amer. Math. Soc. & London Math. Soc., 2010). Translated from the French.

I. CHAVEL, *Riemannian Geometry – A Modern Introduction* (Cambridge Univ. Press, 1983).

E. CODDINGTON, *An Introduction to Ordinary Differential Equations* (Dover Publications, 1989).

H.P. DE SAINT-GERVAIS, *Uniformisation des surfaces de Riemann. Retour sur un théorème centenaire* (ENS éditions, 2010). *Uniformization of Riemann Surfaces. Revisiting a hundred-year-old theorem*, to appear in the Heritage of European Mathematics series, European Mathematical Society. Henri Paul de Saint-Gervais is the collective name of a group of mathematicians, primarily from the École Normale Supérieure de Lyon. They realized a very stimulating book (AN).

O. DEBARRE, *Complex Tori and Abelian Varieties*. SMF/AMS Texts Monogr., vol. 11 (Amer. Math. Soc., 2005). Translated from the French.

J.-P. DEMAILLY, *Complex analytic and differential geometry* (2012). This book is – and will be – available as an "OpenContentBook", see *https://www-fourier.ujf-grenoble.fr/~demailly/manuscripts/agbook.pdf* .

M. DEMAZURE, *Bifurcations and Catastrophes: Geometry of Solutions to Nonlinear Problems*. Universitext (Springer, 2000).

J. DIEUDONNÉ
▶ *Treatise on Analysis III*. Pure Appl. Math. (Amst.), vol. 10 (Academic Press, 1972). Translated from the French. Austere, but contains many technical results that I have never seen elsewhere. There are numerous instructive exercises of reasonable difficulty (AN).
▶ *A History of Algebraic and Differential Topology 1900–1960* (Birkaüser, 1988). A good reference to help understand the motivations which drive the notions of algebraic and differential topology; very practical place to find the reference to "well known" results (AN).

M. DO CARMO
▶ *Differential Geometry of Curves and Surfaces* (Prentice Hall Inc., 1976).
▶ *Riemannian Geometry* (Birkhaüser, 1992).

A. DOUADY and R. DOUADY, *Algèbre et théories galoisiennes*, 2nd ed. (Cassini, 2005). Notable for a very conceptual exposition of covering theory (AN).

B. DOUBROVINE, S. NOVIKOV and A. FOMENKO
▸ *Modern Geometry. Methods and Applications II*. Grad. Texts in Math., vol. 104 (Springer, 1985). Translated from the Russian.
▸ *Modern Geometry. Methods and Applications III*. Grad. Texts in Math., vol. 124 (Springer, 1990). Translated from the Russian.
▸ *Modern Geometry. Methods and Applications I*, 2nd ed. Grad. Texts in Math., vol. 104 (Springer, 1992). Translated from the Russian.

J. DUGUNDJI, *Topology* (Allyn and Bacon, Inc., 1965). A complete exposition of point set topology, followed by the basics of homotopy theory. A little old, but it is hard to do better! (AN).

J.J. DUISTERMAAT and J.A. KOLK, *Lie Groups*. Universitext (Springer, 1999).

H.M. FARKAS and I. KRA, *Riemann Surfaces*, 2nd ed. Grad. Texts in Math., vol. 71 (Springer, 1991).

R. FEYNMAN, R. LEIGHTON and M. SANDS, *The Feynman Lectures on Physics*, The new millenium ed. (Basic books, 1963).

W. FULTON, *Algebraic Topology; A First Course*. Grad. Texts in Math., vol. 153 (Springer, 1995). A little harder than Greenberg-Harper 81, but much more focused toward manifolds (AN).

S. GALLOT, D. HULIN and J. LAFONTAINE, *Riemannian Geometry* (Springer, 2005).

R. GODEMENT, *Introduction à la théorie des groupes de Lie* (Springer, 2005).

M. GOLUBITSKY and V. GUILLEMIN, *Stable Mappings and Their Singularities*. Grad. Texts in Math., vol. 14 (Springer, 1973).

A. GRAY, *Tubes*. Progr. Math., vol. 221 (Birkhaüser, 2004).

M.J. GREENBERG and J.R. HARPER, *Lectures on Algebraic Topology. A First Course* (Benjamin/Cumming, 1981). Singular homology and cohomology, multiplicative structure, duality. Much more of a textbook than a reference book like [Bredon 94] (AN).

W. GREUB, S. HALPERIN and R. VAN STONE, *Connections, Curvature and Cohomology* (Academic Press, 1976). To see a systematic implementation of differential forms (AN).

P. GRIFFITHS and J. HARRIS, *Principles of Algebraic Geometry* (John Wiley & Sons, Inc., 1994).

V. GUILLEMIN and A. POLLACK, *Differential Topology* (Prentice Hall Inc., 1974).

B. HALL, *Lie groups, Lie Algebras and Representations.* Grad. Texts in Math., vol. 222 (Springer, 2003).

R.-S. HAMILTON, "The Inverse Function Theorem of Nash and Moser", Bull. Amer. Math. Soc. **7**, pp. 65–222 (1982). A powerful discussion of the conditions of the Banach inverse function theorem, and a motivated introduction to the "Nash-Moser" machinery (AN).

G. HECTOR and U. HIRSCH, *Introduction to the Differential Geometry of Foliations.* Aspects Math., vol. 1 (Friedr. Vieweg & Sohn, 1981).

S. HELGASON, *Differential Geometry, Lie Groups and Symmetric Spaces.* Pure Appl. Math. (Amst.), vol. 80 (Academic Press, 1978).

Y. HELLEGOUARCH, *Invitation to the Mathematics of Fermat-Wiles* (Academic Press, 2001). Translated from the French.

M. HIRSCH, *Differential Topology.* Grad. Texts in Math., vol. 33 (Springer, 1976).

M. HIRSCH, S. SMALE and R. DEVANEY, *Differential Equations, Dynamical Systems & an Introduction to Chaos* (Academic Press, 2003).

H. HOPF
▶ "Über die Abbildungen von Sphären auf Sphären niedrigerer Dimension", Fundam. Math. **25**, pp. 427–440 (1935).
▶ *Differential Geometry in the Large.* Lecture Notes in Math., vol. 1000 (Springer, 1983).

L. HÖRMANDER, *The Analysis of Linear Partial Differential Operators I.* Classics Math. (Springer, 1990).

M. KAROUBI and C. LERUSTE, *Algebraic Topology* via *Differential Geometry.* London Math. Soc. Lecture Note Ser., vol. 99 (Cambridge Univ. Press, 1987). Translated from the French.

A. KATOK and B. HASSELBLATT, *Introduction to the Modern Theory of Dynamical Systems*, with a supplement by Anatole Katok and Leonardo Mendoza (Cambridge Univ. Press, 1995).

R. KULKARNI and U. PINKALL (eds.), *Conformal Geometry.* Aspects Math., vol. 12 (Friedr. Vieweg & Sohn, 1988).

S. LANG, *Undergraduate Analysis.* Undergrad. Texts Math. (Springer, 1986).

B. LAWSON, *The Theory of Gauge Fields in Four Dimensions.* CBMS Reg. Conf. Ser. Math., vol. 58 (Amer. Math. Soc., 1985).

J. LEE, *Introduction to Smooth Manifolds.* Grad. Texts in Math., vol. 218 (Springer, 2003).

S. MacLane, *Categories for the Working Mathematician*. Grad. Texts in Math., vol. 5 (Springer, 1971).

J. Martinet, *Perfect Lattices in Euclidean Spaces*. Grundleheren Math. Wiss., vol. 327 (Springer, 2003).

W. Massey, *Algebraic Topology: An Introduction*. Grad. Texts in Math., vol. 56 (Springer, 1977). Notable for, amongst other things, the topological classification of compact surfaces (AN).

D. McDuff and D. Salamon, *Introduction to Symplectic Topology* (Oxford Univ. Press, 1998).

J. Merker, *Sophus Lie, Friedrich Engel et le problème de Riemann-Helmholtz* (Hermann, 2010).

J. Milnor
► *Morse Theory* (Princeton Univ. Press, 1963).
►"Whitehead Torsion", Bull. Amer. Math. Soc. **72**, pp. 351–326 (1966).
► *Topology from the Differentiable Viewpoint* (Princeton Univ. Press, 1997). Reprint of a 1965 edition.

J. Milnor and J. Stasheff, *Characteristic Classes* (Princeton Univ. Press, 1974).

C. Misner, K. Thorne and J. Wheeler, *Gravitation* (W.H. Freeman & Co., 1973).

D. Montgomery and L. Zippin, *Topological Transformation Groups* (Interscience Publishers, 1955).

J.M. Munkres, *Topology*, 2nd ed. (Prentice Hall Inc., 2000).

A.L. Onishchik and E.B. Vinberg, *Lie Groups and Algebraic Groups* (Springer, 1990). Translated from Russian.

M. Postnikov, *Lie Groups and Lie Algebras* (Mir, 1994).

A. Pressley and G. Segal, *Loop Groups*. Oxford Math. Monogr. (Clarendon Press, 1986).

J. Robbin, "On the Existence Theorem for Differential Equations", Proc. Amer. Math. Soc. **19**, pp. 1005–1006 (1968).

P. Samuel, *Algebraic Theory of Numbers* (Dover Publications, 2008). Translated from the French.

L.A. Santaló, "Integral Geometry and Geometric Probability", in *Encyclopedia of Mathematics and Its Applications I* (Addison-Wesley Publishing Co., 1976).

J.-P. SERRE, *A Course in Arithmetic*. Grad. Texts in Math., vol. 7 (Springer, 1996).

M. SPIVAK, *Differential Geometry* (Publish or Perish, 1979).

J. STALLING, "The Piecewise-Linear Structure of Euclidean Space", Proc. Cambridge Phil. Soc. **58**, pp. 481–488 (1962).

N. STEENROOD, *The Topology of Fiber Bundles* (Princeton Univ. Press, 1951).

J. STILLWELL, *Naive Lie Theory*. Undergrad. Texts Math. (Springer, 2008).

S. TABACHNIKOV, *Billiards*. Panor. Synthèses, vol. 1 (Soc. Math. France, 1995).

H. WEYL, "On the Volume of Tubes", Amer. J. Math. **61**, pp. 461–472 (1939).

H. WHITNEY, *Geometric Integration Theory* (Princeton Univ. Press, 1957).

J.A. WOLF, *Spaces of Constant Curvature* (Amer. Math. Soc., 1984).

Index

F

fibration, 64–66, 113, 164, 173
flow of a vector field, 122
flux, 256
free action, 76
Frobenius theorem, 140, 142
functions of class C^p, 10
fundamental theorem of algebra, 60–61, 292–295

G

Gauss map, 344
Gauss-Bonnet formula, 325
Gaussian curvature, 334–335
germ, 103–108, 145
Girard's formula, 325
global derivation, 108–111
good covering, 321
gradient, 127, 195, 196, 227, 290
Grassmannian, 175, 193
group action, 74, 93–94, 173, 183

H

Haar measure, 268–269, 372
Hadamard inequality, 35
Hadamard lemma, 105, 145
hairy ball theorem, 246
Hessian, 118–119
holomorphic, 7, 48, 97, 145, 183, 229, 292–294, 354, 365
homogeneous coordinates, 62, 70
homotopy, 81, 349
Hopf fibration, 65, 89, 319

I

identity component, 150, 165
immersion, 18, 22, 26, 69, 88, 133–134
index of a vector field, 289–291, 336–344
infinitesimal generator, 123, 153
interior product, 206, 229
inversion, 12, 54, 91, 357
isotopy, 134, 316

J

Jacobi identity, 112
Jacobian, 7
Jacobian matrix, 6

K

Künneth formula, 327
Klein bottle, 267, 318

L

Lagrange multipliers, 34
Lens spaces, 93
Lie bracket, 111
Lie derivative, 204
Lie subgroup, 150
local diffeomorphism, 17, 94
Lorentz group, 150

M

manifold with boundary, 321, 349
maximal atlas, 54
Möbius group, 91, 357
Möbius strip, 50, 80, 93, 117, 236, 244
model fiber, 65
modulus (of a Lie group), 268
monodromy theorem, 83
morphism (Lie algebra), 154
morphism (of a Lie group), 150
morphism (of a vector bundle), 115
Morse lemma, 45, 131, 352
Moser's theorem, 321
Moser's trick, 129, 232, 321, 370
moving frames, 331

N

normal bundle, 142, 362

O

one-parameter subgroup, 29–32, 152–153, 158
orbit, 75, 162
orientation covering, 242–244
oriented atlas, 237
oriented manifold, 238
orthogonal group, 23, 24, 32, 43, 58, 90, 149

Printed in the United States
By Bookmasters